Cell Physiology and Genetics of Higher Plants

Volume II

Author

A. Rashid
Reader (Associate Professor)
Department of Botany
University of Delhi
Delhi, India

CRC Press
Taylor & Francis Group
Boca Raton London New York

CRC Press is an imprint of the
Taylor & Francis Group, an **informa** business

First published 1988 by CRC Press
Taylor & Francis Group
6000 Broken Sound Parkway NW, Suite 300
Boca Raton, FL 33487-2742

Reissued 2018 by CRC Press

© 1988 by CRC Press, Inc.
CRC Press is an imprint of Taylor & Francis Group, an Informa business

No claim to original U.S. Government works

Library of Congress Cataloging-in-Publication Data

Rashid, A. (Abdur)
    Cell physiology and genetics of higher plants.

    Includes bibliographies and index.
    1. Plant cells and tissues. 2. Plant cytogenetics.
I. Title.
QK725.R33 1988   582'.087   87-9349
ISBN 0-8493-6051-X (set)
ISBN 0-8493-6062-5 (v. 1)
ISBN 0-8493-6063-3 (v. 2)

A Library of Congress record exists under LC control number: 87009349

Publisher's Note
The publisher has gone to great lengths to ensure the quality of this reprint but points out that some imperfections in the original copies may be apparent.

Disclaimer
The publisher has made every effort to trace copyright holders and welcomes correspondence from those they have been unable to contact.

ISBN 13: 978-1-315-89137-8 (hbk)
ISBN 13: 978-1-351-07047-8 (ebk)

Visit the Taylor & Francis Web site at http://www.taylorandfrancis.com and the
CRC Press Web site at http://www.crcpress.com

PREFACE

In the beginning of this century, Haberlandt performed his pioneering experiments on the culture of plant cells. Simultaneously he had visualized the potential applications of this technique, and thus was born plant cell biotechnology. In the intervening period, the technique of plant cell culture has been perfected, but plant cell biotechnology remains in its infancy. The reason for the slow progress is inadequate information about physiology and genetics of higher plant cells. In this series a synthesis of the concepts of cell physiology and genetics, the basic disciplines for cell biotechnology and genetic engineering, is presented. A broad and informative view of these disciplines will be appealing to plant physiologists, geneticists, cell biologists, plant breeders, and biotechnologists.

Volume I comprises four chapters. Initial chapters devoted to fundamental aspects (Cell Multiplication and Cell Differentiation) concerning cell physiology are basic to the theme of biotechnology of higher plants. Included in the chapter on Cell Differentiation is an account of biosynthetic potential of higher plant cells. The next chapter, Cell Totipotency, highlights the regeneration potential of higher plant cells, for micropropagation of plants. Following this is the chapter entitled Induction of Haploid Plant/Cell. Developments in this field have led to the beginning of a new era in crop improvement.

Volume II comprises five chapters. Of these, four chapters are devoted to cell genetics: Protoplast — Isolation and Cell Regeneration, Cell Modification, Cell Fusion, and Cell Transformation. These chapters concern genetic engineering of plants. The final chapter, Cell Preservation, for germplasm storage, concludes the series.

This work is an outcome of the freedom a teacher of this department enjoys in framing a course and offering it to students in lieu of an elective paper for an M.Sc. Degree in Botany. I acknowledge the indirect contribution of students for stimulating discussions. Two of my former research students, now teachers of botany, Dr. Shashi Tyagi (nee Bharal) of Gargi College and Dr. Paramjeet Khurana (nee Gharyal) of Khalsa College, were requested to go through these chapters at the first draft stage and offer critical comments. I am appreciative of the help received from them. Thanks are also due to my research students, Mr. M. A. Mallick, Miss Rajni Dua, and Miss Manju Talwar, for their help in completing literature surveys.

I would also like to acknowledge the much needed encouragement and invaluable help received from my wife, Dr. Zakia Anjum, in undertaking and completing this work.

A. Rashid

THE AUTHOR

A. Rashid, Ph.D., is Reader in Botany at the University of Delhi. Dr. Rashid received his degree from the University of Delhi in 1968 and since then he has been a member of the faculty. Earlier, his research interests were restricted to developmental physiology of lower plants. Later, in 1972, he was attracted towards higher plant cell physiology and completed a postdoctoral fellowship in the laboratory of Professor H. E. Street at the University of Leichester, U.K. From 1980 to 1982, he was also awarded the Alexander von Humboldt Fellowship to collaborate with Professor Dr. J. Reinert, at the Freie Universität Berlin for the induction of haploids in culture of isolated pollen grains.

Dr. Rashid has published 60 research papers and a few review papers. His primary interest is cell physiology, employing pollen and protoplast as systems.

CELL PHYSIOLOGY AND GENETICS OF HIGHER PLANTS

Volume I
Cell Multiplication
Cell Differentiation
Cell Totipotency
Induction of Haploid Plant/Cell

Volume II
Protoplast — Isolation and Cell Regeneration
Cell Modification
Cell Fusion
Cell Transformation
Cell Preservation

TABLE OF CONTENTS

Chapter 1

PROTOPLAST — ISOLATION AND CELL REGENERATION

I. DEFINITION AND SIGNIFICANCE

A plant protoplast is a cell without a wall. Due to the absence of a cell wall and consequent exposure of the plasma membrane, the protoplast becomes a very fragile structure. It is disrupted easily, unless it is maintained in an osmoticum. In an osmoticum, due to partial plasmolysis, the protoplast is a perfect circular structure (Figure 1A). The delicate nature of a protoplast makes it a relatively difficult material to handle as compared to a cell. However, a plant protoplast readily regenerates a wall and is a fully totipotent structure, with all the attributes of a cell.

The absence of a cell wall around the protoplast makes it a suitable system for many studies — fundamental as well as applied — which are not possible with an intact cell, within the confines of a cell wall. The more significant of the potential applications of plant protoplasts is hybridization of two protoplasts for crop improvement. It is a novel method, often described as parasexual hybridization, to expand the pool of genetic variability needed for crop improvement. In parasexual fusion it is possible to combine distant genes due to the absence of crossability barriers, often encountered in gametic fusion, particularly when distant hybridization is programed. It can also bring about selective gene transfer such as incorporation of cytoplasmic genes. Further, protoplasts can be useful in plant improvement by way of cell transformation, i.e., incorporation into a protoplast of purified foreign genetic material, an organelle, or even a microorganism.

The protoplast is an ideal free-cell system suitable for fundamental studies such as cell wall regeneration and isolation of mutants, especially auxotrophs where intercell feeding is undesirable and is a hindrance. Haploid protoplasts are particularly well suited for the raising of mutants. A haploid protoplast is a higher plant-equivalent of a microbe. Other uses of protoplasts are the study of membrane (transport and cotransport of molecules across a membrane) and studies on hormone action, mechanism of fungal and bacterial infection, and viral multiplication, which can ultimately find application in testing for drug and disease resistance. Due to the absence of a cell wall a protoplast can be easily disrupted and this property can be employed for the isolation of organelles and macromolecules without the damaging effects of shearing forces required to break the cell wall.

For the use of protoplasts for these diverse purposes a basic requirement is the isolation of viable protoplasts in high frequency which are capable of regeneration into cells and ultimately into plants. This chapter concerns these fundamental aspects of protoplast technology.

II. ISOLATION

The key operation in the release of a plant protoplast is the removal of cell wall in a way that it does not damage the protoplast and impair its ability to regenerate into a cell.

A prerequisite for the release of a protoplast from the confines of the cell wall is a suitable osmoticum of proper osmolarity. In the absence of an osmoticum, an immediate lysis of a protoplast is a certainty due to the lack of wall pressure usually exerted by the cell wall. Use of an osmoticum of either too high or too low osmotic potential results in irreversible damage to the protoplast. Deviation of the isolation medium from the isotonic concentration can cause the protoplast to shrink or swell. This process may influence membrane physiology and viability of the protoplast.

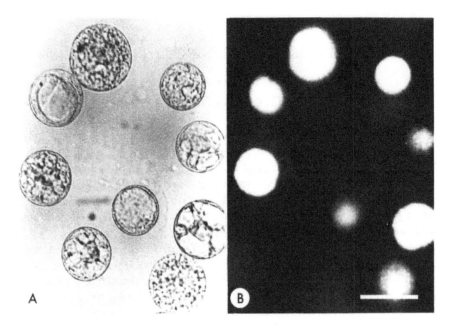

FIGURE 1.   Protoplasts from cell suspension culture of *Nicotiana tabacum*, stained with fluorescein isothiocyanate: (A) bright field illumination and (B) epifluorescence illumination. (From Galbraith, D. W., Afonso, C. L., and Harkins, K. R., *Plant Cell Rep.*, 3, 151, 1984. With permission.)

A. Methods of Isolation

The isolation of plant protoplasts — removal of the cell wall in the presence of a suitable osmoticum — can be accomplished by the following two methods.

1. Mechanical Isolation

The pioneering[154] attempt to isolate protoplasts from a higher plant was by a mechanical method. The method involved a preplasmolysis of tissue followed by its random sectioning which resulted in the release of a few undamaged protoplasts along with several broken cells. Using this method protoplasts could be obtained only from highly vacuolated cells of tissues such as onion bulb scale, mesocarp of cucumber, and roots of radish and beet. In this method it is essential that a protoplast contracts away from cell wall. Therefore, it cannot be applied to meristematic cells. This method is not recommended for routine operations because of (1) poor yield, (2) restricted applicability to tissues which are vacuolated and easy to plasmolyze, and (3) tedium of the operation.

More recently, in a new approach[23] the material (cross sections of petiole of *Saintpaulia ionantha*) was grown on a medium enriched with a relatively high level of auxin, 2,4-D. The thin-walled cells so formed when teased apart with dissecting needles readily resulted into protoplasts, in an osmotic medium.

2. Enzymatic Isolation

Large-scale isolation of protoplasts[59] was possible when enzymatic digestion of cell wall was demonstrated from root cells of tomato. To digest the cell wall the enzyme (cellulase) source was a culture filtrate of the fungus *Myrothecium verucaria*.

Following this, commercial preparations of enzymes[244,276] employed for the isolation of plant protoplasts are macerozyme and cellulase. The former is rich in pectinase and brings about cell maceration and the latter digests the cell wall, releasing protoplasts.

a. Steps in Enzymatic Isolation

In an enzymatic isolation of protoplasts the tissue is subjected to pectin- and cellulose-dissolving enzymes (macerozyme and cellulase) either in two steps (sequentially) or in one step (simultaneously).

In the two-step or sequential method,[276] macerozyme and cellulase are employed one after another. To begin with, there is cell maceration by macerozyme which results in the separation of cells; this is followed by the transformation of cells into protoplasts, effected by cellulase. In this method the cells remain in contact with the enzymes for a shorter period than in the one-step process.

The one-step method[227] is a simplification of the two-step process. Here the tissue is subjected to an enzyme mixture comprising pectinase as well as cellulase. This method minimizes the chances of microbial contamination, as the operation is reduced to a minimum. Higher yields are possible with this method, particularly from leaf tissue because mesophyll as well as palisade cells are transformed into protoplasts. However, in other tissues, such as pods of soybean,[326] no difference either in protoplast yield or viability was recorded with either the one- or two-step method.

A modification of the isolation procedure is an intermediate method[223,304] between the one- and two-step methods. The tissue is treated first with macerozyme to loosen the cells, and before free cells are formed, the tissue is transferred to cellulase.

Yet another modification[126,174,253] is a combination of mechanical and enzymatic methods. Here, a mechanical separation of cells is brought about by homogenizing the tissue, followed by cellulase treatment.

B. Factors Affecting Isolation

A standard practice for isolation of protoplasts is by employing enzymes, pectinase, and cellulase. However, each material has to be investigated for optimal conditions favoring release of a maximum number of viable protoplasts. Various factors, given below, affect the process of protoplast formation.

1. Source Material

It has been possible to isolate protoplasts from plants belonging to different groups. In angiosperms protoplasts have been isolated from almost every part of the plant. Protoplasts are possible from cells having primary walls, without changes such as lignification. As for tissue specificity, the mesophyll cells are commonly used because leaf is the most convenient and abundant material.[38] Also, relatively higher yield and genetic uniformity of protoplasts is possible from mesophyll cells. Another good source is a callus or cell suspension culture. However, protoplasts from mesophyll cells are likely to have a greater degree of genetic uniformity than cells maintained in culture where chromosome aberrations are known to occur.

a. Mesophyll Tissue

In early work, leaves were taken from plants grown either in the field or greenhouses and hence surface sterilization was a prerequisite. However, surface sterilization may be injurious, due to the toxic nature of the sterilizer. Also, at times the tissue, particularly from soft leaves after sterilization, may become too brittle[305] to handle. The problem of surface sterilization can be avoided by taking leaves from plants grown in vitro.

In order to facilitate the penetration of enzymes into the leaf tissue, the lower epidermis is peeled off[276] or partially injured by the use of carborundum.[14,176,258] For peeling the epidermis, when leaves from plants grown in vivo are the source material, restriction of water supply before excision of leaf is reported to be helpful.[276] However, peeling is a tedious process and instead it is convenient to slice a leaf into thin strips. This is a standard practice.

The recovery of viable protoplasts from mesophyll tissue has been found to be critically dependent on the age of the leaf in barley,[250] bean,[220] tobacco,[305] and pea.[7] From fully expanded, mature leaves only 10 to 15% protoplast yield was possible in soybean, whereas this increased to 20 to 30% with young, expanding leaves.[167] Leaf age also affects the regeneration of protoplasts. Protoplasts from younger leaves of alfalfa[149] and apple[303] divided sooner than older leaves. Seasonal variations[53] have also been recorded in protoplast viability obtained from tobacco plants grown in a greenhouse. During winter months protoplast viability could be increased by feeding calcium nitrate and calcium chloride, which increased the membrane calcium. However, feeding treatments were not helpful during spring and summer months.

Growth conditions of donor plants determine the protoplast yield. In particular, light intensity, photoperiod, and relative humidity of the environment affect the yield of protoplasts. Of these, intensity and duration of light are most important for the production of viable protoplasts in good number. Generally, a low light intensity is favorable for tobacco,[27] henbane,[170] and alfalfa.[140,149] Barley[134] and tobacco[163] initially grown under low light intensity gave a higher yield of protoplasts on transfer to dark for 2 to 3 days, prior to isolation of protoplasts. The preparations also contained a smaller number of broken cells.

The transfer of donor plants to dark decreases the starch content of cells, thus allowing an easier protoplast release in barley[134] and giving a better survival rate during centrifugation in tobacco,[163] pea,[62] and grape.[314] Instead of transfer to dark, a reduction in the photoperiod[273] of plants of *Lycopersicon esculentum* from 16 to 9 hr also helped to increase the frequency of regeneration of isolated protoplasts. However, this had no effect in *L. peruvianum*. The beneficial effect of transfer of plants to short-day condition (8-hr light) has also been recorded in *Solanum pennillii*.[130] The influence of starch on protoplast yield is also seen in isolations from pea root. Inclusion of 1 mg/ℓ of GA$_3$ in isolation medium increased protoplast yield.[162] Mitotic activity was, however, reduced in protoplasts obtained from gibberellin-treated tissue.

For raising source plants, a high humidity (60 to 70%) and a moderate temperature (20 to 25°C) are optimal. However, a combination of high humidity (80% RH) and low temperature (20°C) was favorable for isolation and subsequent survival of protoplasts in tomato.[279]

Protoplasts isolated from leaves of plants grown in vitro show greater viability and are endowed with a greater regeneration capacity. This has been observed in many members of Solanaceae, *Petunia*,[25] *Nicotiana*,[27] *Solanum* spp.,[28,32] *Hyoscyamus*,[170] and *S. tuberosum*.[121] It is possibly due to the physiological uniformity of the growth conditions and conditioning of the tissue to in vitro growth. On the other hand, glasshouse-grown plants require a careful supply of inorganic nutrients in addition to the essential nitrogen fertilizers[52,164,305] and other conditions such as light. Hence, in certain cases, such as *Solanum nigrum*,[205] when it was not possible to have viable protoplasts from greenhouse-grown plants, the alternative was axenic shoot culture. Also, in *Pisum sativum*[7] and *Brassica rapa*[287] this problem was circumvented by transfer of plants from greenhouse to growth cabinets in the culture room.

For raising plants in vitro, one need not always germinate seeds; instead, advantage can be taken of the high regenerative capacity of shoots. In plants such as *Petunia*,[26] *Nicotiana*,[27] species of *Solanum*,[28,32,205] and *Hyoscyamus*[170] multiple shoots are formed on culture of shoot tips. Shoot cultures have provided a suitable material for the isolation and regeneration of protoplasts not only in herbaceous plants but also in woody plants[268] such as *Betula*, *Rhododendron*, and *Malus domestica*.[303] A modification of shoot culture medium is further helpful for protoplast isolation and viability. Inclusion of a low concentration of benzylaminopurine to shoot culture medium of *Solanum brevidens*[207] and methionine for *Malus domestica*[303] helped increase the frequency of stable protoplasts and their division.

b. Callus and Cell Suspension

Although callus or cell suspension was the source material for early work on protoplast isolation,[76,243] this choice was relegated in favor of mesophyll as a source tissue. Nevertheless, growing cells from suspension cultures of sugarcane,[178] atropa,[114] and tobacco[69,286] are increasingly employed for protoplast isolation. The main advantage of a callus and a cell suspension is their physiological uniformity due to the controlled conditions, physical and chemical, in which they are grown. Further, these cells are conditioned to in vitro growth. This is clearly seen in a comparative study of the isolation and regeneration of protoplasts from mesophyll cells and callus cells of *Medicago coreulea* and *M. glutinosa.*[5] Conditions of plant growth, plant age, protoplast isolation, and protoplast culture were critical for sustained division of mesophyll protoplasts, whereas they were less stringent for the division of protoplasts from cell suspension cultures.

The yield from callus cultures grown on agar is, however, relatively poor. This may be due to slow growth, diversity of age, and cytological status of cells. Callus cultures are, therefore, unsuitable and suspension cultures are preferred.

The yield from a cell suspension is critically dependent on its age. An increase in the complexity of a cell wall with age is quite expected. Cultures at cell expansion or stationary phase failed to respond to enzyme treatment in cotton[81] and tomato.[281] Hence, a rapidly growing cell line which is regularly subcultured at short intervals (every 3 to 4 days) is desirable. Protoplasts from cell suspension of *Pinus contorta*[122] and *Spinacea oleracea*[204] in a rapidly dividing phase could only divide and form colonies. The efficiency of protoplast isolation from cell suspension of *Haplopappus*[302] could be increased by inducing changes in the nature of the cell wall by increasing the level of auxin in the medium. Also adding sulfur amino acid (e.g., L-cysteine or L-methionine) and sulfydryl compounds such as 2-mercaptoethanol enhanced protoplast yield. Similarly, a better yield of protoplasts from suspension cultures of several plants could be obtained if the cells were cultured in a medium containing ammonium and L-glutamine as the sole nitrogen source.[94] Nitrogen starvation for 3 days also enhanced yield.

Lysis of protoplasts during release and subsequent operations, due to the presence of large starch granules, can be remedied by a preculture of cells on a medium with a low concentration of sucrose or on a sucrose-free medium.[266,302] An interesting case of change in the physiological state of certain cells in response to growth condition has been recorded. Protoplasts from a pantothenate-requiring cell line of *Datura* failed to divide in a medium containing 0.5 M mannitol if the cells from which they were derived had been cultured in liquid medium more than four times. However, divisions could be induced[193] either by changing the concentration of 2,4-D from 1 ppm to 2.5 to 5 ppm or by adding a low concentration of benzyladenine (0.1 to 1 mg/ℓ) along with 1 ppm 2,4-D or by diluting mannitol to 0.2 M immediately after the isolation of protoplasts. Similar physiological changes did not occur in the case of wild-type, adenine-requiring, or isoleucine-valine-requiring cells of *Datura*.

c. Pollen Grains

The isolation of protoplasts from pollen grains remains a formidable task because it has not been possible to find enzymes which can hydrolyze sporopollenin. Instead, it is possible to isolate protoplasts from pollen tetrads[20] while they are without sporopollenin and are within the callose wall. The callose could be digested by helicase which is rich in β-1-3 gluconase. Unfortunately, this work has not been pursued. It has been possible, however, to isolate protoplasts from the pollen of some gymnosperms[71] which shed their exine on hydration.

Recently, it has been possible to obtain protoplasts from pollen,[168] referred to as sporoplasts, of lily, pear, and walnut, employing 4-methylmorpholine *N*-oxide monohydrate (MMNO·H_2O), a potent solvent for cellulose and other cell wall polysaccharides.

2. *Osmoticum*

A prerequisite for the isolation of protoplasts is the determination of a correct level of osmoticum. At a low level of osmoticum multinucleate protoplasts are likely to result due to coalescence and spontaneous fusion of protoplasts formed, whereas a higher level of osmoticum may result in irreversible damage to protoplasts. The osmotic potential of a tissue is also dependent on the environmental conditions of donor plants. Osmotic potential is a critical determinant of protoplast viability.[44] Deviation of isolation media from isoosmotic concentration can cause the protoplasts to shrink and swell. This process may influence membrane physiology and long-term survival of protoplasts.

A simple and direct method for the determination of the correct level of osmoticum is the observation of cells incubated in a medium of different osmotic concentrations. However, it is not simple to determine the required osmotic concentration of a tissue because cell types within a single leaf vary markedly in osmotic concentration.[268,269]

In the earliest[154] work on the isolation of protoplasts from *Stratoides aloides* by the mechanical method, calcium chloride and calcium nitrate salts were used as an osmoticum. However, with the introduction of the enzymatic method, the use of sugars and sugar alcohols has become prevalent. In most investigations, mannitol is the osmoticum of choice. This may be due to its inert nature and its slow diffusion into protoplast, ensuring a constant osmolarity. The change from salt to mannitol became necessary due to the longer incubation period required in enzymatic isolation, as compared to the mechanical method. Salts are relatively more penetrating and are likely to be more harmful. Salts are also known to reduce the cell wall degrading property of enzymes employed for the isolation of protoplasts. Nevertheless, for the isolation of protoplasts from carrot root, a mixture of calcium chloride and potassium chloride was more suitable than mannitol,[142] and reproducible results in tobacco were possible when a combination of potassium chloride and magnesium sulfate was employed as osmoticum.[183] Also, in *Hyoscyamus* salts were employed as an osmoticum.[156]

A determinate effect of osmoticum on plasma membrane and the possible future behavior of protoplasts is seen in the bursting of *Avena* protoplasts.[242] The plasma membrane disintegrated into tiny particles when protoplasts prepared with mannitol were made to burst in distilled water. When sucrose was the osmoticum, the plasma membrane formed interconnecting strands during disintegration, and if salts were employed, the plasma membrane was torn apart into pieces.

Instead of mannitol, sorbitol (an isomer of mannitol) has been employed satisfactorily as an osmoticum either alone[7,13,80,113,114,236] or in combination with mannitol.[70,104,145,150,172,205,239]

It is not essential that the osmotic stabilizer be metabolically inert. For isolation of protoplasts from tobacco cell suspension, various soluble carbohydrates (glucose, fructose, galactose, sorbitol, and mannitol) were equally effective.[286] A combination of sugars and mannitol is more desirable because the gradual decline in overall osmotic potential due to utilization of sugars during the period of early growth of protoplasts helps to avoid a sudden change when regenerated cells are to be transferred to osmoticum-free medium for continued growth.

The isolation of protoplasts has also been done at a relatively low osmotic potential. This was possible by inclusion of a polymer,[260] 2% polyvinyl polypyrrolidone (PVP), along with 0.2 *M* sucrose.

Preplasmolysis of the tissue can help overcome the harmful effects of toxins present in cell wall-dissolving enzymes. When protoplasts prepared from turgid cells were subjected to enzymes associated with soft rot disease, there was a marked increase in the permeability of protoplasts to electrolytes. This resulted in the death of protoplasts. However, when the protoplasts were prepared from preplasmolyzed cells they were resistant to enzymes.[124]

Although plasmolysis is essential for an effective isolation of protoplasts by the enzymatic method, direct microscopic observations of process are often lacking and may reveal the reasons for negative results. Mesophyll cells of *Glycine max* were difficult to plasmolyze in mannitol.[165] This depended on plant age and cell type. Cells of the second palisade layer were more difficult to plasmolyze than those of the first layer of spongy parenchyma. No plasmolysis occurred from podding stage plants (16 weeks old). However, brief treatment of cells with 0.3 mM octylguanidine (OG) in 0.4 M mannitol decreased plasma membrane adherence to the cell wall and enhanced plasmolysis in all cell types except starch-filled palisade cells. Light starvation and OG treatment of tissue or 0.75 M CaCl$_2$ as plasmolyticum improved plasmolysis. These results indicate that tissues difficult to plasmolyze are either due to the strong adherence of the plasma membrane to the cell wall or the mechanical resistance to contraction as a consequence of starch accumulation in chloroplasts.

3. Enzymes

The plant cell wall is a complex structure comprising cellulose fibrils coated with a monolayer of hemicellulose which is xyloglucan in dicots and arabinoxylan in monocots. To degrade such a complex structure, a mixture of enzymes is required with cellulase as the principal component. In nature, the degradation of vegetable matter comprised essentially of cellulose is due to the activity of microorganisms. The enzymes employed for the isolation of protoplasts are also of microbial origin.

A wide range of commercial enzyme preparations are available, under different trade names. For example, the commonly used fungal cellulases and hemicellulases are cellulase Onozuka R-10, cellulysin, driselase, Meicelase P, Rhozyme HP-150, and hemicellulase. It is possible to isolate protoplasts in a relatively short period with driselase. This enzyme is obtained from basidiomycetes and is rich in cellulase and pectinase. Cellulase R-10 also contains a good amount of pectinase.[216,267] Cellulase-RS, a new cellulase from Onozuka, is more potent[194] for the release of bundle sheath protoplasts in C$_4$ plants.

Among the pectinases, Macerozyme R-10 is commonly employed. Pectolyase Y-23 has recently been introduced. This is more potent,[200] and used along with cellulase, it was possible to isolate mesophyll protoplasts of tobacco within 25 min. Pectolyase Y-23 contains an additional maceration factor not found in other macerozyme preparations. This enzyme has proved to be very effective for the isolation of mesophyll protoplasts from *Vicia*[253] and several species of *Medicago*.[139]

The first reaction in enzymatic isolation of plant protoplasts is the maceration of plant tissue which results in cell separation. This is possible due to the breakdown of intercellular cementing material. Endotypes of polygalactouronase and pectin lyase, which randomly cleave the α-1,4-galactouronide linkages in pectic polysaccharides, are the primary enzymes responsible for tissue maceration. These results led to the conclusion that pectic polysaccharides are the principal binding agents of higher plant cells. This is true for many dicot tissues, but not always for graminaceous tissue. For many graminaceous tissues it has been shown that pectic polysaccharides are a very minor component of the cell wall. Therefore, cellulase preparation alone was found to be effective for the release of protoplasts from tissues of Gramineae,[250] suggesting that cellulolytic activity may be all that is required for isolation of protoplasts. Thus, it is not known what kinds of cell wall components play a role in cementing cells in Gramineae. To an extent, this was resolved when treatment of grass leaves with purified pectin lyase from *Aspergillus japonicus* and a purified xylanase from *Trichoderma viride*, which is specific for methyl-galactouronide linkages, led to the isolation of some single cells. A mixture of the two was found to be synergistic. Analysis of components released from oat cell wall by these enzymes indicated[136] that both homo-galactouronans with a high degree of esterification and a kind of glucuronoarabinoxylan with a ferulic acid ester may play a role in cell wall cementing of monocots.

Due to the complex nature of the cell wall and rather impure preparations of commercially available enzymes, it is advantageous to use a mixture of enzymes for the isolation of protoplasts. A combination of four enzymes (cellulase, pectinase, driselase, and rhozyme) was employed for the isolation of cereal protoplasts.[172,295] However, a mixture of enzymes is not always helpful. Pectinases inhibited the release of protoplasts from coleoptiles of *Avena*.[243] A new trend is to have a mixture of enzymes[239] or new enzyme combinations.[218] This is supported by an example of the isolation of protoplasts from *Glycine max*.[171] A mixture of 4% Meicelase, 2% Rhozyme, and 0.3% Macerozyme R-10, found suitable for isolation of protoplasts from seedling cotyledons of several genera, failed to release protoplasts from soybean seedlings, but released protoplasts from immature soybean seeds. These results suggest some basic difference in the composition of walls of cotyledon cells of immature seeds and seedlings of soybean and between soybean seedlings and other plants.

For many workers, commercial preparations of cellulase and pectinase are not enough to obtain protoplasts from all tissues. For example, aleurone cells of barley when treated with cellulase still retained a thin wall which could be dissolved by glusulase.[274] Glusulase is a commercially available snail enzyme preparation. Another enzyme preparation from snail is helicase, rich in β-1-3 gluconase which can digest callose wall. Using this enzyme, protoplasts were obtained from pollen tetrads[20] while they were still enclosed in callose wall and did not have sporopollenin.

Commonly employed enzyme preparations Onozuka cellulase and macerozyme are crude preparations. Of these, cellulase could be resolved into three fractions, one of which could be further resolved into five peaks showing cellulase activity.[283] In addition, the cellulase preparation contains cellobiase, xylanase, gluconase, pectinase, lipase, phospholipases, and various nucleases. Traces of β-1-3 gluconase and chitinase activity can also be seen. Because of these enzyme complexes, Onozuka cellulase is a potent cell wall-degrading enzyme. Concentration[44] of enzymes employed for isolation of protoplasts is a critical determinant of protoplast viability.

Partial purification of enzymes by elution through Sephadex G-25 is helpful. It shortens the incubation period, increases the viability of protoplasts, and enhances regeneration frequency.[118,146,219,246,251,257] Highly purified preparation of cellulase is, however, less potent than the commercial preparation and is unable to digest the complex cell wall which contains hemicellulase, pectin, lipid, protein, and cellulose. Enzyme complexes purified of toxic substances, such as nucleases and impurities,[124] are therefore more effective. Highly purified preparation[7] resulted in fewer surviving protoplasts.

Desalting of enzyme preparation (Onozuka cellulase) resulted in a tenfold increase in activity.[146] For this, about 15 g of cellulase was suspended in 0.5 M NaCl solution overnight. To remove the undissolved material, the preparation was centrifuged. The enzyme solution was loaded onto a Bio-gel® P6 column and eluted with water. Proteins elute in void volume and salts in approximately twice the void volume. The fractions which precede the salts are bulked together and freeze dried and can also be preserved. Desalting of driselase also resulted in increased frequency of protoplasts from sugarbeet.[18]

4. Isolation Medium

Although protoplasts can be isolated in an aqueous solution of osmoticum and enzymes, the need for a large number of intact and viable protoplasts has resulted in many additions to the isolation mixture.

A positive influence on yield and stability of protoplasts has resulted from inclusion of calcium (2 to 6 mM)[31,67,86,146,170,241,243] in the isolation mixture. In addition to being an essential constituent of the cell wall, calcium preserves the membrane integrity of protoplasts. An effect similar to that of Ca^{++} is possible with Mg^{++}.[241,280] Many workers prefer to dissolve enzymes in a special salt solution or even a nutrient medium. Phosphate[105] is also beneficial for the preservation of protoplast viability. Optimal yield of mesophyll protoplasts from grape[314] was possible on inclusion of 2,4-D (5 ppm), BAP (1 ppm), and 1/10 strength MS salts in isolation mixture.

In the earliest studies[201,202] on isolation and regeneration of protoplasts, potassium dextran sulfate was included in the isolation mixture. It is supposed to remove the deleterious proteins from the enzyme sample and improve the yield of protoplasts. Subsequent workers have also recorded beneficial effect of potassium dextran sulfate.[26,160,196,212,232]

Another substance found to be beneficial is PVP. It enhances the yield of protoplasts and helps release them at lower osmotic potential.[260,261] It is very likely that PVP is helpful by means of adsorption of phenolics present in enzyme mixture or released from tissue during isolation of protoplasts. However, PVP is harmful for petunia[26] protoplasts. For the isolation of photosynthetically active protoplasts it is essential to have 10 mM EGTA or EDTA[270] in isolation medium. Another substance to be added to this list is 2-mercaptoethanol.[313] Recently, glycine[217] is said to have a stabilizing effect on leaf protoplasts of *Cucumis sativus*.

As for the volume of isolation medium, about 10 mℓ/g of tissue is sufficient.

5. Incubation Conditions

In an enzymatic isolation of protoplasts, the most important conditions affecting release and survival of protoplasts are temperature and pH of the incubation medium.

The optimal pH values for commonly employed Onozuka enzyme preparations Cellulase R-10 and Macerozyme as given by the manufacturer are 5 to 6 and 4 to 5, respectively. Therefore, a range from 4.8 to 7.2 has been employed by different workers.[117,196,301] However higher pH value, above 6.00,[220] has been recommended because of the better survival of protoplasts. A possible explanation for the promotive effect of higher pH may be the reduced activity of deleterious enzymes released during digestion of tissue or present in the isolation mixture. To avoid changes in pH, inclusion of 2(*N*-morpholino)-ethanesulfonic acid (MES) buffer (3 mM) is recommended.[35,99,172,261,295] MES buffer was also essential for obtaining higher yield and increased viability of grape mesophyll protoplasts.[263]

The optimal temperature for Onozuka enzymes is 40 to 50°C. This range is too high for plant cells. In general, the time of incubation for release of protoplasts is inversely proportional to temperature. However, protoplasts isolated in a shorter period at high temperature are not suitable for culture. They turn brown and tend to burst. Therefore, the temperature[99] generally employed is 23 to 32°C. However, lower temperature has been used — 3 to 5°C for tobacco[68] and 12°C for *Solanum melongena*.[325] Some workers have used a combination of low and high temperature.[226,294,295]

Incubation is normally done in the dark, but in certain cases exposure to light has been found to be beneficial.[26,28,205]

During incubation, occasional shaking at low speed facilitates the penetration of enzymes.

6. Tissue Pretreatment

The beneficial effect of preplasmolysis of tissue has already been described. Recently, certain other tissue pretreatments affecting isolation and subsequent behavior of protoplasts that have come to light are hormone pretreatment and cold conditioning of tissue.

A preculture of tissue (leaves of *Vicia narbonensis*[66] and *Medicago*[149] and cotyledons of some gymnosperms[64,152]) in hormone medium for 1 to 2 days greatly increased the regeneration of isolated protoplasts. Protoplasts isolated from precultured cotyledons of *Brassica oleracea* cv. *botrytis* divided to form colonies, whereas protoplasts from mesophyll cells from plants grown in vivo or in vitro failed to divide.[298] The preculture may condition the cells for more efficient isolation and increased viability of protoplasts.

However, more interesting is cold conditioning of potato leaves at 4°C for 16 to 24 hr which enhanced the stability of protoplasts.[261] This effect was perceived even by isolated protoplasts. In *Lycopersicon*, chilling of protoplasts at 7°C stimulated their frequency of division.[197]

Of the two pretreatments, cold conditioning of cotyledons of *Cyamopsis*[248] at 10°C for

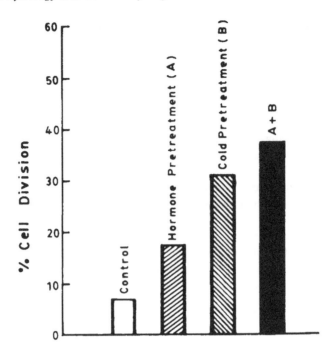

FIGURE 2.   Effect of different pretreatments on division frequency of protoplasts of *Cyamopsis tetragonoloba*. (From Saxena, P. K., Gill, R., Rashid, A., and Maheshwari, S. C., *Z. Pflanzenphysiol.*, 106, 277, 1982. With permission.)

48 hr was more effective than preculture on hormone medium. The protoplasts isolated from cold-conditioned cotyledons were more stable and divided quickly (Figure 2). A combination of hormone pretreatment and cold conditioning slightly improved the response. The effect of cold conditioning of material for increased regenerative capacity of protoplasts should be investigated in greater detail. It is of interest to know that cold treatment[273] of entire plants of *L. esculentum* and *L. peruvianum* significantly increased the frequency of protoplast regeneration. Cold conditioning of cotyledons of *Dalbergia sisso*[247] also increased the frequency of division.

III. HARVEST, PURIFICATION, AND VIABILITY

Protoplasts are harvested by removing the undigested tissue through filtration, using either miracloth, nylon filter, or stainless steel.

The protoplasts are also to be washed of enzyme solution. Enzyme is removed by centrifugation at low speed. In order to handle a dense protoplast suspension with ease, in place of centrifuge tubes, milk test Babcock bottles have been recommended,[260,261,312] but others have found narrow graduated centrifuge tubes to be better than Babcock bottles.[35] Complete removal of enzymes requires repeated washings, employing centrifugation, and this results in a heavy loss in yield. To overcome this, gravity sedimentation[63] and discontinuous density gradient centrifugation[134] have been tried. Centrifugation can be avoided for washing protoplasts if a fleshy tissue such as stem or hypocotyl is the source material.[17] The digested tissue, while still intact, is first washed off the enzyme by repeated addition of new medium and is removed by Pasteur pipette. Protoplasts are then released by gentle teasing of tissue and harvested by filtration.

Purification of intact protoplasts from broken cells and debris is possible by (1) flotation on dense sucrose,[79,107,116,227] (2) fractionation in percoll discontinuous gradient,[100,255] (3) banding between the two-phase aqueous system,[143,164,267] and (4) isoosmotic gradient.[129]

Flow cytometry and cell-sorting techniques, employed for animal systems, offer a new system for characterizing, identifying, and selecting plant protoplasts regardless of their origin. Individual cells can be analyzed at a very high rate by passing them in a liquid stream through a laser beam. In principle, the operation of a flow cytometer cell sorter involves the passage of cells through a flow chamber containing an orifice of 70 to 100 μm. The resultant precise fluid stream intersects the focus of a turnable laser. The cells, in passing through the focus, absorb light and, if they contain fluorochrome, remit the light in the form of fluorescence. The degree of fluorescence within an individual cell can be quantified. Cell sorting involves the operation of a piezoelectric crystal attached to a flow chamber. This causes the fluid stream to break into precise droplets. On analysis of fluorescent signals generated, it is possible to program the microprocessor to recognize only those cells that display a desired degree of fluorescence. In this way individual cells can be separated from a large heterogenous population. Up to 2000 intact protoplasts of *Euphorbia lathyrus*[235] could be sorted and recovered within 1 hr. These protoplasts produced callus and regenerated shoots indicating nontoxicity of fluorochromes. Protoplasts of tobacco were also sorted employing a flow cytometer cell sorter. These protoplasts readily regenerated. The procedure was based either on chlorophyll autofluorescence[127] or staining with fluorescein[96] isothio-cyanate. Flow sorting has been utilized for selection of fused protoplasts and also for selection of protoplasts having alkaloid content.[42]

A new method[272] introduced for purification, separating of contaminating cells from protoplasts, is passage through a Sepharose 6 MB cyanogen-bromide-activated macrobead column coupled with cellulase-RS. In a single passage through this column, contamination of protoplasts by cells possessing partial to complete walls was reduced from 25% to near zero. Coupled macrobeads were easily recovered, washed free of cells, and stored for repeated use. Corn protoplasts appeared undamaged and rapeseed protoplasts, which were passed through the column, divided and formed colonies.

The viability of protoplasts can be tested by the dye exclusion method, using either Evan's blue[134,143] or phenosafranin.[134,151,241,287] The living protoplasts remain unstained, whereas dead and broken protoplasts take up the stain. This method is suitable for protoplasts obtained from cell suspension. Another method to test the viability is to use the vital stain fluorescein diacetate (FDA). On hydrolysis of FDA, possible by esterases from protoplasts, the fluorescein moiety released accumulates in a viable protoplast due to its intact membrane and causes it to fluoresce in UV light (Figure 1B). This method is especially suitable for mesophyll protoplasts. Instead of FDA, fluorescein isothiocyanate (FITC) can also be used.[35]

The main factors affecting viability of protoplasts are osmotic potential of the medium and enzyme concentration.[44]

IV. PROTOPLAST TO CELL — A TRANSITION

The formation of protoplasts from cells is a process of dedifferentiation. It is an interesting study from a fully differentiated to a dedifferentiated state. Protoplasts are different from cells in a number of ways: (1) protoplasts are without a cell wall, (2) protoplasts are under osmotic stress, (3) protoplasts are removed from their original cell environment and cell to cell contact through plasmodesmata is snapped, (4) protoplasts are in a new solute environ-ment, (5) protoplasts lack the proteins present at the interface of the plasma membrane and cell wall, and (6) protoplasts are likely to be damaged by toxic effects of enzymes used for isolation.

At the subcellular level, changes in chloroplast structure, presence of paracrystalline arrays, multimembrane inclusions, and other malformations are reported in freshly isolated protoplasts. Biosynthesis of large subunit of fraction I protein, which is chloroplast coded,

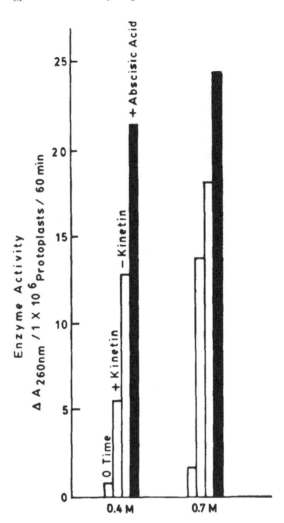

FIGURE 3. Ribonuclease level in protoplasts of tobacco at the time of completion of isolation (0 time). After 48 hr with or without kinetin, ABA, at different osmolarities of medium. (From Premecz, G., Olah, T., Guylas, A., Nyitrani, A., Palfi, G., and Farkas, G. L., *Plant Sci. Lett.*, 9, 195, 1977. With permission.)

is not detectable in isolated protoplasts of tobacco.[85] Studies on proteins synthesized in vitro by messenger RNA extracted from tobacco protoplasts showed that the changes in protein synthesis and especially the lack of ribulose 1,5 bis-phosphate carboxylase did not result from a failure of translation.[83] It is not certain whether these changes are due to the osmoticum employed for isolation of protoplast or are a result of isolation. It is, however, very clear that there is a reduction in ribosomes and organelles and there is also a disappearance of polysomes. The cytoplasm that is clear at isolation is later filled with polysomes and other cytoplasmic constituents during division.[106,245]

Protoplasts are under osmotic stress. In a freshly isolated protoplast there is an increase of RNAse,[221] irrespective of whether the protoplast is isolated by a mechanical or enzymatic method. The level of RNAse activity in tobacco protoplasts could be reduced by cycloheximide and kinetin (Figure 3), indicating thereby that it is due to stress-induced synthesis of new proteins.[229] In fact, protoplasts of *Nicotiana sylvestris* do produce osmotic shock-

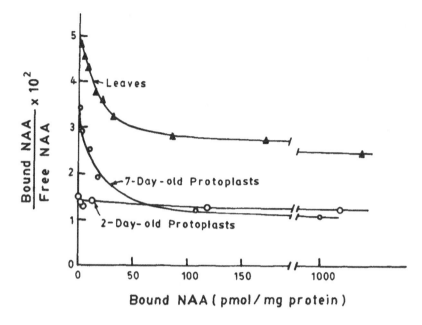

FIGURE 4.   Auxin binding by tobacco protoplasts, Scatchard plot of NAA binding by particulate fraction from leaves, 2- and 7-day-old protoplasts. (From Vreugdenhil, D., Harkes, P. A. A., and Libbenga, K. R., *Planta*, 150, 9, 1980. With permission.)

induced stress proteins.[84] During culture of protoplasts there is steady incorporation of leucine, uridine, and thymidine, indicative of active protein and nucleic acid syntheses, but these are strongly reduced by stress.[230] Also, the protoplasts from *Centaurea* cell culture differed from cells in several respects, such as poly A$^+$ RNA profiles, level of rRNA synthesis, and rRNA processing.[161]

In brief, it can be said that isolation of protoplasts, a process of dedifferentiation, is marked with changes in RNA pattern and content and there is also a change in ribosomal content. Even the protoplasts from oat, which did not divide, showed an increased incorporation of precursors into protein, RNA, and DNA, and this was promoted by polyamines and to an extent by kinetin which basically inhibits senescence.[98]

There is also a basic difference between protoplasts and cells in respect to auxin-binding sites. In freshly harvested protoplasts, no specific binding site for NAA could be detected[299] in vitro (Figure 4), whereas it was present in a particulate fraction from tobacco leaves. From this it can be concluded that NAA binding sites are probably located at the external face of the plasma membrane and are destroyed during protoplast isolation by proteolytic enzymes in the protoplast isolation mixture. On culture of protoplasts for 3 to 4 days, first cell division was observed and at the same time specific NAA binding sites were detectable (Figure 5).

V. CULTURE

The problems encountered in the culture of protoplasts differ in three important respects from those of cells.

First, protoplasts require an osmotic stabilizer, in the culture medium, until a wall is synthesized. The presence of an osmotic stabilizer for protoplasts is, however, the cause of stress, and hence the concentration of osmotic stabilizer should be as low as to maintain the stability of protoplasts.

Second, the plasma membrane of protoplasts is relatively more leaky than that of cells.

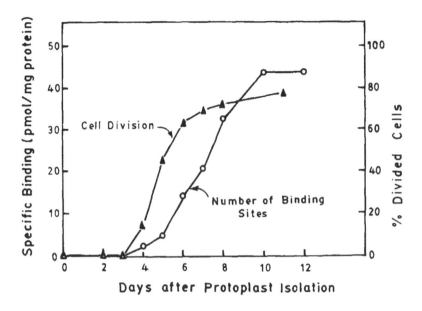

FIGURE 5.   Auxin binding by tobacco protoplasts, cell division, and number of binding sites as a function of time. (From Vreugdenhil, D., Harkes, P. A. A., and Libbenga, K. R., *Planta*, 150, 9, 1980. With permission.)

This is inferred from an experiment on culture of protoplasts requiring additional substances than those needed for cells from which the protoplasts are derived. Crown gall cells of *Parthenocissus tricuspidata* do not require any growth regulator, but protoplasts obtained from these cells required growth regulators, and this requirement ceased soon after the synthesis of cell wall.[254]

Third, due to the delicate nature of plasma membrane, protoplasts are fragile structures, requiring careful handling during culture. The protoplasts are to be transferred from one medium to another with a Pasteur pipette.

A. Methods of Culture

The methods employed for the culture of protoplasts (Figure 6) are basically similar to those employed for culture of cells. However, some minor modifications are necessary in view of the delicate nature of protoplasts. The choice of a particular method is dependent upon (1) the object of the experiment and (2) supply of protoplasts. Protoplasts can be cultured in one of the following ways.

1. Suspension Culture

In the original work[201] which demonstrated the regeneration potential of plant protoplasts, cultures were raised in the form of a suspension in a small volume of liquid medium at a density of 10^5 protoplasts per milliliter in an Erlenmeyer flask. Since then, this procedure has been adopted either in the same or in a modified way. Instead of flasks, 2 to 5 mℓ of protoplast suspension is dispensed in a petri dish of appropriate size (3- to 9-cm diameter) so as to maintain a shallow layer, approximately (0.1 mm). This procedure is routinely employed and has proved to be successful in regeneration of protoplasts of many plants such as *Petunia*,[26,223] *Datura*,[252] *Nicotiana* spp.,[10] *Hyoscyamus*,[170] and a number of other dicot plants.[30,31]

Liquid culture of protoplasts in a petri dish can be regularly monitored for divisions. Liquid cultures are advantageous because it is possible to change: (1) osmotic potential of the medium (a gradual decrease after 3 to 5 days promotes rapid proliferation of cells), (2)

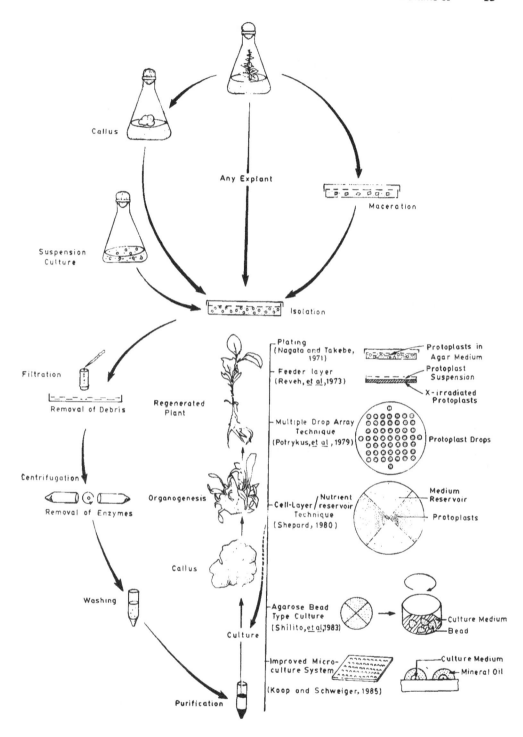

FIGURE 6. Summary diagram of enzymatic isolation, purification, methods of culture, and regeneration of protoplasts into plants.

the medium itself, in the eventuality of any toxicity produced in it due to death of a large number of protoplasts, and (3) the density of cells by incorporating fresh medium. Occasionally, it has been found that protoplasts of some species divide only in liquid medium[113,114,308] and fail to do so in agar medium.

A serious limitation is the bursting of protoplasts in liquid medium in a petri dish, particularly when a plastic dish is used. This can be alleviated by addition of a low concentration (0.04 to 0.02%) of Tween-80®.[58,184,185] However, in tobacco[48] the detergent was found to be inhibitory for the growth of protoplasts.

Using liquid medium, the protoplasts can also be cultured on a fabric support or on filter paper.[152] In addition to the benefit of changing the medium, the filter paper culture of protoplasts is better for growth. It was considered to be one of the key factors responsible for regeneration of protoplasts of the legumes *Medicago sativa*[246] and *Trifolium resupinatum*.[211]

For the liquid culture of protoplasts in a flask or petri dish, a large volume of protoplast suspension is required. However, liquid culture using a small volume is possible in one of the following ways.

a. Single Drop or Multiple Drops

Hanging drop culture of protoplasts is possible by placing a small single drop[147] or multiple drops[225] of protoplast suspension on the inner side of the lid of a petri dish which when inverted brings the culture drops in hanging position. This is also possible by using a cavity slide.[13] In this procedure a drop of protoplast suspension is placed on a sterile coverglass. This is in turn inverted onto a cavity slide and sealed with sterile paraffin oil. In this way, progress of an individual protoplast can be followed. Instead of hanging drops, a simpler way is to use standing drops[5,146,234,257,291] of protoplast suspension.

The drop size can range from 40 to 100 $\mu\ell$ to as small as 20 $\mu\ell$.[225] Using multiple drops of small volume, many media combinations can be tested at one time with a few million protoplasts. Drop culture is better for aeration of protoplasts because of an increase in surface-to-volume ratio.

b. Microchamber Culture

Regeneration of either an individual protoplast or a small number of protoplasts of tobacco[292,293] and petunia[72] could be followed in microchambers which were set up in a manner essentially similar to the one described for culture of cells.

c. Microdroplet Culture

In droplet culture, the smallest volume previously employed was 25 $\mu\ell$.[147] However, it has been possible to culture protoplasts in a volume as small as 0.25 to 0.50 $\mu\ell$[110] using a cuprak dish which has built-in wells capable of accommodating the required volume of culture medium. The regeneration frequency of tobacco protoplasts on culture in microdroplets was comparable to that obtained in a macroculture. Microdroplet culture with a cuprak dish is especially suitable for the culture of an individual protoplast or selectively isolated cell hybrid. In principle, it is the culture of a few protoplasts in a very small volume and is comparable to a large number of protoplasts in a large volume.

More recently,[159] an improved microdroplet culture system (Figure 7) has been devised for culture of individual protoplasts. Each microdroplet (20 to 80 $\mu\ell$) is contained within a separate drop of mineral oil (1 $\mu\ell$); 50 such microdroplets could be placed on a cover glass. In this way, complete physical separation of the individual protoplast is possible, thus excluding diffusion of substances between aqueous and lipophilic phases. This method is an improvement over the method practiced for culture of individual cells (see Volume I, Chapter 1 of this book). Details for the preparation of microculture of protoplasts are given in Figure 7B.

2. Plating of Protoplasts

This method is essentially similar to that employed for culture of cells.[15] The plating technique was adopted for culture of mesophyll protoplasts of tobacco.[202] Protoplast sus-

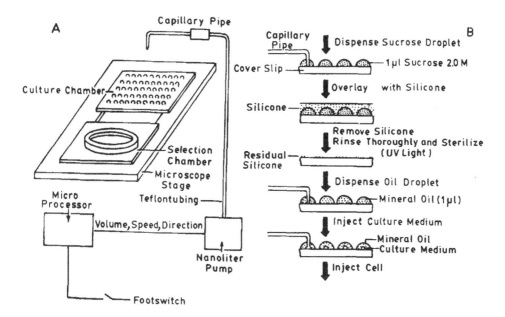

FIGURE 7. Diagrammatic representation of set up for transfer of single cells/protoplasts and method for preparing medium microdroplet for culture of individual protoplasts. (From Koop, H-U. and Schweiger, K-G., *J. Plant Physiol.*, 121, 245, 1985. With permission.)

pension at an appropriate density is mixed gently with an equal volume of nutrient medium having 1.2% agar kept molten at 45°C and then dispensed in petri dishes, forming a thin layer of agar. The dishes are sealed with parafilm or cellotape and kept upside down for incubation at appropriate conditions for regeneration of protoplasts into cells.

The plating of protoplasts enables not only the growth of a large population, but it is also possible to mark and observe an individual protoplast. However, in agar medium it is not possible to change the medium during the growth of protoplasts. This limitation can be overcome when, instead of agar, agarose (Sea-Prep) is employed; it can facilitate an efficient transfer of calli.[1] Using agarose it was possible to culture tobacco protoplasts at a lower density than on agar. Protoplasts of *Nicotiana tabacum* developed into colonies from lower initial population densities in agarose than in agar or liquid. Protoplasts from *Hyoscyamus muticus*,[262] which did not divide in agar, formed colonies in agarose at higher efficiency than in liquid medium. Also, the protoplasts of *Glycine canescens* had to be cultured on agarose medium for initiation of division and were then transferred to agar medium.[210]

In the conventional plating technique, described above, using either agar or agarose, protoplasts are inadvertantly exposed to a temperature shock. Therefore, instead of agar/agarose as a jelling agent, sodium alginate[180] has been recommended. Sodium alginate, a stabilizing colloid used in the ice cream industry, can be gelled by the addition of Ca^{++} and again liquified with a chelate (sodium citrate). This can be performed at room temperature and thus the procedure, besides avoiding temperature shock, also eases the recovery of regenerants as compared to the agar/agarose medium wherein the colonies are embedded. On alginate medium colony formation by protoplasts of tobacco and petunia was comparable to agar medium.

Innovation of the plating technique permits the growth of protoplasts even at a very low density. In one such instance, protoplasts were cultured at a low density of 10^2 protoplasts per milliliter provided they are grown on a feeder layer of X-irradiated nondividing protoplasts.[233,289,290] With this technique it was possible to culture protoplasts even at a lower density, as low as 5 to 50 protoplasts per milliliter.[232] Irradiated carrot cell suspension,[54] when employed as feeder layer, could support the growth of *Nicotiana* protoplasts/cells. Nurse culture[181] was also used to clone somatic hybrids and for the culture of protoplasts of oil palm, *Elaesis guineensis*.[12] Use of a feeder layer for the culture of protoplasts at low density can be made more effective when feeder cells are embedded in agar medium and it is overlaid with cellophane membrane,[264] on which cells to be nursed are plated. This method allows for transfer of cells without disturbance and washing. Being transparent, the membrane allows for an easy microscopic follow-up of feeder layer and cells to be nursed. More convenient is the use of cellulose nitrate filter.[173] On pipetting 0.2 mℓ of protoplast suspension of maize over such a filter, placed on top of an agarose-solidified medium in which suspension cells were embedded, it was possible to show colony formation by 10% of protoplasts. This was 100- to 1000-fold higher than the conventional plating method.

Another innovation of the plating technique is the reservoir system.[258] For this, in a specially designed petri dish, the protoplasts are plated in a cell layer medium in diagonally opposite sectors; the other two sectors serve as reservoir for medium of different composition. This method permits an exchange of media components through the reservoir layer and has been found suitable for the culture of protoplasts by many workers.[22,120,256,259] When activated charcoal was included in reservoir medium in x-plates, the regeneration ability of protoplasts of *Solanum tuberosum* was significantly increased.[51]

Yet another innovation of the plating technique is to culture protoplasts in twin layers of agar-liquid medium.[175] Employing this technique only, it was possible to culture protoplasts isolated from cells of *Lithospermum erythrorhizon*.

A combination of gel and liquid medium seems to be more effective for the culture of protoplasts (Figure 6). Gel-embedded protoplasts[262] on suspension in a large volume of liquid medium divided at an increased frequency in *Lycopersicon* and *Crepis* and enabled sustained proliferation of protoplasts of *Brassica* and *Petunia* which had not previously developed beyond a few-cell stage. This method was found effective for increasing the frequencies of division of protoplasts of *Alnus*,[284] *Medicago*,[132] and *Cucumis*.[217]

VI. CELL REGENERATION

The two critical steps of morphogenesis occurring during cell regeneration from protoplasts are cell wall formation and cell division.

A. Cell Wall Formation

On culture, a viable protoplast soon starts regenerating a new cell wall. In fact, it has been suggested that wall synthesis initiates immediately after the removal of enzymes. It has been demonstrated as early as 10 min after culture in *Vicia hajastana*,[309] whereas in other genera, such as mesophyll protoplasts of tobacco, cell wall deposition is preceded by a lag period of at least 7 to 10 hr. It may be prolonged to 16 to 24 hr.[46,119,310] However, the lag may be longer, 45 hr in *Antirrhinum majus*.[47] The lag period in the initiation of cell wall seems to be due to the adjustment of protoplasts to the new medium. However, once the fiber formation begins, a dense mat is possible within 2 to 4 hr.

Protoplasts isolated from cells in culture regenerate cell wall more readily than those isolated from a differentiated tissue such as mesophyll.[291] After the synthesis of cell wall, protoplasts lose their spherical shape.

To begin with, there is deposition of cellulose microfibrils[87] at the surface of the plasma membrane, which are loosely organized and later become organized to form a typical wall. Studies on incorporation of labeled precursors have shown that protoplasts synthesize cellulose and a large number of polymer components, some of which may be liberated into the medium.[9,125,278] The onset and accumulation of radioactivity into cellulose coincides with the appearance of fibrils on the surface of protoplasts, as seen in an electron microscope.[153]

However, there is evidence to the contrary that the cell wall of regenerating protoplast comprises mainly noncellulosic polysaccharides with glucose predominating (65%) and only a small fraction of cellulose (5%),[34] whereas the cell wall of an isolated mesophyll cell comprises 60% cellulose.

The role of specific organelles in cell wall synthesis is not clear. In protoplasts of *Skimmia japonica*,[238] only the endoplasmic reticulum is considered to be involved and the role of Golgi apparatus is ruled out. The plasma membrane regulates the transport between apoplast and symplast and participates in the deposition of the cell wall. A technique developed for examining the inner surface of an algal plasma membrane has been applied to higher plant protoplasts for the study of associated organelles which may be important in wall formation. In particular, coated vesicles and patches of coat material are seen on the inner surface of the plasma membrane of mesophyll protoplasts of tobacco.[90] The frequency of coated vesicles in mesophyll protoplasts was relatively less than that seen in protoplasts from cell suspension culture. Coated pits and coated vesicles are distinctive ultrastructural features of plant protoplasts. Coated membranes are associated with the plasmalemma and dictyosome cisternae. They have also been observed in association with smooth membrane structure. It has also been possible to isolate coated vesicles from protoplasts of soybean.[182] Coated pits and vesicles play an important role in the receptor-mediated endocytosis of specific macromolecules in animal cells. Other roles currently envisaged for these structures are in membrane recycling and intracellular transport. Similar inference can be derived for plant cells.

A simple method[137,201] for monitoring the synthesis of the cell wall is the staining of protoplasts with a cellulose specific stain, calcofluor white-ST (American Cyamide, N.J.). It is an optical brightener that binds to cellulose and fluoresces in UV light. The protoplasts are incubated for about 5 min in 0.1% solution of calcofluor in an osmoticum and then washed to remove the excess stain. The preparation is then examined in UV light, using a fluorescence microscope. Freshly isolated unwalled protoplasts do not show any fluorescence, whereas protoplasts forming a cell wall show different degrees of fluorescence (Figure 8). Instead of calcofluor, Tinopal Bopt (Geigy, U.K.) or Ranipal (Surhid-Geigy, India) can be used. For confirmation of cell wall, subcellular techniques of silver hexamine staining,[88] freeze-etching,[311] and platinum-palladium replica[309] have been employed.

A microfluorimetric procedure[95] using calcofluor white has been developed for measurement of cellulose biosynthesis on protoplasts of *Nicotiana*. The intensity of fluorescence emitted following calcofluor treatment is a specific measurement of the cell wall cellulose level. The procedure is relatively easy and sensitive up to the picogram range, as compared to the conventional method of cellulose estimation by anthrone reagent on exhaustive extraction of developing cell wall.

For the regeneration of cell wall only a carbon source is required; in its absence, cell wall synthesis did not occur on protoplasts of *Convolvulus*.[133] For quick and uniform deposition of cell wall, polyethylene glycol (PEG 1500) was helpful for protoplasts from carrot[301] cell suspension.

The physical environment of the culture is an important factor affecting cell wall formation. Compared to liquid medium, cell division was stimulated when protoplasts of *Vinca rosea* were cultured on agar[278] and their regenerated cell walls had a composition similar to that in nature.

The nature of the osmoticum specifically affects the regeneration of the cell wall. When salts were used as osmotic stabilizer instead of mannitol, protoplasts failed to form a rigid cell wall and this was also not accompanied by sustained cell division.[184,185] Therefore, there may be a relationship between the ability to synthesize a rigid cell wall and the ability to divide. However, an inhibitor of wall synthesis, 2,6-dichlorobenzonitrile (DB) (a weed killer), inhibited production of cellulose, monitored by quantitative fluorescence microscopy, but did not inhibit DNA synthesis or protein accumulation by protoplasts of tobacco.[97,190]

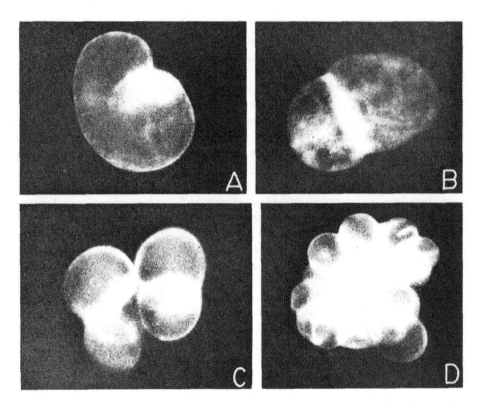

FIGURE 8.   Cell regeneration from protoplasts of *Capsicum annuum* monitored on staining with calcofluor white-ST: (A) regeneration of cell wall and initiation of division, (B, C) bicellular regenerants, and (D) multicellular colony. (From Saxena, P. K., Gill, R., Rashid, A., and Maheshwari, S. C., *Protoplasma*, 108, 357, 1981. With permission.)

Continuous culture of protoplasts in DB resulted in the accumulation of increased capacity of cellulose synthesis, expressed following the removal of inhibitor (Figure 9).

B. Cell Division

Wall formation is followed by the loss of spherical shape of protoplast and the result is an elongate cell. Before division, increased metabolic activity is noticeable in terms of cytoplasmic streaming, respiration, and synthesis of macromolecules. There is also an increase in the number of cell organelles. In mesophyll protoplasts there is rearrangement of chloroplasts, along the cytoplasmic strands, which become scant and yellow. Protoplasts of mesophyll origin are slower to divide than protoplasts from cells in culture.[293] In general, there is no synchrony in division of protoplasts.

The first mitotic division normally occurs within 2 to 5 days. However, the division is delayed up to 14 days in rye[306] and in mesophyll protoplasts of tomato var. Hilda 72 and Rutgers.[196] It can also be 21 days in protoplasts of cotton[21] and protoplasts from lily callus.[266] However, protoplasts of *Gossypium klotzschianum*[81] divided after 3 days. After initiation, due to sustained divisions, within 7 to 20 days multicellular colonies visible to the unaided eye are possible.

At times, multinucleate structures are formed when karyokinesis is not accompanied by cytokinesis.[76,89,195,237]

C. Factors Affecting Division

Apart from a number of factors affecting isolation and viability of protoplasts and cell

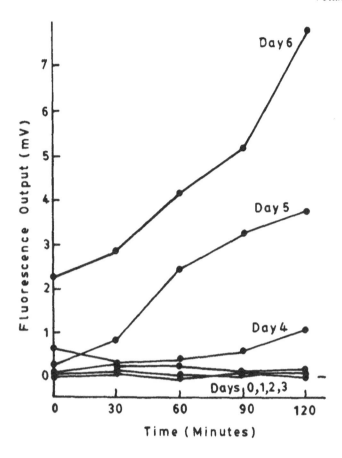

FIGURE 9.　Effect of DB on inhibition of cell wall formation monitored in terms of changes in intensity of calcofluor fluorescence following transfer of protoplasts to medium lacking DB. The protoplasts were precultured in the presence of DB for 0 to 6 days. (From Galbraith, D. W. and Shields, B. A., *Physiol. Plant.*, 55, 25, 1982. With permission.)

wall regeneration, certain other factors that affect cell division are basically the chemical and physical environments of culture and the genetic nature of cells.

1. Physical Environment
The critical determinants of the physical environment are density of protoplasts, light, and temperature.

a. Density
Protoplasts fail to divide if cultured below a certain minimal number. Mesophyll protoplasts of tobacco did not divide at a density below 10^3 units/mℓ.[202] For a good regeneration frequency the protoplasts are generally cultured at a density of 5×10^3 to 5×10^5 units/mℓ.[78,114] A large population is suggested to help detoxify the deleterious substances present either in the culture medium or released during culture of protoplasts.[148]

b. Light
In general, incubation of protoplasts, after culture, in the dark for about a week has been found to be beneficial for wall formation and cell regeneration.[67,81,92,108,170,205,325] Protoplasts of *Nicotiana plumbaginifolia*[108] required darkness for cell regeneration. Maximum plating efficiency could be achieved by keeping the cultures in dark instead of light or a dark/light

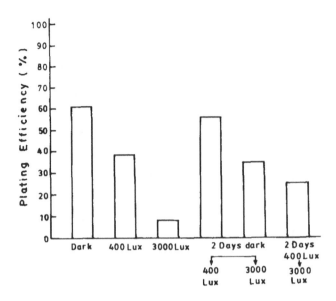

FIGURE 10.   Plating efficiency of protoplasts of *Nicotiana plumbagin-ifolia* in response to different light conditions. (From Gill, R., Rashid, A., and Maheshwari, S. C., *Protoplasma*, 106, 351, 1981. With permission.)

sequence. Only 2 days of dark (Figure 10) prior to illumination resulted in appreciable plating efficiency. The specific role of light, if any, in regeneration of protoplasts remains to be resolved. For mesophyll protoplasts of tobacco[202] high light intensity of 2300 lux was better than 700 lux. In another cultivar of tobacco, preincubation of cultures at low intensity (400 lux) for 48 hr followed by transfer to light at 3000 lux enhanced plating efficiency as compared to that obtained at 3000 lux.[74] The variation in light requirement of different cultivars of tobacco *(N. tabacum* cv. Xanthi[202] and cv. Samsun[74]) indicates that response to light is probably a genetic trait. This is evidenced from work on different species of *Nicotiana.*[10] Protoplasts of *N. tabacum* and *N. sylvestris* were light tolerant and those of *N. otophora* were light sensitive and grew better in the dark. A comparison of F$_1$ hybrids in relation to light revealed that protoplasts of the hybrid *(N. tabacum* × *N. otophora)* were light tolerant and of *N. sylvestris* × *N. otophora* were light insensitive, although overall growth was better in the dark.

Some workers obtained regeneration on illuminating the cultures at 1000 to 4000 lux.[25,211,306] An improvement in division of protoplasts of flax[102] in light was seen. Interestingly, protoplasts of *Vicia faba* were insensitive to light intensity at 800 to 2000 lux,[29] and in *Arachis hypogea* light as well as dark were equally effective.[211]

c. Temperature

Normally protoplasts are cultured at 22 to 25°C; higher and lower temperatures are considered to be harmful. However, protoplasts of *Lycopersicon esculentum* and *L. peruvianum* divided at a maximum frequency at 27 and 29°C. No colony formation occurred at 25°C in *L. peruvianum* and a simple change to 27°C made the difference.[322] Therefore, a detailed study of the effect of temperature is desirable.

More interesting is the chilling effect on the conditioning of protoplasts for division. Chilling of tomato protoplasts at 7°C for 12 hr in nutrient medium triggered a high frequency of division.[197] During chilling there is elaboration of some factor(s) into the medium by protoplasts which improve their viability and regeneration. When protoplasts were washed off the medium in which they were chilled, the effect of chilling was lost. Even cold conditioning of tissue from which protoplasts are isolated enhances the viability and regeneration of protoplasts.[247,248,261] For details, see Section II.B.6.

2. Chemical Environment

Critical chemical factors affecting regeneration of protoplasts are nutrients, osmoticum and carbon source, growth regulators, and pH of the nutrient medium.

a. Nutrient Medium

Protoplasts have special nutrient requirements. Metabolites are easily lost to the external environment, possibly due to exposure of the plasma membrane, and the loss is to be compensated for by active growth and division. Nutrient formulations employed for culture of protoplasts are generally similar to those for cell culture. In most studies, the media employed are formulations or modifications of high salt, Murashige and Skoog (MS),[199] or low salt B_s medium.[101] However, it is essential to have a correct nutrient formulation for the culture of protoplasts because occasionally different results are possible in the same plant. For instance, the Nagata and Takebe (NT)[202] formulation employed for culture of mesophyll protoplasts of tobacco was found to be unsuitable by subsequent workers.[184,185,286] Another example is the protoplasts of *Lycopersicon* spp.[196] The protoplasts of *L. peruvianum* divided, but protoplasts of *L. esculentum* cv. Hilda 72 and Rutgers divided only when the major elements of the nutrient medium were reduced to one half the original concentration and mannitol as osmoticum was replaced by glucose, whereas protoplasts of cv. Retina failed to divide in 15 different modifications of 5 media. At times, one must resort to the use of nurse medium-containing cells to obtain regeneration of protoplasts, as has been done for protoplasts from cell cultures of oil palm.[12]

As for the individual ions affecting regeneration, common modifications have been alteration in the level of Ca^{++} and ammonium. Calcium, an essential component of the cell wall, is considered to preserve the structural and functional integrity of protoplasts. An increase in concentration of $CaCl_2$ increased the frequency of dividing protoplasts of *Vicia hajastana* and *Bromus inermis*[145] and *Brassica oleracea*.[93] However, supplementing the medium with 20 mM NH_4NO_3 reduced the frequency of division in *Vicia* and *Bromus*. The inhibitory effect of ammonium has also been recorded in *N. tabacum*,[183] *Solanum tuberosum*,[261,288] *Salpiglossis sinuata*,[39] and *L. esculentum*.[322] The reduction of ammonium concentration and the use of glucose as carbon source were essential for sustained cell division and colony formation by protoplasts of *Broussonetia kazinoki*, paper mulberry.[214] Contrary to these results, inclusion of ammonium nitrate in the medium has been found to be beneficial for *Pisum sativum*,[7] *Gossypium hirsutum*,[21] and *Solanum nigrum*.[208] To lessen the toxic effect and improve utilization of ammonium, the addition of organic acid such as succinate is recommended.[105]

When ammonium nitrate was replaced by glutamine and serine, and these served as the sole source of nitrogen, the protoplasts of *S. sinuata*[40] divided rapidly. Also, leaf protoplasts of *Chicorium intybus* did not show sustained divisions on MS medium and replacement of ammonium and nitrate by glutamine as the sole source of nitrogen helped increase plating efficiency.[61] Replacement of ammonium nitrate is not required for the regeneration of protoplasts of moth bean *Vigna aconitifolia*, but inclusion of glutamine and asparagine is helpful.[257] Also, for protoplasts of *Rehmannia glutinosa* a mixture of amino acids comprising glutamine, arginine, glycine, and aspartic acid promoted sustained division.[317] Glutamine alone significantly promoted the plating efficiency of protoplasts of *S. brevidens*.[208] In a comparative study of the effect of ammonium on protoplast culture of members of Asteraceae (*Artemisia vulgaris* and *Chrysanthemum indicum*) and *N. tabacum*, this ion was inhibitory for protoplasts of Asteraceae but promotory for tobacco (Figure 11).[215]

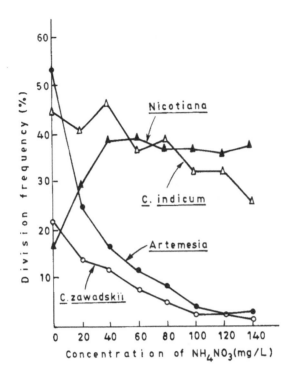

FIGURE 11.   Division frequency of protoplasts of different plants as a function of increasing ammonium level. Protoplasts of *Nicotiana tabacum* are not affected but those of *Artemisia* decreased markedly, whereas those of *Chrysanthemum indicum* and *C. zawadskii* decreased gradually. (From Okamura, M., Hayashi, T., and Miyazaki, S., *Plant Cell Physiol.*, 25, 281, 1984. With permission.)

A beginning has been made towards understanding the nitrogen metabolism of regenerating protoplasts. In tobacco, the addition of glycine caused the accumulation of serine in dark cultured protoplasts via the photorespiratory pathway. Activities of enzymes of nitrogen metabolism (glutamate dehydrogenase and glutamine synthetase) were seen in the photorespiratory assimilation of ammonium. On inclusion of 3 mM glycine, uptake and metabolism of glucose and cell divisions were inhibited, suggesting that the accumulation of serine or release of ammonium during serine synthesis had a toxic effect in this system.[206] Further studies would resolve the role of ammonium in protoplast regeneration.

Another ion specifically affecting protoplast regeneration is iron. The level of iron (Fe-EDTA, 10^{-4} M or 100 µM) as recommended in MS formulation, although employed in several investigations, was inhibitory for culture of mesophyll protoplasts of tobacco.[183,185] Also, in *Pisum sativum*[6] only 50 µM of Fe-EDTA gave optimal results. Since Fe-EDTA in a high amount tends to lower the pH of the medium, changing from Fe-EDTA to Fe-EDDHA turned out to be beneficial for protoplasts of *N. plumbaginifolia*.[109]

b. Osmoticum and Carbon Source

Proper choice of an osmoticum and the correct concentration is not only essential for obtaining a large number of protoplasts, but is also critical for their viability during handling and culture. In general, the same compounds are used for culture as for isolation.

In most of the earlier investigations mannitol was employed. However, better results were obtained with sorbitol, either alone[56,113,308] or in combination with mannitol.[146,148,149,295] Some workers have also used sucrose or glucose either alone or in combination with mannitol.[103,144,260,261] This combination is desirable because sugars are metabolized by the tissue, and in this way, osmotic potential is reduced and the tissue is saved from osmotic shock when it has to be transferred to a medium of low osmotic potential for continued proliferation of cells.

Sucrose alone can also serve as an osmoticum and it is beneficial for culture of protoplasts of *Nicotiana*,[286] *P. sativum*,[6] and *N. sylvestris*.[204] However, in several other studies it was inhibitory and best results were obtained with a combination of several sugars and sugar alcohols.[104,115,145] A new trend is to use sucrose or glucose either alone or in combination with mannitol as osmotic stabilizer for culture of protoplasts[103,144,158,260,261] so that when sugars are metabolized, a reduction in the osmotic level of the medium is beneficial for continued cell division. As an alternative, repeated dilution of the medium in which cell clusters have appeared is emphasized for successful regeneration of protoplasts.[31] A gradual reduction of osmotic level of the medium is also possible using a feeder layer without osmoticum.[158]

A comparative study of several sugars and sugar alcohols as osmotic stabilizers for culture of protoplasts revealed that glucose was best for *Vicia* and *Glycine* protoplasts and sucrose for protoplasts of *Bromus*.[191]

Sucrose was the choice as an energy source in earlier investigations. It is suitable for protoplasts of *Nicotiana*[286] and *Pisum*[6] but unfavorable for *Daucus*[115,301] and *Solanum melongena*.[19] Therefore, in recent studies, instead of sucrose, glucose is preferred for regeneration of protoplasts of *Medicago sativa*,[149] *Lycopersicon* spp.,[196,322] *Pennisetum americanum*,[295] and *Panicum maximum*.[172]

c. *Hormones and Growth Regulators*

Type and concentration of growth regulator needed for culture of protoplasts must be determined empirically. Of the growth regulators, an auxin is required from the beginning of culture for wall formation and in some systems auxin alone can support a high frequently of division and wall formation. This is true for carrot[115] and tobacco.[286] The most frequency employed auxin is 2,4-D and it is essential for *Solanum*.[205] Other auxins are NAA and chlorophenoxyacetic acid, used in combination with a cytokinin.[81,128,170,325] Rarely, indoleacetic acid has been used along with 2,4-D as in *Salpiglossis sinuata*[40] and *Trifolium arvense*.[308]

Generally, cytokinin and auxin are employed together. The essentiality of both growth regulators for division is seen for protoplasts of snapdragon,[222] pea,[8] and carrot.[70] Among the cytokinins, benzyladenine, kinetin, N[6]-dimethylallylaminopurine, and zeatin are used.[19,81,128]

Of the hormones, the presence of an auxin is beneficial for cell wall formation and essential for induction of division.[9,189,275] The type and concentration of auxin for culture of protoplasts should be carefully determined. Auxin may lead to rapid bursting of protoplasts either by increasing the volume of existing vacuoles or by formation of new vacuoles.[291]

Protoplasts of tobacco at all stages of culture and in all culture media incorporate a low level of thymidine into their DNA. However, incorporation of a considerable amount of thymidine, indicative of S phase, takes place only after 24 to 30 hr and requires the presence of auxin and cytokinin. In the absence of hormones, the protoplasts do not synthesize a measurable quantity of chloroplast ribosomal RNA, but actively synthesize ribosomal RNA and polyadenylated RNA. These syntheses are essentially related to the process of aging.[60]

Two-dimensional separation of proteins newly synthesized by tobacco mesophyll protoplasts on culture could be resolved reproducibly into 257 spots.[188] This pattern was stable throughout 3 days of culture; only the intensity of 24 spots varied during this time. The absence of cytokinin in the medium did not modify the pattern but prohibited the entry of protoplasts into S phase, indicating that none of the proteins were synthesized in S_1, G_2, or M phases. However, the presence of 2,4-D was necessary for mitotic development; it induced the appearance of one protein, increased the level of another, and reduced the level of eight others. The proteins, synthesized at a reduced level in the presence of auxin, are proline rich. Their synthesis was no longer inhibited by auxin, if dichlorobenzonitrile, a weed killer which inhibits cell wall reformation of tobacco, was added to the culture medium. Also, the synthesis of two proteins that are formed only when protoplasts are cultured in an auxin medium was not modified by dichlorobenzonitrile.[186] These results suggest that proteins reduced by auxin are related to cell wall formation and do not play a role in the induction of cell cycle. By contrast, proteins whose synthesis is stimulated in the presence of auxin are good candidates for a role in induction of cell cycle. When protoplasts previously cultivated in a medium lacking auxin were transferred to hormone-containing medium, the proteins stimulated by auxin were detected after about 30 min and reached a constant level within 2 to 4 hr. By contrast, proteins reduced by auxin were affected after 8 hr of treatment.[187] Also, analysis of in vitro translation products of protoplast RNA showed that the time course of the effect of auxin on protein synthesis and mRNA accumulation was perfectly superimposed. These results indicate that auxin acts by regulating the concentration of auxin-sensitive protein mRNAs.

The hormonal requirement of protoplasts changes soon after the induction of mitotic activity, and an early transfer to a new medium with a relatively low level of hormones supports rapid growth in *Nicotiana* and *Hyoscyamus*.[307] Frequent dilution of culture medium has been emphasized for successful regeneration of protoplasts from a large number of plants,[31] and this was found to be essential for sustained division in protoplast culture of *Lactuca*[16] and *Solanum brevidens*.[208]

The presence of vitamins is beneficial for regeneration of protoplasts.[202] Of the various vitamins, nicotinic acid, pyridoxine, and thiamine are needed for pea protoplasts,[8] whereas inclusion of inositol was obligatory for tobacco[48] and beneficial for mung bean[318] protoplasts. The inclusion of undefined substances such as casein hydrolysate and coconut milk has been recommended by various workers.

Specificity of certain amino acids has been found in many systems. Glutamine promoted division in protoplasts of *Asparagus*,[43] *Ranunculus*,[67] and *S. brevidens*.[208] Asparagine, glutamine, and serine were benificial for protoplasts of *Vicia narbonensis*,[66] whereas only glutamine and serine were enough for protoplasts of *Salpiglossis sinuata*,[40] and glutamine as well as asparagine promoted plating efficiency in *Vigna aconitifolia*.[257] Arginine stimulated division of protoplasts of *Prunus dulcis*.[316]

Polyamine-induced stimulation of protoplast division is also on record. Ornithine (0.025 mM) and putrescine (0.05 and 0.01 mM) stimulated cell division and colony formation in protoplasts of *Alnus glutinosa* and *A. indica*;[135] by contrast, spermine and spermidine were inhibitory. Polyamines were also stimulatory for division of *Prunus*[316] protoplasts.

d. pH

In addition to affecting the yield and viability of protoplasts, pH also markedly affects regeneration of protoplasts. In general, pH in the range of 5 to 5.8 is used but pH values above 6 have been found to be stimulatory for division of *Asparagus*,[174] *Pisum*,[104] and *Vigna*,[17] protoplasts. In order to monitor pH changes during culture of protoplasts, inclusion of an indicator dye, bromocresol purple, is helpful.[240] It does not affect colony formation. The color of this dye is very sensitive to a change in pH from 5.5 to 8.0. This dye is also stable on autoclaving.

3. Genetic Nature

The ability of protoplasts to divide is also dependent on genetic nature. This is inferred from the behavior of relatively recalcitrant types among the highly regenerative genera of Solanaceae. The division frequency of protoplasts of *N. sylvestris* is much less than *N. tabacum* and *N. otophora*.[10,58] However, F₁ hybrids of *N. sylvestris* × *N. otophora* gave protoplasts which regenerated at the same frequency as *N. otophora*. Similarly, in petunia,[228] it is more difficult to regenerate protoplasts of *Petunia parodii* than *P. hybrida*, but protoplasts from F₁ hybrid *P. parodii* × *P. hybrida* regenerated at the same frequency as *P. hybrida*. Varietal differences in regeneration of protoplasts are seen in many crop plants such as alfalfa,[140] potato,[86,259,282] and white clover.[2]

VII. CULTURE AT LOW DENSITY

In view of various uses of plant protoplasts in fundamental as well as applied research, it is essential to devise a nutrient formulation which can be routinely employed for culture of protoplasts at low density or even for an individual protoplast.

Culture of *Vicia faba* protoplasts at low density was possible on a complex medium.[148] This medium, in addition to mineral salts, comprised 14 vitamins, auxins and cytokinins, various inorganic acids, 10 sugars and sugar alcohols, 21 amino acids, 6 nucleic acid bases, casein hydrolysate, and coconut milk.

However, culture of protoplasts at low density on a defined medium is more desirable. Purification of enzyme[219] is described as essential for the culture of single protoplasts of *P. parodii* in microdroplets. It has been possible to culture protoplasts at low density (Figure 12) and surprisingly the requirements are very simple.[48] Protoplasts of *N. tabacum* at high density (1 to 4 × 10⁴ units/mℓ) required 3 mg/ℓ of NAA for 30 to 40% plating efficiency. In this medium, inositol was the only vitamin required stringently. However, after 4 days of culture at high-density conditioning of medium, cell density could be lowered to 1 to 4 cells/mℓ, provided that the concentration of NAA in the medium was lowered to 0.3 mg/ℓ.[48] Experiments with ¹⁴C-labeled NAA indicated that after 4 days of the presence of protoplasts at high density, the amount of free NAA in the medium was drastically reduced. Therefore, conditioning of the medium by the presence of high-density protoplasts in fact lowers the auxin level. Later it was found that auxin is in fact a critical determinant in low-density culture of protoplasts of *N. tabacum*, *N. sylvestris*, *N. plumbaginifolia*, and *Petunia*.[198] Auxin, NAA, is inhibitory at a concentration above 0.3 mg/ℓ; at the same level, IAA could support the growth of low-density protoplasts, whereas it is unable to support the growth of protoplasts at high density. The optimal level of 2,4-D was higher and was pH dependent. Interestingly, picloram induced low-density growth of protoplasts over a wide concentration range.

The induction of growth by 2,4-D and picloram which was not affected by cell density was explained on the basis of auxin conjugation data[49] from tobacco mesophyll protoplasts. A comparison of ¹⁴C-labeled auxins by protoplasts showed that IAA and NAA were rapidly accumulated and conjugated unlike 2,4-D and picloram. Therefore, there may be a relationship between the conjugation process and auxin cytotoxicity.

Protoplasts of tobacco cultured at low density on medium containing 0.05 micromolar NAA were sensitive to toxicity of individual amino acids.[179] However, cytotoxicity of alanine, aspartic acid, asparagine, glutamic acid, glutamine, glycine, lysine, proline, and valine was reduced when the concentration of NAA was increased to 1 μM. This selective modification of amino acid toxicity by NAA could not be correlated with modification of the uptake rate or incorporation of these amino acids into proteins or amino acid auxin conjugates.

In brief, the strategies adopted for culture of protoplasts at low density are culture on (1) complex medium,[148] (2) feeder layer,[232] (3) optimized defined medium, after initial culture of cells at high density,[48] and (4) preculture of protoplasts at high density followed by culture

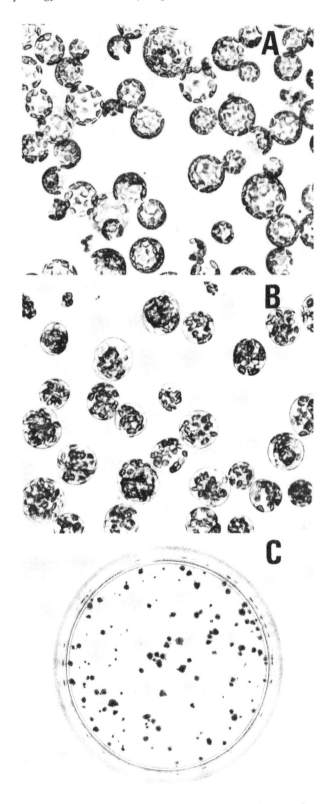

FIGURE 12.   Culture of protoplasts of *Nicotiana tabacum* at low density: (A) protoplasts at the time of isolation, (B) protoplasts in liquid medium showing division, and (C) protoplast-derived cell colonies. (From Caboche, M., *Planta*, 149, 7, 1980. With permission.)

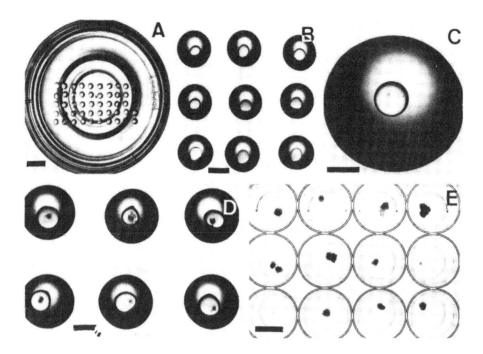

FIGURE 13.   Culture of individual protoplasts of tobacco. (A) Culture droplets on a cover glass in a moist chamber. (B) Assay of oil droplets containing culture medium. (C) Single protoplast in a microdroplet of culture medium covered by an oil droplet. (D) Microcalli developed within 2 weeks on culture of individual protoplasts. (E) Calli on agarose medium. (From Koop, H. U. and Schweiger, K. G., *J. Plant Physiol.*, 121, 245, 1985. With permission.)

of individual protoplasts in microdroplets.[110] More recently,[159] the technique for culture of individual (Figure 6) protoplasts has been substantially improved (Figure 7), representing the utility of this method in terms of reproducibility and efficiency. For individual protoplasts of tobacco (Figure 13), plant regeneration frequencies ranged from 50 to 60%. The methods also work well for *Brassica napus*.[271] Critical determinants of microculture of single protoplasts are composition and volume of culture medium, pH, buffering system osmolarity, and plant genotype.

VIII. PROBLEMS IN PROTOPLAST CULTURE

A. General

In many of the successful protoplast systems the individual protoplasts respond differently.[45] Some never divide, while others undergo early division but stop dividing at the microcolony stage and eventually degenerate. Therefore, frequency of colony formation is much lower than frequency of protoplasts undergoing division (Figure 14). This problem has been analyzed in protoplasts of *Vicia hajastana*.[265] Freshly prepared protoplasts of this plant showed mitoses and a high frequency of binucleate cells, which probably resulted from failure of cytokinesis. In early divisions many mitoses showed metaphase chromosomes with kinetochore microtubules, but lacking polar microtubules. This aberrant microtubular organization can result in genetic abnormalities and it may be possible to manipulate the frequency of abnormalities by controlling the onset of the first division in protoplast culture.

B. Protoplast Culture of Monocot Plants

Although it has been possible to regenerate protoplasts into cells in a number of dicoty-

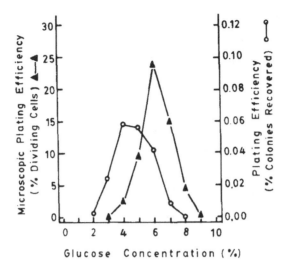

FIGURE 14.   Percent dividing cells and percent colonies recovered from protoplast cultures of *Malus*. (From Konider, M., Hauptmann, R., Widholm, J. M., Skirvin, R. M., and Korban, S. S., *Plant Cell Rep.*, 3, 142, 1984. With permission.)

ledonous plants, the protoplasts of monocots and particularly of cereals are relatively refractory to regeneration. The isolation of protoplasts from mesophyll tissue of cereals is least problematic, but regeneration of these protoplasts has yet to be accomplished. This is essential for realization of the potential of protoplast technology.

The factors ascribed to account for failure of regeneration of mesophyll protoplasts of cereals vary from genetic inability of the tissue to conditions of stress resulting in synthesis of cell division inhibitors such as abscisic acid[315] and ethylene.[75] The higher content of nucleases is ascribed to result in failure of protoplast regeneration in oat. The level of nucleases could be reduced by antisenescence compounds, such as polyamine, but it does not favor the division of oat protoplasts.[3] Therefore regeneration of mesophyll protoplasts of monocots remains a challenge in protoplast technology. In this context, reported regeneration of mesophyll protoplasts of sugarcane[77] and sorghum[151] must be mentioned. Of these, sorghum protoplasts regenerated if they were isolated from plants kept in the dark for 3 days and cultured on a medium preconditioned by growth of cells of the same plant. A mitotic frequency of 36% was recorded, but no sustained divisions were seen.

Faced with the failure of regeneration of mesophyll protoplasts of cereals, attempts were made to regenerate protoplasts obtained from callus tissues of *Hordeum vulgare*,[155] *Oryza sativa*,[65] *Triticum monococcum*,[209] and *Zea mays*,[224] but without a measure of success. Occasional divisions recorded could not be substantiated by subsequent workers.

However, sustained divisions of protoplasts derived from cell suspensions of monocots have been reported. Divisions were sporadic in *Sorghum bicolor*[41] but sustained divisions were reported in *Pennisetum americanum*,[295] *Panicum maximum*,[172] and *Pennisetum purpureum*[296] (Figure 15). Along with these reported successes are reports of regeneration of protoplasts of suspension cell origin from other monocots, *Lolium multiflorum*,[141] *Z. mays*,[57,55,173] *Panicum miliaceum*[131] and *O. sativa*.[321] These reports are consistent with the suggestion that protoplast regeneration is possible provided they are isolated from an actively growing cell suspension.

That the nature of tissue is important for successful regeneration of protoplasts is evidenced in the ready regeneration of protoplasts obtained from roots of *Phaseolus aureus*,[318] *Glycine max*,[319] *Medicago sativa*, and *Trigonella foenum graecum*.[320] Also, a high rate of regeneration

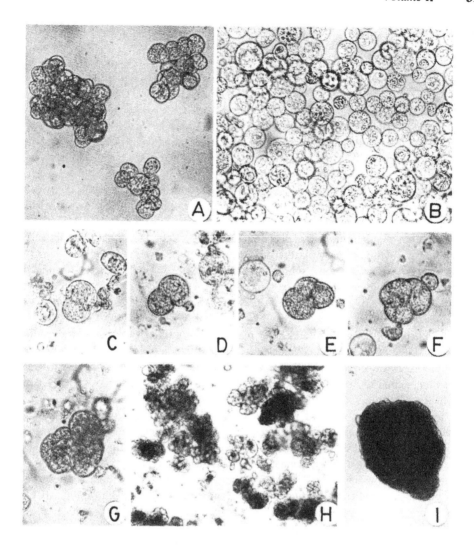

FIGURE 15. Tissue differentiation and embryo formation from protoplasts of *Pennisetum purpureum:* (A) cells from a suspension culture. (B) protoplasts at the time of isolation, (C-H), initial division and tissue formation from protoplasts, and (I) an embryo from protoplast-derived tissue. (From Vasil, V., Wang, D-Y., and Vasil, I. K., *Z. Pflanzenphysiol.*, 111, 235, 1983. With permission.)

capacity is seen in protoplasts obtained from hypocotyl of *Vigna sinensis*[17] and *Brassica* spp.[11,111] The mesophyll protoplasts of the plants listed above are relatively refractory to regeneration. Therefore, in *B. napus* the protoplasts were regenerated from secondary embryos differentiating from pollen-derived plants.[157] When mesophyll protoplasts were refractory to regeneration it was possible to have regeneration from cotyledons of *Cyamopsis*,[248] *Dalbergia*,[247] *Hedysarum*,[6] and *Gossypium*.[82]

IX. PROTOPLAST TO PLANT

Regeneration of plants from protoplast-derived tissue is a prerequisite for the realization of potentials of protoplast technology in plant improvement. Totipotency of higher plant protoplasts was first demonstrated in *N. tabacum*[202] and other species,[203] and since then, it has been possible to have plants from protoplasts of a number of crop plants. However, many of these are dicots, a few legumes, and a few grasses. Unfortunately, this has not been possible in important seed legumes and cereals — our staple crops. Therefore, much remains to be learned and achieved in protoplast technology.

X. MICROPLASTS/CYTOPLASTS AND THEIR UTILITY

The formation of subprotoplasts or microplasts, which are basically cytoplasmic masses without nuclei (enucleated protoplasts), has been recorded by many previous workers in various tissues both at the time of isolation of protoplasts as well as during their culture. These enucleated protoplasts are thought to be of significance in the transfer of cytoplasmic genes, resulting in the formation of cybrids.

Separation of enucleated protoplasts from normal protoplasts was accomplished by density gradient centrifugation of protoplast suspension.[169] Employing percoll, it was possible to isolate about 10% of the subprotoplasts.

The formation of subprotoplasts is described to be dependent on osmolarity of the enzyme mixture during isolation of protoplasts from cotyledons of *Brassica*.[297] The more hypertonic the enzyme mixture was, the more subprotoplasts were formed. This is ascribed to an increase in surface area to volume ratio in hypertonic solution.

A simple method for obtaining microplasts is through mechanical disruption of thin-walled callus cells developed on an increased auxin medium. When auxin-induced, highly vacuolated callus cells of *Saintpaulia* and other plants were ruptured with dissecting needles, it resulted in a large number of subcellular units, most of which were enucleate.[24] On culture, these microplasts formed a cell wall and showed budding. However, it remains to be seen whether this method of obtaining cytoplasts holds true for other systems.

The factor affecting subprotoplast formation is the presence of elongated cells.[4] In *N. glauca* fewer cytoplasts occurred in isolates from young leaves than from mature fully expanded leaves which have more elongated cells. Suspension cultures of *N. debneyi* which were isodiametric produced no cytoplasts, whereas suspension cultures derived from *N. glutinosa* which contained many elongated cells produced cytoplasts.

Of late, a technique has been described for mass production of cytoplasts and miniprotoplasts on density gradient centrifugation of suspension culture-derived protoplasts of *Solanum nigrum*.[166] The vigor of the donor cell suspension culture was the critical factor in obtaining a high frequency of enucleated protoplasts. It depended upon the age of culture; optimal yield was possible from 3-day-old suspensions. The protoplasts from such a suspension culture was enucleated by centrifugation through a stepwise mannitol sucrose gradient. Two bands were routinely obtained. A minor band at the 6.4%/18.2% mannitol border contained highly vacuolate cytoplasts which were 95% enucleate. The major band was at the 18.2% mannitol/33% sucrose border. It contained the bulk of cytoplasts, of which 90% were enucleate.

XI. PROTOPLAST — APPLICATIONS IN FUNDAMENTAL RESEARCH

The potential applications of protoplasts for fundamental research were highlighted at the beginning of this chapter; in this section some details are given, citing specific examples.

A. Isolation of Cell Components

For understanding the structure and function of individual plant cell components, the main problem is their isolation from cells, due to the presence of the rigid cell wall. This is not encountered when they are isolated from protoplasts. Employing protoplasts, various cell components that have been isolated are plasmalemma, chloroplasts, mitochondria, nuclei, and chromosomes.

Since plant protoplasts bind to lectin concanavalin-A (con-A), plasmalemma was isolated from protoplasts of carrot with this marker.[36] Feeding of ^{14}C-acetyl con-A before homogenization and subsequent separation on discontinuous (10 to 38%) or continuous (5 to 50%) renographin gradients resulted in a labeled fraction which peaked at about 1 to 14 g cm^{-3} density and had enzyme characteristics of plasmalemma.

Large-scale isolation of chloroplasts was possible by osmotic rupture of protoplasts in phosphate buffer containing Mg^{++} and dithiothreitol. The chloroplasts were separated on discontinuous sucrose gradients and were found to be photochemically acive.[300] A relatively simpler method[231] of chloroplast isolation was osmotic rupture of protoplasts in a buffered osmoticum containing bovine serum albumin and dithiothreitol and passing the suspension through a 20-μm net.

Along with chloroplasts, mitochondria can also be isolated when 0.05 to 0.1% bovine serum albumin is included and sucrose density gradients are preformed of disrupted protoplasts. The mitochondria will form a band at 1.18 g cm^{-3} and can be identified by fumarase activity, whereas chloroplasts form a band at 1.22 g cm^{-3}.

Prior to routine isolation of protoplasts there was no satisfactory method for isolating vacuoles from cells, since they comprise a very fragile cell component. Now, large-scale isolation of vacuoles is possible by disruption of protoplasts in a protoplast buffer.[300] The suspension is centrifuged at low speed and purified by layering over 5% Ficoll containing 0.55 M sorbitol and 1 μM tris MES buffer.

Nuclei can also be isolated from protoplasts by lysing with Triton X-100® and further purification of nuclei on sucrose or sorbitol.[213] Isolation of nuclei is also possible without detergent.[280] The protoplasts are suspended in hypertonic solution. This treatment results in nuclei with an intact double membrane. An efficient and rapid method[249] for the isolation of nuclei from plant protoplasts requires disruption of protoplasts in 10 mM MES buffer supplemented with a very low concentration of 0.01% detergent, Triton X-100®. The pH of buffer (5.3) is a critical factor in the recovery of stable nuclei in large numbers. Supplementing buffer with spermine (0.1 mM), dithiothreitol (2.5 mM), EDTA (2.5 mM), and NaCl and KCl (each 10 mM) improved nuclear yield and quality. With this method it is possible to routinely recover 95% nuclei (Figure 16) from protoplasts within 30 min.

Not only nuclei, but also chromosomes can be isolated from plant protoplasts. The procedure involves a gentle lysis of protoplasts from synchronized cells with a low osmotic strength and low detergent concentration. Mitotic chromosomes were isolated from synchronized cell suspension cultures of *N. tabacum* and *Lycopersicon esculentum* with maximum yield of 7 and 12%, respectively, whereas 50% yield of meiotic chromosomes was possible for naturally synchronized meiocytes of *Lilium* and *Hemerocallis*.[177] A method for purification of chromosomes isolated from protoplasts included differential centrifugation and repeated layering on top of a dense sucrose solution.[123] In case the chromosomes are sticky, as in *Vicia*, they can be separated by filtration through polycarbonate filter.[192] Isolated chromosomes were comparable morphologically to those in intact cells.

B. Photosynthesis

Due to the difference in density, it was possible to isolate and separate bundle sheath protoplasts from mesophyll protoplasts in C$_4$ plants with Kranz anatomy. This enabled the identification of enzymatic activity specific to these cells. The carboxylation phase of the C$_4$ pathway is located in mesophyll cells, whereas the decarboxylative phase of this pathway, as well as the carboxylative phase of the Calvin-Benson pathway, is located in bundle sheath cells. Ribulose-1,5-bisphosphate carboxylase was found only in bundle sheath cells and not in mesophyll cells.[73] Further, subtle differences were revealed in enzymic activities and transport mechanism between mesophyll and bundle cells.

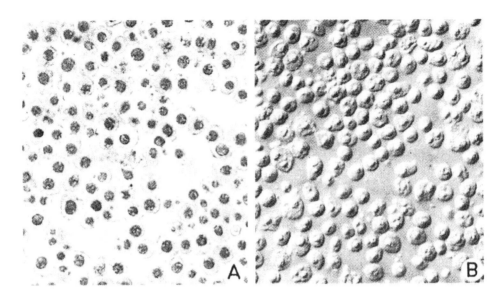

FIGURE 16.   Nuclei isolated from protoplasts of *Brassica nigra:* (A) stained with toluidine blue and (B) unstained, examined with Normarksi differential interference contrast optics. (After Saxena, P. K., Fowke, L. C., and King, J., *Protoplasma,* 128, 184, 1985. With permission.)

C. Stomatal Physiology

The role of blue light in the opening of stomata was strengthened when protoplasts from guard cells, and not from epidermal cells, were stimulated to swell by blue light.[327] This was possible due to a microchamber in which enzymatic digestion of cell wall could be monitored. Protoplasts from stomatal guard cells swell in response to light and shrink in response to abscisic acid. Guard cells differ in biochemical characteristics from mesophyll cells.[33] This could be resolved by taking protoplasts from guard cells and mesophyll cells of *Commelina communis.*

D. Environmental Monitoring

The membrane integrity of plasmalemma and changes in fluidity can be utilized to estimate cell damage. A positive correlation was found between bursting of protoplast and ozone[50] concentration.

E. Cell Wall Formation

Protoplasts represent a state of dedifferentiation in plant cells. The formation of the cell wall, a state of redifferentiation, has already been described.

When protoplasts of tobacco were maintained in the absence of hormones for 96 hr, but in presence of coumarin, an inhibitor of wall synthesis, the protoplast resumed division on washing and transfer to hormone medium.[324] Therefore, it is very likely that during hormone starvation, an inhibitor related to wall formation accumulated in the medium.

F. Herbicide Physiology

Employing protoplasts, the mode of action of the herbicide paraquat (1,1'-dimethyl(4,4'-bipyridinium ions) was revealed.[37] To begin with, it caused segregation of cytoplasm into isolated areas near the plasmalemma and ultimately led to rupture of this membrane.

G. Circadian Rhythm

Nyctinastic plants, many of the Leguminoseae, spread their paired leaflets during daytime and fold them at night. This movement is controlled by light and darkness, but continues under a constant free-running condition since it is controlled by an internal circadian rhythm. The movement is accomplished by changes in the curvature of a cylindrical motor organ or pulvinus at the base of each leaflet. Cortical cells on opposing sides of the pulvinus swell and shrink to generate changes in the relative lengths of the two sides that are required for bending. Extensor cells swell as the leaflet opens in daytime and oppositely placed flexor cells swell as the leaflet closes at night. The turgor changes responsible for swelling and shrinking are caused by the uptake and release of large amounts of solutes. Protoplasts from pulvinus cells[112] might aid in finding answers about pulvinar physiology and ion transport that are difficult to approach in intact cells. They should also enable us to approach the basic question about circadian rhythmality, i.e., do rhythms persist in individual cells isolated from a multicellular rhythmic system.

H. Nitrogen Metabolism

The mesophyll of soybean leaf consists of two major cell types: photosynthetic mesophyll cells (palisade and spongy and paraveinal mesophyll cells) and an essentially nonphotosynthetic cell layer, which is specialized for nitrogen metabolism and compartmentation. These cells are larger than mesophyll cells and contain very few chloroplasts. This gives rise to a considerable difference in buoyant density and, taking advantage of it, the protoplasts from these two types of cells could be separated.[91] Further research will unravel the metabolic machinery of these interesting cell types.

I. Hormone Physiology

By depriving the protoplasts of hormones, the specific role of hormones in cell division can be unraveled. In the absence of hormones, metabolic activity of tobacco protoplasts in terms of incorporation of labeled leucine and uridine was not impaired.[323] The growth regulators NAA and BAP were required during early hours of culture. When added after 10 hr, a reduction of 50% in cell division was recorded. If they were not given up to 15 hr of culture, 95% reduction was recorded.

In another study[189] on tobacco protoplasts it was found that division was normal if protoplasts were cultured on auxin medium alone for 24 hr, but a further delay of BAP drastically reduced cell division. The inhibitory effect was possibly due to the release of an inhibitory factor by protoplasts in the absence of hormone. Washing of protoplasts followed by reculture in hormone medium restored division capacity.

Protoplasts can also serve as a useful system for the study of GA_3 action in barley aleurone, because in this way many disadvantages of the barley aleurone layer (that cell walls are very thick and tough, making cell fractionation very difficult) are removed. Also, the cell wall acts as a barrier for the release of α-amylase secreted by gibberellic acid, making the interpretation of enzyme secretion studies difficult. Such difficulties are likely to be alleviated by the use of protoplasts. Recently, gibberellic acid-sensitive protoplasts have been prepared from mature aleurone of barley[138] and interesting data on the action of gibberellin are expected.

REFERENCES

1. **Adams, T. L. and Townsend, J. A.,** A new procedure for increasing efficiency of protoplast plating and clone selection, *Plant Cell Rep.,* 2, 165, 1983.
2. **Ahuja, P. S., Lu, D. Y., Cocking, E. C., and Davey, M. R.,** An assessment of the cultural capabilities of *Trifolium repens* (white clover) and *Onobrychis viciifolia* (Sainfoin) mesophyll protoplasts, *Plant Cell Rep.,* 2, 269, 1983.

3. **Altmann, A., Kaur Sawhney, R., and Galston, A. W.,** Stabilization of oat leaf protoplasts through polyamine-mediated inhibition of senescence, *Plant Physiol.,* 60, 570, 1977.

4. **Archer, E. K., Landgren, C. R., and Bonnett,** Cytoplast formation and enrichment from mesophyll tissues of *Nicotiana* spp., *Plant Sci. Lett.,* 25, 175, 1982.

5. **Arcioni, S., Davey, M. R., Dos Santos, A. V. P., and Cocking, E. C.,** Somatic embryogenesis in tissues from mesophyll and cell suspension protoplasts of *Medicago coerulea* and *M. glutinosa, Z. Pflanzenphysiol.,* 106, 105, 1982.

6. **Arcioni, S., Mariotti, D., and Pezzotti, M.,** *Hedysarum coronarium,* in vitro conditions for plant regeneration from protoplasts and callus of various explants, *J. Plant Physiol.,* 121, 144, 1985.

7. **S. Von Arnold, and Eriksson, T.,** Factors influencing the growth and division of pea mesophyll protoplasts, *Physiol. Plant.,* 36, 193, 1976.

8. **S. Von Arnold, and Eriksson, T.,** A revised medium for growth of pea mesophyll protoplasts, *Physiol. Plant.,* 39, 257, 1977.

9. **Asamizu, T. and Nishi, A.,** Regenerated cell wall of carrot protoplasts isolated from suspension cultured cells, *Physiol. Plant.,* 48, 207, 1980.

10. **Banks, M. S. and Evans, P. K.,** A comparison of the isolation and culture of mesophyll protoplasts from several *Nicotiana* species and their hybrids, *Plant Sci. Lett.,* 7, 409, 1976.

11. **Barsby, T. L., Yarrow, S. A., and Shepard, J. F.,** A rapid and efficient alternative procedure for the regeneration of plants from hypocotyl protoplasts of *Brassica napus, Plant Cell Rep.,* 5, 101, 1986.

12. **Bass, A. and Hughes, W.,** Conditions for isolation and regeneration of viable protoplasts of oil palm (*Elaeis guineensis*), *Plant Cell Rep.,* 3, 169, 1984.

13. **Bawa, S. B. and Torrey, J. G.,** Budding and nuclear division in cultured protoplasts of corn, *Convolvulus,* and onion, *Bot. Gaz.,* 132, 240, 1971.

14. **Beier, H. and Bruening, G.,** The use of an abrasive in the isolation of cowpea leaf protoplasts which support the multiplication of cowpea mosaic virus, *Virology,* 64, 272, 1975.

15. **Bergmann, L.,** Growth and division of single cells of higher plants, in vitro, *J. Gen. Physiol.,* 43, 841, 1960.

16. **Berry, S. F., Lu, D. Y., Pental, D., and Cocking, E. C.,** Regeneration of plants from protoplasts of *Lactuca sativa, Z. Pflanzenphysiol.,* 108, 31, 1982.

17. **Bharal, S. and Rashid, A.,** Isolation of protoplasts from stem and hypocotyl of the legume *Vigna sinensis* and some factors affecting their regeneration, *Protoplasma,* 102, 307, 1980.

18. **Bhat, S. R., Ford-Lloyd, B. Y., and Callow, J. A.,** Isolation of protoplasts and regeneration of callus from suspension cultures of cultivated beets, *Plant Cell Rep.,* 4, 348, 1985.

19. **Bhatt, D. P. and Fassuliotis, G.,** Plant regeneration from mesophyll protoplasts of egg plant, *Z. Pflanzenphysiol.,* 104, 81, 1981.

20. **Bhojwani, S. S. and Cocking, E. C.,** Isolation of protoplasts from pollen tetrads, *Nature (London) New Biol.,* 239, 29, 1972.

21. **Bhojwani, S. S., Power, J. B., and Cocking, E. C.,** Isolation, culture and division of cotton callus protoplasts, *Plant Sci. Lett.,* 8, 85, 1977.

22. **Bidney, D. L. and Shepard, J. F.,** Colony development from sweet potato petiole protoplasts and mesophyll cells, *Plant Sci. Lett.,* 18, 335, 1980.

23. **Bilkey, P. C. and Cocking, E. C.,** A non-enzymatic method for the isolation of protoplasts from callus of *Saintpaulia ionantha* (African violet), *Z. Pflanzenphysiol.,* 105, 285, 1982.

24. **Bilkey, P. C., Davey, M. R., and Cocking, E. C.,** Isolation, origin and properties of enucleate plant protoplasts, *Protoplasma,* 110, 147, 1982.

25. **Binding, H.,** Cell cluster formation by leaf protoplasts from axenic cultures of haploid *Petunia hybrida, Plant Sci. Lett.,* 2, 185, 1974.

26. **Binding, H.,** Regeneration von haploiden und diploiden Pflanzen aus protoplasten von *Petunia hybrida, Z. Pflanzenphysiol.,* 74, 327, 1974.

27. **Binding, H.,** Reproducibly high plating efficiencies of isolated mesophyll protoplasts from shoot cultures of tobacco, *Physiol. Plant.,* 35, 225, 1975.

28. **Binding, H. and Nehls, R.,** Regeneration of isolated protoplasts to plants in *Solanum dulcamara, Z. Pflanzenphysiol.,* 85, 279, 1977.

29. **Binding, H. and Nehls, R.,** Regeneration of isolated protoplasts of *Vicia faba, Z. Pflanzenphysiol.,* 88, 327, 1978.

30. **Binding, H., Nehls, R., and Kock, R.,** Versuche zur protoplasten regeneration dikotyler pflanzen unterschiedlicher systematischer zugchorigkeit, *Ber. Dtsch. Bot. Ges.,* 83, 667, 1980.

31. **Binding, H., Nehls, R., Kock, R., Finger, J., and Mordhorst, G.,** Comparative studies on protoplast regeneration in herbaceous species of the Dicotyledoneae class, *Z. Pflanzenphysiol.,* 101, 119, 1981.

32. **Binding, H., Nehls, R., Schieder, O., Sopory, S. K., and Wenzel, G.,** Regeneration of mesophyll protoplasts isolated from dihaploid clones of *Solanum tuberosum, Physiol. Plant.,* 43, 52, 1978.

33. **Birkenhead, K. and Willmer, C. M.,** Some biochemical characteristics of guard cell and mesophyll cell protoplasts from *Commelina communis, J. Exp. Bot.,* 37, 119, 1986.
34. **Blaschek, W., Haass, D., Koehler, H., and Franz, G.,** Cell wall regeneration by *Nicotiana tabacum* protoplasts, chemical and biochemical aspects, *Plant Sci. Lett.,* 22, 47, 1981.
35. **Bornman, J. F., Bornman, C. H., and Bjorn, L. O.,** Effects of ultraviolet radiation on viability of isolated *Beta vulgaris* and *Hordeum vulgare* protoplasts, *Z. Pflanzenphysiol.,* 105, 297, 1982.
36. **Boss, W. F. and Ruesink, A. W.,** Isolation and characterization of concanavalin A-labelled plasmamembranes of carrot protoplasts, *Plant Physiol.,* 64, 1005, 1979.
37. **Boulwere, M. A. and Camper, N. D.,** Effects of selected herbicides on plant protoplasts, *Physiol. Plant.,* 26, 313, 1972.
38. **Bourgin, J. P., Chupeau, Y., and Missionier, C.,** Plant regeneration from mesophyll protoplasts of several *Nicotiana* species, *Physiol. Plant.,* 45, 288, 1979.
39. **Boyes, C. J. and Sink, K. C.,** Regeneration of plants from callus-derived protoplasts of *Salpiglossis, J. Am. Soc. Hortic. Sci.,* 106, 42, 1981.
40. **Boyes, C. J., Zapata, F. J., and Sink, K. C.,** Isolation, culture and regeneration to plants of callus protoplasts of *Salpiglossis sinuata, Z. Pflanzenphysiol.,* 99, 471, 1980.
41. **Brar, D. S., Rambold, S., Constabel, F., and Gamborg, G. L.,** Isolation, fusion and culture of Sorghum and corn protoplasts, *Z. Pflanzenphysiol.,* 96, 269, 1980.
42. **Brown, S., Renaudin, J. P., Prevot, C., and Guern, J.,** Flow cytometry and sorting of plant protoplasts: technical problems and physiological results from a study of pH and alkaloids in *Catharanthus roseus, Physiol. Veg.,* 22, 541, 1984.
43. **Bui-Dang-Ha, D. and Mackenzie, I. A.,** The division of protoplasts from *Asparagus officinalis* and their growth and differentiation, *Protoplasma,* 78, 215, 1973.
44. **Burger, D. W. and Hackett, W. P.,** The isolation, culture and division of protoplasts from citrus cotyledons, *Physiol. Plant.,* 56, 324, 1982.
45. **Burgess, J. and Lawrence, W.,** Studies of the recovery of tobacco mesophyll protoplasts from an evacuolation treatment, *Protoplasma,* 126, 140, 1985.
46. **Burgess, J. and Linstead, P. J.,** Coumarin inhibition of microfibril formation at the surface of cultured protoplasts, *Planta,* 133, 267, 1977.
47. **Burgess, J., Linstead, P. J., and Bonsall, V. E.,** Observations on the time course of wall development at the surface of the isolated protoplasts, *Planta,* 139, 85, 1978.
48. **Caboche, M.,** Nutritional requirements of protoplast-derived haploid tobacco cells grown at low cell densities in liquid medium, *Planta,* 149, 7, 1980.
49. **Caboche, M., Arauda, G., Poll, A. M., Huet, J-C., and Leguay, J-J.,** Auxin conjugation by tobacco mesophyll protoplasts, *Plant Physiol.,* 75, 54, 1984.
50. **Cailloux, M., Phan, C. T., and Chung, Y. S.,** Damage by ozone to the mechanical integrity of the protoplast plasmalemma, *Experientia,* 34, 730, 1978.
51. **Carlberg, I., Glimelius, K., and Eriksson, T.,** Improved culture ability of potato protoplasts by use of activated charcoal, *Plant Cell Rep.,* 2, 223, 1983.
52. **Cassells, A. C. and Barlass, M.,** A method for the isolation of stable mesophyll protoplasts from tomato leaves throughout the year under standard conditions, *Physiol. Plant.,* 42, 36, 1978.
53. **Cassells, A. C. and Cocker, F. M.,** Seasonal and physiological aspects of the isolation of tobacco protoplasts, *Physiol. Plant.,* 56, 69, 1982.
54. **Cella, R. and Galun, E.,** Utilization of irradiated carrot cell suspension as feeder layer for cultured *Nicotiana* cells and protoplasts, *Plant Sci. Lett.,* 19, 243, 1980.
55. **Chang, Y. F.,** Plant regeneration in vitro from leaf tissues derived from cultured immature embryos of *Zea mays, Plant Cell Rep.,* 2, 183, 1983.
56. **Chin, J. C. and Scott, K. J.,** A large scale isolation procedure for cereal mesophyll protoplasts, *Ann. Bot.,* 43, 23, 1979.
57. **Chourey, P. S. and Zurawski, D. B.,** Callus formation from protoplasts of a maize cell cultures, *Theor. Appl. Genet.,* 59, 341, 1981.
58. **Chupeau, Y., Bourgin, J. P., Missonier, C., Dorion, N., and Morel, G.,** Preparation et culture de protoplasts de divers *Nicotiana, C.R. Acad. Sci. Ser. D,* 278, 1565, 1974.
59. **Cocking, E. C.,** A method for isolation of plant protoplasts and vacuoles, *Nature (London),* 187, 927, 1960.
60. **Cooke, R. and Meyer, Y.,** Hormonal control of tobacco protoplast nucleic acid metabolism during in vitro culture, *Planta,* 152, 1, 1981.
61. **Crepy, L., Chupeau, M. C., and Chepeau, Y.,** The isolation and culture of leaf protoplasts of *Chicorium intybus* and their regeneration in white plants, *Z. Pflanzenphysiol.,* 107, 123, 1982.
62. **Crowder, A. J., Landgren, C. R., and Rockwood, L. L.,** Cultivar differences in starch content and protoplasts yields from root cortical explants of *Pisum sativum, Physiol. Plant.,* 46, 85, 1979.

63. **Davey, M. R., Bush, E., and Power, J. B.,** Cultural studies of a dividing legume leaf protoplast system, *Plant Sci. Lett.,* 3, 127, 1974.

64. **David, A. and David, H.,** Isolation and callus formation from cotyledon protoplasts of pine (*Pinus pinaster*), *Z. Pflanzenphysiol.,* 94, 173, 1979.

65. **Deka, P. C. and Sen, S. K.,** Differentiation in calli originated from isolated protoplasts of rice (*Oryza sativa*) through plating technique, *Mol. Gen. Genet.,* 145, 239, 1976.

66. **Donn, G.,** Cell division and callus regeneration from leaf protoplasts of *Vicia narbonensis, Z. Pflanzen-physiol.,* 86, 66, 1978.

67. **Dorion, N., Chupeau, Y., and Bourgin, J. P.,** Isolation, culture and regeneration into plants of *Ranunculus sceleratus* leaf protoplasts, *Plant Sci. Lett.,* 5, 325, 1975.

68. **Dorokhov, Y. L. and Alexandrova, N. M.,** Isolation of tobacco mesophyll protoplasts at low temperature and analysis of polyribosomes during their regeneration, *Soviet Plant Physiol.,* 28, 1151, 1981.

69. **Douglas, G. C., Keller, W. A., and Setterfield, G.,** Somatic hybridization between *Nicotiana rustica* and *N. tabacum.* 1. Isolation and culture of protoplasts and regeneration of plants from cell cultures of wild type and chlorophyll deficient strains, *Can. J. Bot.,* 59, 208, 1981.

70. **Dudits, D., Kao, K. N., Constabel, F., and Gamborg, O. L.,** Embryogenesis and formation of tetraploid and hexaploid plants from carrot protoplasts, *Can. J. Bot.,* 54, 1063, 1976.

71. **Duhonx, E.,** Protoplasts isolation of Gymnosperm pollen, *Z. Pflanzenphysiol.,* 99, 210, 1980.

72. **Durand, J., Potrykus, I., and Donn, G.,** Plantes issues de protoplastes de *Petunia, Z. Pflanzenphysiol.,* 69, 26, 1973.

73. **Edwards, G. E. and Huber, S. C.,** Usefulness of isolated cells and protoplasts for photosynthetic studies, in *Proc. 4th Int. Congr. Photosynthesis,* Hall, D. O., Coobs, J., and Goodwin, T. W., Eds., Biochemical Society, London, 1978, 95.

74. **Enzmann-Becker, G.,** Plating efficiency of protoplasts of tobacco in different light conditions, *Z. Natur-forsch.,* 28C, 470, 1973.

75. **Eriksson, T., Bonnett, H., Gliemlius, K., and William, A.,** Technical advances in protoplast isolation, culture and fusion, in *Tissue Culture and Plant Science,* Street, H. E., Ed., Academic Press, New York, 1974.

76. **Eriksson, T. and Jonasson, K.,** Nuclear division in isolated protoplasts from cells of higher plants grown *in vitro, Planta,* 89, 85, 1969.

77. **Evans, D. A., Crocomo, O. J., and De Carrotho, M. T. V.,** Protoplast isolation and subsequent callus regeneration in sugarcane, *Z. Pflanzenphysiol.,* 98, 355, 1980.

78. **Evans, P. K. and Cocking, E. C.,** The techniques of plant cell culture and somatic cell hybridization, in *New Techniques in Biophysics and Cell Biology,* Pain, R. H. and Smith, B. J., Eds., John Wiley & Sons, London, 1975, 127.

79. **Evans, D. A., Keates, A. G., and Cocking, E. C.,** Isolation of protoplasts from cereal leaves, *Planta,* 104, 173, 1972.

80. **Facciotti, D. and Pilet, P. E.,** Plants and embryoids from haploid *Nicotiana sylvestris* protoplasts, *Plant Sci. Lett.,* 15, 1, 1979.

81. **Finer, J. J. and Smith, R. H.,** Isolation and culture of protoplasts from cotton (*Gossypium klotzschianum*) callus cultures, *Plant Sci. Lett.,* 26, 147, 1982.

82. **Firoozabady, E. and DeBoer, D. L.,** Isolation, culture and cell division in cotyledon protoplasts of cotton *Gossypium hirsutum* and *G. barbadense, Plant Cell Rep.,* 5, 127, 1986.

83. **Fleck, J., Durr, A., Fritsch, C., Lett., M. C., and Hirth, L.,** Comparison of proteins synthesized in vitro and in vivo by mRNA from isolated protoplasts, *Planta,* 148, 453, 1980.

84. **Fleck, J., Durr, A., Fritsch, C., Vernet, T., and Hirth, L.,** Osmotic shock "stress proteins" in protoplasts of *Nicotiana sylvestris, Plant Sci. Lett.,* 26, 159, 1982.

85. **Fleck, J., Durr, A., Lett, M. C., and Hirth, L.,** Changes in protein synthesis during the initial stage of life of tobacco protoplasts, *Planta,* 145, 279, 1979.

86. **Foulger, D. and Jone, M. G. K.,** Improved efficiency of genotype dependent regeneration from protoplasts of important potato cultivars, *Plant Cell Rep.,* 5, 72, 1986.

87. **Fowke, L. C.,** Ultrastructure of isolated and cultured protoplasts, in *Frontiers of Plant Tissue Culture,* Thorpe, T. A., Ed., University of Calgary Press, Canada, 1978, 223.

88. **Fowke, L. C., Bech-Hanson, C. W., and Gamborg, O. L.,** Electron microscopic observations of cell regeneration from cultured protoplasts of *Ammi visnaga, Protoplasma,* 79, 235, 1974.

89. **Fowke, L. C., Bech-Hanson, C. W., Gamborg, O. L., and Constabel, F.,** Electron microscopic observations of mitosis and cytokinesis in multinucleate protoplasts of soybean, *J. Cell Sci.,* 18, 491, 1975.

90. **Fowke, L. C., Rennie, P. J., and Constabel, F.,** Organelles associated with plasma membrane of tobacco leaf protoplasts, *Plant Cell Rep.,* 2, 292, 1983.

91. **Franceschi, V. R., Ku, M. S. B., and Wittechbach, V. A.,** Isolation of mesophyll and paraveinal mesophyll protoplast from soybean leaves, *Plant Sci. Lett.,* 36, 181, 1984.

92. **Frearson, E. M., Power, J. B., and Cocking, E. C.,** The isolation, culture and regeneration of *Petunia* leaf protoplasts, *Dev. Biol.,* 33, 130, 1973.

93. **Fu, Y-Y., Jia, S-R., and Lin, Y.,** Plant regeneration from mesophyll protoplast culture of cabbage, *Brassica oleracea* var. *capitata, Theor. Appl. Genet.,* 71, 495, 1985.

94. **Fukunaga, Y. and King, J.,** Effect of different nitrogen sources in culture media on protoplast release from plant cell suspension cultures, *Plant Sci. Lett.,* 11, 241, 1978.

95. **Galbraith, D. W.,** Microfluorimetric quantitation of cellulose biosynthesis by plant protoplasts using calcofluor white, *Physiol. Plant.,* 53, 111, 1981.

96. **Galbraith, D. W., Afonso, C. L., and Harkins, K. R.,** Flow-sorting and culture of protoplasts: conditions for high frequency recovery growth and morphogenesis from sorted protoplasts of suspension cultures of nicotiana, *Plant Cell Rep.,* 3, 151, 1984.

97. **Galbraith, D. W. and Shields, B. A.,** The effects of inhibitors of cell wall synthesis on tobacco protoplast development, *Physiol. Plant,* 55, 25, 1982.

98. **Galston, A. W., Altman, A., and Kaur-Sawhney, R.,** Polyamines, ribonuclease and the improvement of oat leaf protoplasts, *Plant Sci. Lett.,* 11, 69, 1978.

99. **Gamborg, O. L.,** Plant protoplast isolation, culture and fusion, in *Cell Genetics in Higher Plants,* Dudits, D., Farkas, G. L., and Maliga, P., Eds., Akademici Kiado, Budapest, 1976, 107.

100. **Gamborg, O. L., Davis, B. P., and Stahlhut, R. W.,** Cell division and differentiation in protoplasts from cell cultures of *Glycine* species and leaf tissue of soybean, *Plant Cell Rep.,* 2, 213, 1983.

101. **Gamborg, O. L., Miller, R. A., and Ojima, K.,** Nutrient requirements of suspension cultures of soybean root cells, *Exp. Cell Res.,* 50, 151, 1968.

102. **Gamborg, O. L. and Shyluk, J. P.,** Tissue culture, protoplasts and morphogenesis in flax, *Bot. Gaz.,* 137, 301, 1976.

103. **Gamborg, O. L., Shyluk, J. P., Fowke, L. C., Wetter, L. R., and Evans, D.,** Plant regeneration from protoplasts and cell cultures of *Nicotiana tabacum* sulfur mutant (Su/Su), *Z. Pflanzenphysiol.,* 95, 255, 1979.

104. **Gamborg, O. L., Shyluk, J. P., and Kartha, K. K.,** Factors affecting the isolation and callus formation in protoplasts from the shoot apices of *Pisum sativum, Plant Sci. Lett.,* 4, 285, 1975.

105. **Gamborg, O., Shyluk, J. P., and Shahin, E. A.,** Isolation, fusion and culture of plant protoplasts, in *Plant Tissue Culture: Methods and Applications in Agriculture,* Thorpe, T. A., Ed., Academic Press, New York, 1981, 115.

106. **Gigot, C., Phillips, C., and Hirth, L.,** Evenements biochimiques accompagnant les etapes ¢hapes initiales de la dedifferentiation des protoplastes de tabacos en culture, in *Memoires Origulaux,* Morel, G., Ed., Masson, Paris, 1976, 186.

107. **Gill, R., Rashid, A., and Maheshwari, S. C.,** Regeneration of plants from mesophyll protoplasts of *Nicotiana plumbaginifolia, Protoplasma,* 96, 375, 1978.

108. **Gill, R., Rashid, A., and Mesheshari, S. C.,** Dark requirement for cell regeneration and colony formation by mesophyll protoplasts of *Nicotiana plumbaginifolia, Protoplasma,* 106, 351, 1981.

109. **Gill, R., Saxena, P. K., Rashid, A., and Maheshwari, S. C.,** Factors affecting enhancement in plating efficiency of mesophyll protoplasts of *Nicotiana plumbaginifolia, J. Indian Bot. Soc.,* 61, 295, 1982.

110. **Gleba, Y. Y.,** Microdroplet cultures: tobacco plants from single mesophyll protoplasts, *Naturwissenschaften,* 65, 158, 1978.

111. **Glimelius, K.,** High growth rate and regeneration capacity of hypocotyl protoplasts in some Brassicaceae, *Physiol. Plant.,* 61 38, 1984.

112. **Gorton, H. L. and Satter, R. L.,** Extensor and flexor protoplasts from *Samanea pulvini, Plant Physiol.,* 76, 680, 1984.

113. **Gosch, G., Bajaj, Y. P. S., and Reinert, J.,** Isolation, culture and fusion studies on protoplasts from different species, *Protoplasma,* 85, 321, 1975.

114. **Gosch, G., Bajaj, Y. P. S., and Reinert, J.,** Isolation, culture and induction of embryogenesis in protoplasts from culture suspensions of *Atropa belladonna, Protoplasma,* 86, 405, 1975.

115. **Grambow, A. J., Kao, K. N., Miller, R. A., and Gamborg, O. L.,** Cell division and plant development from protoplasts of carrot cell suspension cultures, *Planta,* 103, 348, 1972.

116. **Gregory, D. W. and Cocking, E. C.,** The large scale isolation of protoplasts from immature tomato fruit, *J. Cell Biol.,* 24, 143, 1965.

117. **Gregory, D. W. and Cocking, E. C.,** Studies on isolated protoplasts and vacuoles. I. General properties, *J. Exp. Bot.,* 17, 57, 1966.

118. **Gresshoff, P. M.,** In vitro cultures of white clover: callus suspension protoplast culture, and plant regeneration, *Bot. Gaz.,* 141, 157, 1980.

119. **Grout, B. W. W.,** Cellulose microfibril deposition on the plasmalemma surface of regenerating tobacco mesophyll protoplasts. A deep etch study, *Planta,* 123, 275, 1975.

120. **Gunn, R. E. and Shepard, J. F.,** Regeneration of plants from mesophyll-derived protoplasts of British potato (*Solanum tuberosum*) cultivars, *Plant Sci. Lett.,* 22, 97, 1981.

121. **Haberlach, G. T., Cohen, B. A., Reichert, N. A., Baer, M. A., Towill, L. E., and Helgeson, J. P.,** Isolation, culture and regeneration of protoplasts from potato and several related *Solanum* species, *Plant Sci.*, 39, 67, 1985.

122. **Hackman, I. C. and Arnold, S. V.,** Isolation and growth of protoplasts from cell suspensions of *Pinus contorta*, *Plant Cell Rep.*, 2, 92, 1984.

123. **Hadlaczky, Gy., Bisztray, Gy., Praznovszky, T., and Dudits, D.,** Mass isolation of plant chromosomes and nuclei, *Planta*, 157, 278, 1983.

124. **Hall, J. A. and Wood, R. K. S.,** Plant cell killed by softrot parasites, *Nature (London)*, 227, 1266, 1970.

125. **Hanke, D. E. and Northcote, D. H.** Cell wall formation by soybean callus protoplasts, *J. Cell Sci.*, 14, 29, 1974.

126. **Harada, H.,** A new method for obtaining protoplasts from mesophyll cells, *Z. Pflanzenphysiol.*, 69, 97, 1973.

127. **Harkins, K. R. and Galbraith, D. W.,** Flow sorting and culture of plant protoplasts, *Physiol. Plant.*, 60, 43, 1984.

128. **Harms, C. T., Lorz, H., and Potrykus, I.,** Multiple drop array (MDA) technique for the large scale testing of culture media variations in hanging microdrop cultures of single cell system. II. Determination of phytohormone combinations for optimal division responses in *Nicotiana tabacum* protoplasts cultures, *Plant Sci. Lett.*, 14, 237, 1979.

129. **Harms, C. T. and Potrykus, I.,** Enrichment for heterokaryocytes by the use of iso-osmotic density gradients after plant protoplast fusion, *Theor. Appl. Genet.*, 53, 49, 1978.

130. **Hassanpour-Estahbanati, A. and Demarly, Y.,** Plant regeneration from protoplasts of *Solanum pennellii*. Effect of photoperiod applied to donor plants, *J. Plant Physiol.*, 121, 171, 1985.

131. **Heyser, J. W.,** Callus and shoot regeneration from protoplasts of proso millet (*Panicum miliaceum*), *Z. Pflanzenphysiol.*, 113, 293, 1984.

132. **Holbrook, L. A., Reich, T. J., Iyer, V. N., Haffner, M., and Miki, B. L.,** Induction of efficient cell division in alfalfa protoplasts, *Plant Cell Rep.*, 4, 229, 1985.

133. **Horine, R. K. and Ruesink, A. W.,** Cell wall regeneration around protoplasts isolated from *Convolvulus* tissue culture, *Plant Physiol.*, 50, 438, 1972.

134. **Huges, B. G., White, F. G., and Smith, M. A.,** Effect of plant growth isolation and purification condition on barley protoplast yield, *Biochem. Physiol. Pflanz.*, 172, 67, 1978.

135. **Huhtinen, O., Honkanen, J., and Simola, L. K.,** Ornithine and putreseine-supported division and cell colony formation in leaf protoplasts of alder (*Alnus glutinosa* and *A. incana*), *Plant Sci. Lett.*, 28, 3, 1983.

136. **Ishii, S.,** Cell wall cementing materials of grass leaves, *Plant Physiol.*, 76, 959, 1984.

137. **Itoh, T., O'Neil, R. M., and Brown, R. M.,** Interference of cell wall regeneration of *Boergesenia forbesii* protoplasts by tinopal LPW, a fluorescent brightening agent, *Protoplasma*, 123, 174, 1984.

138. **Jacobson, J. V., Zwar, J. A., and Chandler, P. M.,** Gibberellic acid responsive protoplasts from mature aleurone of Himalaya barley, *Planta*, 163, 430, 1985.

139. **Johnson, L. B., Stuteville, D. L., Higgins, R. K., and Douglas, H. L.,** Pectolyase Y-23 for isolating mesophyll protoplasts from several *Medicago* species, *Plant Sci. Lett.*, 26, 135, 1982.

140. **Johnson, L. B., Stuteville, D. L., Higgins, R. K., and Skinner, D. Z.,** Regeneration of alfalfa plants from protoplasts of selected clones, *Plant Sci. Lett.*, 20, 297, 1981.

141. **Jones, M. G. K. and Dale, R. J.,** Reproducible regeneration of callus from suspension culture protoplasts of the grass *Lolium multiflorum*, *Z. Pflanzenphysiol.*, 106, 267, 1982.

142. **Kameya, T. and Uchimiya, H.** 1972, Embryoids derived from isolated protoplasts of carrot, *Planta*, 103, 356, 1972.

143. **Kanai, R. and Edwards, G. E.,** Purification of enzymatically isolated mesophyll protoplasts from C_3, C_4 and crassulacean acid metabolism plants using an aqueous dextran polyethylene glycol two phase system, *Plant Physiol.*, 52, 484, 1973.

144. **Kao, K. N.,** Chromosomal behaviour in somatic hybrids of soybean *Nicotiana glauca*, *Mol. Gen. Genet.*, 150, 225, 1977.

145. **Kao, K. N., Constabel, F., Michayluk, M. R., Keller, W. A., and Miller, R. A.,** The effects of sugars and inorganic salts on cell regeneration and sustained division in plant protoplasts, in *Protoplastes et Fusion de Cellules Sometiques Vegetales (Colloq. Int. C.N.R.S.)*, No. 212, 207, 1973.

146. **Kao, K. N., Gamborg, O. L., Miller, R. A., and Keller, W. A.,** Cell divisions in cells regenerated from protoplasts of soybean and *Haplopappus gracilis*, *Nature (London) New Biol.*, 232, 124, 1971.

147. **Kao, K. N., Keller, W. A., and Miller, R. A.,** Cell division in newly formed cells from protoplasts of soybean, *Exp. Cell Res.*, 62, 338, 1970.

148. **Kao, K. N. and Michayluk, M. R.,** Nutrient requirements for growth of *Vicia hajastana* cells and protoplasts at a very low population density in liquid media, *Planta*, 126, 105, 1975.

149. **Kao, K. N. and Michayluk, M. R.,** Plant regeneration from mesophyll protoplasts of alfalfa, *Z. Pflanzenphysiol.*, 96, 135, 1980.

150. **Kartha, K. K., Michayluk, M. R., Kao, K. N., Gamborg, O. L., and Constabel, F.** Callus formation and plant regeneration from mesophyll protoplasts of rape plants (*Brassica napus* cv. Zephyr.), *Plant Sci. Lett.*, 3, 265, 1974.

151. **Karunaratne, S. M. and Scott, K. J.**, Mitotic activity in protoplasts isolated from *Sorghum bicolor* leaves, *Plant Sci. Lett.*, 23, 11, 1981.

152. **Kirby, E. G. and Cheng, T. Y.**, Colony formation from protoplasts derived from douglas fir cotyledons, *Plant Sci. Lett.*, 14, 145, 1979.

153. **Klein, A. S., Montezinos, D., and Delmer, D. P.**, Cellulose and 1,3-glucan synthesis during early stages of wall regeneration in soybean protoplasts, *Planta*, 152, 105, 1981.

154. **Klercker, J. A.**, Eine methode zur isolierung lebender protoplasten, *Swerdk. Vel. Kad. Forh. (Stockholm)*, 9, 463, 1892.

155. **Koblitz, H.**, Isolierung and kultvierung von protoplasten aus calluskulturen der Gerste, *Biochem. Physiol. Pflanzen*, 170, 287, 1976.

156. **Kohlenbach, H. W. and Bohnke, E.**, Isolation and culture of mesophyll protoplasts of *Hyoscyamus niger* var. *annuus*, *Experientia*, 31, 1281, 1975.

157. **Kohlenbach, H. W., Wenzel, G., and Hoffmann, F.**, Regeneration of *Brassica napus* plants in cultures from isolated protoplasts of haploid stem embryos as compared with leaf protoplasts, *Z. Pflanzenphysiol.*, 105, 131, 1982.

158. **Konider, M., Hauptmann, R., Widholm, J. M., Skirvin, R. M., and Korban, S. S.**, Callus formation from *Malus* × *domestica* cv. Jonathan protoplasts, *Plant Cell Rep.*, 3, 142, 1984.

159. **Koop, H.-U. and Schweiger, H.-G.**, Regeneration of plants from individually cultivated protoplasts using an improved microculture system, *J. Plant Physiol.*, 121, 245, 1985.

160. **Krishnamurthy, M.**, Isolation, fusion and multiplication of sugarcane protoplasts and comparison of sexual and parasexual hybridization, *Euphytica*, 25, 145, 1976.

161. **Kulikowski, R. R. and Mascarenhas, J. P.**, RNA synthesis in whole cells and protoplasts of *Centaurea* — a comparison, *Plant Physiol.*, 61, 575, 1978.

162. **Landgren, C. R.**, Gibberellin enhancement of the enzymic release of *Pisum* root cell protoplasts, *Physiol. Plant.*, 52, 349, 1981.

163. **Landgren, C. R. and Bonnett, H. T.**, The culture of albino tobacco protoplasts treated with polyethylene glycol to induce chloroplast incorporation, *Plant Sci. Lett.*, 16, 15, 1979.

164. **Larkin, P. J.**, Purification and viability determination of plant protoplasts, *Planta*, 128, 213, 1976.

165. **Lee-Stadelmann, O., Chung, I., and Stadelmann, E. J.**, Plasmolysis of *Glycine max* mesophyll cells: the use of octylguanidine and its implications in protoplast isolation, *Plant Sci.*, 38, 1, 1985.

166. **Lesney, M. S., Callow, P. W., and Sink, K. C.**, A technique for bulk production of cytoplasts and mini protoplasts from suspension culture-derived protoplasts, *Plant Cell Rep.*, 5, 115, 1986.

167. **Lin, W.**, Isolation of mesophyll protoplasts from mature leaves of soybeans, *Plant Physiol.*, 73, 1067, 1983.

168. **Loewus, F. A., Baldi, B. G., Franceschi, V. R., Meinert, L. D., and McCollum, J. J.**, Pollen sporoplasts: dissolution of pollen walls, *Plant Physiol.*, 78, 652, 1985.

169. **Lörz, H., Paszkowski, J., Dierks-Ventling, C., and Potrykus, I.**, Isolation and characterization of cytoplasts and miniprotoplasts derived from protoplasts of cultured cells, *Physiol. Plant.*, 53, 385, 1981.

170. **Lörz, H., Wernicke, W., and Potrykus, I.**, Culture and plant regeneration of *Hyoscyamus* protoplasts, *Planta Medica*, 36, 21, 1979.

171. **Lu, D. Y., Cooper-Bland, S., Cocking, E. C., and Davey, M. R.**, Isolation and sustained division of protoplasts from cotyledons of seedlings and immature seeds of *Glycine max*, *Z. Pflanzenphysiol.*, 111, 389, 1983.

172. **Lu, D. Y., Vasil, V., and Vasil, I. K.**, Isolation and culture of protoplasts of *Panicum maximum* (Guinea Grass): somatic embryogenesis and plantlet formation, *Z. Pflanzenphysiol.*, 104, 311, 1981.

173. **Ludwig, S. R., Somers, D. A., Peterson, W. L., Pohlman, R. F., Zarowitz, M. A., Gengenbach, B. G., and Messing, J.**, High-frequency callus formation from maize protoplasts, *Theor. Appl. Genet.*, 71, 344, 1985.

174. **Mackenzie, I. A., Bui-Dang, Ha, D., and Davey, M. R.**, Some aspects of the isolation, fine structure and growth of protoplasts from *Asparagus officinalis*, in *Protoplastes et Fusion de Cellules Somatiques Vegetales (Colloq. Int. C.N.R.S.)*, No. 212, 291, 1973.

175. **Maeda, Y., Fujita, Y., and Yamada, Y.**, Callus formation from protoplasts of cultured *Lithospermum erythrorhizon* cells, *Plant Cell Rep.*, 2, 179, 1983.

176. **Magnien, E., Daleschaert, X., Roumengous, M., and Devreux, M.**, Improvement of protoplast isolation and culture technique from axenic plantlets of wild *Nicotiana* species, *Acta Genet. Sin.*, 7, 231, 1980.

177. **Malemberg, R. L. and Greisbach, R. J.**, The isolation of mitotic and meiotic chromosomes from plant protoplasts, *Plant Sci. Lett.*, 17, 141, 1980.

178. **Maretzki, A. and Nickell, L. G.,** Formation of protoplasts from sugarcane cell suspensions and the regeneration of cell cultures from protoplasts, in *Protoplastes et Fusion de Cellules Somatiques Vegetales (Colloq. Int. C.N.R.S.),* No. 212, 51, 1973.

179. **Marion-Poll, A. and Caboche, M.,** Relationship between auxin and amino acid metabolism of tobacco protoplast derived cells, *Plant Physiol.,* 75, 1048, 1984.

180. **Mbanaso, E. N. A. and Roscoe, D. H.,** Alginate an alternative to agar in plant protoplast culture, *Plant Sci. Lett.,* 25, 61, 1982.

181. **Menczel, L., Lazar, G., and Maliga, P.,** Isolation of somatic hybrids by cloning *Nicotiana* heterokaryons in nurse cultures, *Planta,* 143, 29, 1978.

182. **Mersey, B. G., Griffing, L. R., Rennie, P. J., and Fowke, L. C.,** The isolation of coated vesicles from protoplasts of soybean, *Planta,* 163, 317, 1985.

183. **Meyer, Y.,** Isolation and culture of tobacco mesophyll protoplasts using a saline medium, *Protoplasma,* 81, 363, 1974.

184. **Meyer, Y. and Abel, W. O.,** Importance of the wall for cell division and in the activity of the cytoplasm in cultured tobacco protoplasts, *Planta,* 123, 33, 1975.

185. **Meyer, Y. and Abel, W. O.,** Budding and cleavage division of tobacco mesophyll protoplasts in relation to pseudo-wall and wall formation, *Planta,* 125, 1, 1975.

186. **Meyer, Y., Aspart, L., and Chartier, Y.,** Auxin-induced regulation of protein synthesis in tobacco mesophyll protoplasts cultivated in vitro. I. Characteristics of auxin sensitive proteins, *Plant Physiol.,* 75, 1027, 1984.

187. **Meyer, Y., Aspart, L., and Chartier, Y.** II. Time course and level of auxin control, *Plant Physiol.,* 75, 1034, 1984.

188. **Meyer, Y. and Chartier, Y.,** Hormonal control of mitotic development in tobacco, *Plant Physiol.,* 68, 1273, 1981.

189. **Meyer, Y. and Cooke, R.,** Time course of hormonal control of the first mitosis in tobacco mesophyll protoplasts cultivated in vitro, *Planta,* 147, 181, 1979.

190. **Meyer, Y. and Herth, W.,** Chemical inhibition of cell wall formation and cytokinesis, but not of nuclear division in protoplasts of *Nicotiana tabacum* cultivated in vitro, *Planta,* 142, 253, 1978.

191. **Michayluk, M. R. and Kao, K. N.,** A comparative study of sugars and sugar alcohols on cell regeneration and sustained cell division in plant protoplasts, *Z. Pflanzenphysiol.,* 75, 181, 1975.

192. **Mii, M., Saxena, P. K., Fowke, L. C., and King, J.,** Isolation of chromosomes from cell suspension cultures of *Vicia hajastana, Cytologia,* in press.

193. **Mii, M., Seeni, S., Fowke, L. C., and King, J.,** The isolation and cultivation of protoplasts from cell suspensions of a pantothenate-requiring auxotroph of *Datura, Can. J. Bot.,* 63, 779, 1985.

194. **Moore, B., Ku, M. S. B., and Edwards, G. E.,** Isolation of bundle sheath protoplasts from C_4 dicot species and intracellular localization of selected enzymes, *Plant Sci. Lett.,* 35, 127, 1984.

195. **Motoyoshi, F.,** Protoplasts isolated from callus cells of maize endosperm, formation of multinucleate protoplasts and nuclear division, *Exp. Cell Res.,* 68, 452, 1971.

196. **Muhlbach, H. P.,** Different regeneration potentials of mesophyll protoplasts from cultivated and a wild species of tomato, *Planta,* 148, 89, 1980.

197. **Muhlbach, H. P. and Thiele, H.,** Response to chilling of tomato mesophyll protoplasts, *Planta,* 151, 399, 1981.

198. **Muller, J. F., Missionier, C., and Caboche, M.,** Low density growth of cells derived from *Nicotiana* and *Petunia* protoplasts: influence of the source of protoplasts and comparison of the growth promoting activity of various auxins, *Physiol. Plant.,* 57, 35, 1983.

199. **Murashige, T. and Skoog, F.,** A revised medium for rapid growth and bioassays with tobacco tissue cultures, *Physiol. Plant.,* 15, 473, 1962.

200. **Nagata, T. and Ishii, S.,** A rapid method for isolation of mesophyll protoplasts, *Can. J. Bot.,* 57, 1820, 1979.

201. **Nagata, T. and Takebe, T.,** Cell wall regeneration and cell division in isolated tobacco mesophyll protoplasts, *Planta,* 92, 301, 1970.

202. **Nagata, T. and Takebe, T.,** Plating of isolated tobacco mesophyll protoplasts on agar medium, *Planta,* 99, 12, 1971.

203. **Nagy, J. I. and Maliga, P.,** Callus induction and plant regeneration from mesophyll protoplasts of *Nicotiana sylvestris, Z. Pflanzenphysiol.,* 78, 453, 1976.

204. **Nakagawa, H., Tanaka, H., Oba, T., Ogura, N., and Iizuka, M.,** Callus formation from protoplasts of cultured *Spinacia oleracea* cells, *Plant Cell Rep.,* 4, 148, 1985.

205. **Nehls, R.,** Isolation and regeneration of protoplasts from *Solanum nigrum, Plant Sci. Lett.,* 12, 183, 1978.

206. **Neeman, M., Aviv, D., Degani, H., and Galun, E.,** Glucose and glycine metabolism in regenerating tobacco protoplasts, *Plant Physiol.,* 77, 374, 1985.

207. **Nelson, R. S., Creissen, G. P., and Bright, S. W. J.,** Plant regeneration from protoplasts of *Solanum brevidens, Plant Sci. Lett.,* 30, 355, 1983.

208. **Nelson, R. S., Karp, A., and Bright, S. W.,** Ploidy variation in *Solanum brevidens* plants regenerated from protoplasts using an improved culture system, *J. Exp. Bot.,* 37, 253, 1986.

209. **Nemet, G. and Dudits, D.,** Potentials of protoplast cell and tissue culture in cereal research, in *Use of Tissue Cultures in Plant Breeding,* Novak, F. J., Ed., Czechoslovak Academy of Science, Prague, 1977, 145.

210. **Newell, C. A. and Luu, H. T.,** Protoplast culture and plant regeneration in *Glycine canescens, Plant Cell Tissue Organ Culture,* 4, 145, 1985.

211. **Oelck, M. M., Bapat, V. A., and Schieder, O.,** Protoplast culture of three legumes: *Arachis hypogea, Melilotus officinalis, Trifolium resupinatum, Z. Pflanzenphysiol.,* 106, 173, 1982.

212. **Ohyama, K. and Nitsch, J. P.,** Flowering haploid plants obtained from protoplasts of tobacco leaves, *Plant Cell Physiol.,* 13, 229, 1972.

213. **Ohyama, K., Pelchar, L. E., and Horn, D.,** A rapid, simple method for nuclei isolation from plant protoplasts, *Plant Physiol.,* 60, 179, 1977.

214. **Oka, S. and Ohyama, K.,** Plant regeneration from leaf mesophyll protoplasts of *Broussonetie kazinoki* (paper mulberry), *J. Plant Physiol.,* 119, 455, 1985.

215. **Okamura, M., Hayashi, T., and Miyazaki, S.,** Inhibiting effect of ammonium ion in protoplast culture of some Asteraceae plants, *Plant Cell Physiol.,* 25, 281, 1984.

216. **Okuno, T. and Furusawa, I.,** A simple method for the isolation of intact mesophyll protoplasts from higher plants, *Plant Cell Physiol.,* 10, 917, 1977.

217. **Orczyk, W. and Malepszy, S.,** In vitro culture of *Cucumis sativus.* V. Stabilizing effect of glycine on leaf protoplasts, *Plant Cell Rep.,* 4, 269, 1985.

218. **Patnaik, G. and Cocking, E. C.,** A new enzyme mixture for the isolation of leaf protoplasts, *Z. Pflanzenphysiol.,* 107, 41, 1982.

219. **Patnaik, G., Wilson, D., and Cocking, E. C.,** Importance of enzyme purification for increased plating efficiency and plant regeneration from single protoplasts of *Petunia parodii, Z. Pflanzenphysiol.,* 102, 199, 1981.

220. **Pelcher, L. E., Gamborg, O. L., and Kao, K. N.,** Bean mesophyll protoplasts: production, culture and callus formation, *Plant Sci. Lett.,* 3, 107, 1974.

221. **Pilet, P. E., Prat, R., and Ronald, J. C.,** Morphology, RNAse and transaminase of root protoplasts, *Plant Cell Physiol.,* 13, 297, 1972.

222. **Poirier-Hamon, S., Rao, P. S., and Harda, H.,** Cultures of mesophyll protoplasts and stem segments of *Antirrhinum majus* (snapdragon): growth and organization of embryoids, *J. Exp. Bot.,* 25, 752, 1974.

223. **Potrykus, I. and Durand, J.,** Callus formation from single protoplasts of *Petunia, Nature (London) New Biol.* 237, 286, 1972.

224. **Potrykus, I., Harms, C. T., and Lorz, H.,** Callus formation from stem protoplasts of corn *(Zea mays), Mol. Gen. Genet.,* 156, 347, 1979.

225. **Potrykus, I., Harms, C. T., and Lorz, H.,** Multiple drop array (MDA) technique for the large scale testing of culture media variations in hanging microdrop cultures of single cell systems. I. The technique, *Plant Sci. Lett.,* 14, 231, 1979.

226. **Potrykus, I., Harms, C. T., Lorz, H., and Thomas, E.,** Callus formation from stem protoplasts of corn *(Zea mays L.), Mol. Gen. Genet.,* 156, 347, 1977.

227. **Power, J. B. and Cocking, E. C.,** Isolation of leaf protoplasts: macromolecular uptake of growth substance responses, *J. Exp. Bot.* 21, 64, 1970.

228. **Power, J. B., Frearson, E. M., George, D., Evans, P. H., Berry, S. F., Hayward, C., and Cocking, E. C.,** Isolation, culture and regeneration of leaf protoplasts in the genus *Petunia, Plant Sci. Lett.,* 7, 51, 1976.

229. **Premecz, G., Olah, T., Gulyas, A., Nyitrani, A., Palfi, G., and Farkas, G. L.,** Is the increase in ribonuclease level in isolated tobacco protoplasts due to osmotic stress, *Plant Sci. Lett.,* 9, 195, 1977.

230. **Premecz, G., Ruzieska, P., Olah, T., and Farkas, G. L.,** Effect of osmotic stress on protein and nucleic acid synthesis in isolated tobacco protoplasts, *Planta,* 141, 33, 1978.

231. **Rathnam, C. K. M. and Edwards G. E.,** Protoplasts is a tool for isolating functional chloroplasts from leaves, *Plant Cell Physiol.,* 17, 177, 1976.

232. **Raveh, D. and Galun, E.,** Rapid regeneration of plants from tobacco protoplasts plated at low densities, *Z. Pflanzenphysiol.,* 76, 76, 1975.

233. **Raveh, D., Huberman, E., and Galun, E.,** *In vitro* culture of tobacco protoplasts. Use of feeder techniques to support division of cells plated at low densities, *In Vitro,* 9, 216, 1973.

234. **Razdan, M. K., Cocking, E. C., and Power, J. B.,** Callus regeneration from mesophyll protoplasts of sweet pea *(Lathyrus odoratus), Z. Pflanzenphysiol.,* 96, 181, 1980.

235. **Redenbaugh, K., Ruzin, S., Bartholomew, J., and Bassham, J. A.,** Characterization and separation of plant protoplasts via flow cytometry and cell sorting, *Z. Pflanzenphysiol.,* 107, 65, 1982.

236. **Reid, R. K. and Galston, A. W.,** Experiments on the isolation and cultivation of protoplasts and calli of agriculturally important plants. II. Soybean *(Glycine max), Biochem. Physiol. Pflanzen.,* 168, 473, 1975.

237. **Reinert, J. and Hellman, S.**, Mechanisms of the formation of polynuclear protoplasts from cells of higher plants, *Naturwissenschaften*, 58, 419, 1971.
238. **Robeneck, H. and Peveling, E.**, Ultrastructure of the cell wall regeneration of isolated protoplasts of *Skimmia Japonica*, *Planta*, 136, 135, 1977.
239. **Roper, W.**, Callus formation from protoplasts derived from cell suspension cultures of *Vicia faba*, *Z. Pflanzenphysiol.*, 101, 75, 1981.
240. **Roscoe, D. H. and Bell, G. M.**, Use of pH indicator in protoplast culture medium, *Plant Sci. Lett.*, 21, 275, 1981.
241. **Rose, R. J.**, Factors that influence the yield, stability in culture and cell wall regeneration of spinach mesophyll protoplasts, *Aust. J. Plant Physiol.*, 7, 713, 1980.
242. **Ruesink, A. W.**, Surface membrane properties of isolated protoplasts, in *Protoplasates et Fusion de Cellules Somatiques Vegetales (Colloq. Int. C.N.R.S.)*, No. 212, 41, 1973.
243. **Ruesink, A. W. and Thimann, K. V.**, Protoplasts from the *Avena* coeloptile, *Proc. Natl. Acad. Sci. U.S.A.*, 54, 56, 1965.
244. **Ruesink, A. W. and Thimann, K. V.**, Protoplasts: preparation from higher plants, *Science*, 154, 280, 1966.
245. **Ruzieska, P., Mettrie, R., Dorkhov, Y. L., Premecz, G., Olah, T., and Farkas, G. L.**, Polyribosomes in protoplasts isolated from tobacco leaves, *Planta*, 145, 199, 1979.
246. **Santos, A. V. P., Jr., Outka, D. E., Cocking, E. C., and Davey, M. R.**, Organogenesis and somatic embryogenesis in tissues derived from leaf protoplasts and leaf explants of *Medicago sativa*, *Z. Pflanzenphysiol.*, 99, 261, 1980.
247. **Saxena, P. K.**, Enhancement of protoplast regeneration by cold-conditioning of donor tissue, *J. Plant Physiol.*, 119, 385, 1985.
248. **Saxena, P. K., Gill, R., Rashid, A., and Maheshwari, S. C.**, Colony formation by cotyledonary protoplasts of *Cyamopsis tetragenoloba*, *Z. Pflanzenphysiol.*, 106, 277, 1982.
249. **Saxena, P. K., Fowke, L. C., and King, J.**, An efficient procedure for isolation of nuclei for plant protoplasts, *Protoplasma*, 128, 184, 1985.
250. **Schaskolskaya, N. D., Sacharovskaya, G. W., and Sacharova, E. V.**, The optimal conditions for isolation and incubation of barley mesophyll protoplasts, in *Protoplastes et Fusion de Cellules Somatiques Vegetales (Colloq. Int. C.N.R.S.)*, No. 212, 93, 1973.
251. **Schenk, R. U. and Hildebrandt, A. C.**, Production of protoplasts from plant cells in liquid culture using purified commercial cellulases, *Crop. Sci.*, 9, 629, 1969.
252. **Schieder, O.**, Regeneration von haploiden und diploiden *Datura innoxia*. Mesophyll protoplasten zu pflanzen, *Z. Pflanzenphysiol.*, 76, 462, 1975.
253. **Schwenk, F. W., Pearson, C. A., and Roth, M. R.**, Soybean mesophyll protoplasts, *Plant Sci. Lett.*, 23, 153, 1981.
254. **Scowcroft, W. R., Davey, M. R., and Power, J. B.**, Crown gall protoplasts, isolation culture and ultrastructure, *Plant Sci. Lett.*, 1, 451, 1973.
255. **Scowcroft, W. R. and Larkins, P. J.**, Isolation culture and plant regeneration from protoplasts of *Nicotiana debneyi*, *Aust. J. Plant Physiol.*, 7, 635, 1980.
256. **Shahin, E. A. and Shepard, J. F.**, Cassava mesophyll protoplasts: isolation, proliferation, and shoot formation, *Plant Sci. Lett.*, 17, 459, 1980.
257. **Shekhawat, N. S. and Galston, A. W.**, Isolation, culture and regeneration of moth bean *Vigna aconitifolia* leaf protoplasts, *Plant Sci. Lett.*, 32, 43, 1983.
258. **Shepard, J. F.**, Mutant selection and plant regeneration from potato mesophyll protoplasts, in *Genetic Improvement of Crops, Emergent Techniques*, Rubenstein, I., Gengenbach, B., Phillips, R. L., and Green, C. E., Eds., University of Minnesota Press, Minneapolis, 1980, 185.
259. **Shepard, J. F.**, Cultivar dependent cultural refinements in potato protoplast regeneration, *Plant Sci. Lett.*, 26, 127, 1982.
260. **Shepard, J. F. and Totton, R. E.**, Isolation and regeneration of tobacco mesophyll cell protoplasts under low osmotic conditions, *Plant Physiol.*, 55, 689, 1975.
261. **Shepard, J. F. and Totton, R. E.**, Mesophyll cell protoplasts of potato isolation proliferation and plant regeneration, *Plant Physiol.*, 60, 313, 1977.
262. **Shilito, R. D., Paszkowski, J., and Potrykus, I.**, Agarose plating and a bead type culture technique enable and stimulate development of protoplast derived colonies in a number of plant species, *Plant Cell Rep.*, 2, 244, 1983.
263. **Shimizu, J. I.**, Cell regeneration and division of grape mesophyll protoplasts, *J. Plant Physiol.*, 119, 419, 1985.
264. **Shneyour, Y., Zelcher, A., Izhar, S., and Beckmann, J. S.**, A simple feeder-layer technique for the plating of plant cells and protoplasts at low density, *Plant Sci. Lett.*, 33, 293, 1984.
265. **Simmonds, D. H. and Setterfield, G.**, Aberrant microtubule organization can result in genetic abnormalities in protoplast cultures of *Vicia hajastana*, *Planta*, 167, 468, 1986.

266. **Simmonds, J. A., Simmonds, D. H., and Cumming, B. G.**, Isolation and cultivation of protoplasts from morphogenetic callus cultures of *Lilium*, *Can. J. Bot.*, 57, 512, 1979.

267. **Slabas, A. R., Powell, A. J., and Lloyd, C. W.**, An improved procedure for the isolation and purification of protoplasts from carrot suspension culture, *Planta*, 147, 283, 1980.

268. **Smith, M. A. L. and McCown, B.**, A comparison of source tissue for protoplast isolation from 3 woody plant species, *Plant Sci. Lett.*, 28, 149, 1982.

269. **Smith, M. A. L., Palta, J. P., and McCown, B. H.** The measurement of isotonicity and maintenance of osmotic balance in plant protoplast manipulation, *Plant Sci. Lett.*, 33, 249, 1984.

270. **Sommerville, C. R., Sommerville, S. C., and Ogren, W. L.**, Isolation of photosynthetically active protoplasts and chloroplasts from *Arabidopsis thaliana*, *Plant Sci. Lett.*, 21, 89, 1981.

271. **Spangenberg, G., Koop, H.-U., Lichter, R., and Schweiger, H. G.**, Microculture of single protoplasts of *Brassica napus*, *Physiol. Plant.*, 66, 1, 1986.

272. **Swanson, E. B., Wong, R. S. C., and Kemble, R. J.**, A novel method for the isolation and purification of protoplasts from friable embrogenic corn (*Zea mays*) callus, *Plant Sci.*, 40 137, 1985.

273. **Tabaeizadeh, Z., Bunisset-Bergounioux, C., and Perennes, C.**, Environmental growth conditions of protoplast source plants. Effect on subsequent protoplast division in two tomato species, *Physiol. Veg.*, 22, 223, 1984.

274. **Taiz, L. and Jones, R. L.**, The isolation of barley-aleurone protoplasts, *Planta*, 101, 95, 1971.

275. **Takebe, I. and Nagata, T.**, Culture of isolated tobacco mesophyll protoplasts, in *Protoplastes et Fusion de Cellules Somatiques Vegetales (Colloq. Int. C.N.R.S.)*, No. 212, 175, 1973.

276. **Takebe, I., Otsuki, Y., and Aoki, S.**, Isolation of tobacco mesophyll cells in intact and active state, *Plant Cell Physiol.*, 9, 115, 1968.

277. **Takeuchi, Y. and Komamine, A.**, Composition of the cell wall formed by protoplasts isolated from cell suspension cultures of *Vinca rosea*, *Planta*, 140, 227, 1978.

278. **Takeuchi, Y. and Komamine, A.**, Effects of culture conditions on cell division and composition of regenerated cell walls in *Vinca rosea* protoplasts, *Plant Cell Physiol.*, 23, 249, 1982.

279. **Tal, M. and Watts, J. W.**, Plant growth conditions and yield of viable protoplasts isolated from leaves of *Lycopersicon esculentum* and *L. peruvianum*, *Z. Pflanzenphysiol.*, 92, 207, 1979.

280. **Tallman, G. and Reeck, G. R.**, Isolation of nuclei from plant protoplasts without the use of a detergent, *Plant Sci. Lett.*, 18, 271, 1980.

281. **Tewes, A., Glund, K., Walther, R., and Rinbothe, H.**, High yield, isolation and rapid recovery of protoplasts from suspension cultures of tomato, *Z. Pflanzephysiol.*, 113, 141, 1985.

282. **Thomas, E.**, Plant regeneration from shoot culture-derived protoplasts of tetraploid potato (*Solanum tuberosum* cv. Moris Bard), *Plant Sci. Lett.*, 23, 81, 1981.

283. **Tomita, Y., Suzuki, H., and Nisizawa, K.**, Chromatographic patterns of cellulose components of *Trichoderma viridae* grown on the synthetic and natural media, *J. Ferment. Technol.*, 46, 701, 1968.

284. **Tremblay, F. M., Power, J. B., and Lalonde, M.**, Callus regeneration from *Alnus incana* protoplasts isolated from cell suspension, *Plant Sci.*, 41, 221, 1985.

285. **Uchimiya, H. and Murashige, T.**, Evaluation of parameters in the isolation of viable protoplasts from cultured tobacco cells, *Plant Physiol.*, 54, 936, 1974.

286. **Uchimiya, H. and Murashige, T.**, Influence of the nutrient medium on the recovery of dividing cells from tobacco protoplasts, *Plant Physiol.*, 57, 424, 1976.

287. **Ulrich, T. H., Chowdhury, J. B., and Widholm, J. M.**, Callus and root formation from mesophyll protoplasts of *Brassica rapa*, *Plant Sci. Lett.*, 19, 347, 1980.

288. **Upadhya, M. D.**, Isolation and culture of mesophyll protoplasts of potato (*Solanum tuberosum* L.), *Potato Res.*, 18, 438, 1975.

289. **Vardi, A. and Raveh, D.**, Cross feeder experiments between tobacco and orange protoplast, *Z. Pflanzenphysiol.*, 78, 350, 1976.

290. **Vardi, A., Spiegel-Roy, P., and Galun, E.**, Plant regeneration from citrus protoplasts: viability in methodological requirements among cultivars and species, *Theor. Appl. Genet.*, 62, 171, 1982.

291. **Vasil, I. K. and Vasil, V.**, Isolation and culture of protoplasts, *Int. Rev. Cytol.*, 11B (Suppl.), 1, 1980.

292. **Vasil, V. and Vasil, I. K.**, Growth and cell division in isolated plant protoplasts in microchambers, in *Protoplastes et Fusion de Cellules Somatiques Vegetales (Colloq. Int. C.N.R.S.)*, No. 212, 139, 1973.

293. **Vasil, V. and Vasil, I. K.**, Regeneration of tobacco and *Petunia* plants from protoplasts and cultures of corn protoplasts, *In Vitro*, 10, 83, 1974.

294. **Vasil, V. and Vasil, I. K.**, Isolation and cultures of cereal protoplasts. I. Callus formation from pearl millet (*Pennisetum americanum*) protoplasts, *Z. Pflanzenphysiol.*, 92, 379, 1979.

295. **Vasil, V. and Vasil, I. K.**, Isolation and cultures of cereal protoplasts. II. Embryogenesis and plantlet formation from protoplasts of *Pennisetum americanum*, *Theor. Appl. Genet.*, 56, 97, 1980.

296. **Vasil, V., Wang, D-Y., and Vasil, I. K.**, Plant regeneration from protoplasts of Napier grass *Pennisetum purpurem*, *Z. Pflanzenphysiol.*, 111, 233, 1983.

297. **Vatsya, B. and Bhaskaran, S.**, Production of subprotoplasts in *Brassica oleracea* var *capitata* — a function of osmolarity of the medium, *Plant Sci. Lett.*, 23, 277, 1981.

298. **Vatsya, B. and Bhaskaran, S.**, Plant regeneration from cotyledonary protoplasts of cauliflower (*Brassica oleraceae* var. *botrytis*), *Protoplasma*, 113, 161, 1982.

299. **Vreugdenhil, D., Harkes, P. A. A., and Libbenga, K. R.**, Auxin-binding by particulate fraction from tobacco leaf protoplasts, *Planta*, 150, 9, 1980.

300. **Wagner, G. T. and Siegelman, H. W.**, Large scale isolation of intact vacuoles and isolation of chloroplasts of mature plant tissues, *Science*, 190, 1298, 1975.

301. **Wallin, A. and Eriksson, T.**, Protoplast culture from cell suspensions of *Daucus carota*, *Physiol. Plant.*, 28, 31, 1973.

302. **Wallin, A., Glimelius, K., and Eriksson, T.**, Pretreatment of cell suspensions and a method to increase the protoplast yield of *Haplopappus gracilis*, *Physiol. Plant.*, 40, 307, 1977.

303. **Wallin, A. and Welander, M.**, Improved yield of apple leaf protoplasts from in vitro cultured shoots by using very young leaves and adding L-methionine to shoot medium, *Plant Cell Tissue Organ Culture*, 5, 69, 1985.

304. **Watts, J. W. and King, J. M.**, A method for the preparation of plant protoplasts, *Z. Naturforsch.*, 28C, 231, 1973.

305. **Watts, J. W., Motoyoshi, F., and King, J. M.**, Problems associated with the production of stable protoplasts of cells of tobacco mesophyll, *Ann. Bot.*, 38, 667, 1974.

306. **Wenzel, G.**, Isolation of leaf protoplasts from haploid plants of *Petunia*, rape and rye, *Z. Pflanzenzuecht.*, 69, 58, 1973.

307. **Wernicke, W. and Thomas, E.**, Studies on morphogenesis from isolated plant protoplasts: shoot formation from mesophyll protoplast of *Hyoscyamus muticus* and *Nicotiana tabacum*, *Plant Sci. Lett.*, 77, 401, 1980.

308. **White, D. W. R. and Bhojwani, S. S.**, Callus formation from *Trifolium arvense* protoplast-derived cells plated at low densities, *Z. Pflanzenphysiol.*, 102, 257, 1981.

309. **Williamson, F. A., Fowke, L. C., Weber, G., Constabel, F., and Gamborg, O. L.**, Microfibril deposition on cultured protoplasts of *Vicia hajastana*, *Protoplasma*, 91, 213, 1977.

310. **Willison, J. H. M. and Cocking, E. C.**, Microfibril synthesis at the surface of isolated tobacco mesophyll protoplasts, a freeze-etch study, *Protoplasma*, 84, 147, 1975.

311. **Willison, J. H. M. and Grout, B. W. W.**, Further observations on cell wall formation around isolated protoplasts of tobacco and tomato, *Planta*, 140, 53, 1978.

312. **Wilson, H. M., Styer, D. J., Conrad, P. L., Durbin, R. D., and Helgeson, J. P.**, Isolation of sterile protoplasts from unsterilized leaves, *Plant Sci. Lett.*, 18, 151, 1980.

313. **Wilson, V. M., Haq, N., and Evans, P. K.**, Protoplast isolation culture and plant regeneration in the winged bean *Psophocarpus tetragonolobus*, *Plant Sci.*, 41, 61, 1985.

314. **Wright, D. C.**, Factors affecting isolation of protoplasts from leaves of grapes *(Vitis vinifera)*, *Plant Cell Tissue Organ Culture*, 4, 95, 1985.

315. **Wright, S. T. C. and Hiron, R. W. P.**, Abscisic acid, the growth inhibitor-induced in detached wheat leaves by a period of wilting, *Nature (London)*, 224, 719, 1969.

316. **Wu, S. C. and Kuniyuki, A. H.**, Isolation and culture of almond protoplasts, *Plant Sci.*, 41, 55, 1985.

317. **Xu, Z. H. and Davey, M. R.**, Shoot regeneration from mesophyll protoplasts and leaf explants of *Rehmania glutinosa*, *Plant Cell Rep.*, 2, 55, 1983.

318. **Xu, Z. H., Davey, M. R., and Cocking, E. C.**, Isolation and sustained division of *Phaseolus aureus* (mung bean) root protoplasts, *Z. Pflanzenphysiol.*, 104, 289, 1981.

319. **Xu, Z. H., Davey, M. R., and Cocking, E. C.**, Callus formation from root protoplasts of *Glycine max* (soybean), *Plant Sci. Lett.*, 24, 111, 1982.

320. **Xu, Z. H., Davey, M. R., and Cocking, E. C.**, Organogenesis from root protoplasts of the forage legumes *Medicago sativa* and *Trigonella foenum-graecum*, *Z. Pflanzenphysiol.*, 107, 231, 1982.

321. **Yamada, Y., Zhi-Qi, Y., and Ding-Tai, T.**, Plant regeneration from protoplast derived callus of rice, *Plant Cell Rep.*, 5, 85, 1986.

322. **Zapata, F. J., Sink, K. C., and Cocking, E. C.**, Callus formation from leaf mesophyll protoplasts of three *Lycopersicon* species, *L. esculantum* cv. Walter, *L. pimpinillifolium* and *L. hirsutum* f. *glabratum*, *Plant Sci. Lett.*, 23, 41, 1981.

323. **Zelcher, A. and Galun, E.**, Culture of newly isolated tobacco protoplasts: precursor incorporation into protein, RNA and DNA, *Plant Sci. Lett.*, 7, 331, 1976.

324. **Zelcher, A. and Galun, E.**, Culture of newly isolated tobacco protoplasts: cell division and precursor incorporation following a transient exposure to coumarin, *Plant Sci. Lett.*, 18, 185, 1980.

325. **Zia, J. F. and Potrykus I.**, Mesophyll protoplasts of *Solanum melongena* regenerate to fertile plants, *Plant Cell Rep.*, 1, 71, 1981.

326. **Zieg, R. G. and Outka, D. E.**, The isolation, culture and callus formation of soybean pod protoplasts, *Plant Sci. Lett.*, 18, 105, 1980.

327. **Zieger, E. and Hepler, P. K.**, Light and stomatal function: blue light stimulates swelling of guard cell protoplast, *Science*, 196, 887, 1979.

Chapter 2

CELL MODIFICATION

I. INTRODUCTION

Regeneration of whole plants from an isolated cell or protoplast in an ever-increasing number of species has introduced the possibility of employing these experimental systems for induction and selection of variant cell lines or mutants and, ultimately, the development of new plants for human welfare. The rapid progress in microbial genetics and the consequent utility of microbial systems for industrial purposes can be attributed to many mutants that could be generated for specific biochemical events. This progress helped in the understanding of metabolic pathways and increased the yield of desirable metabolites. Similar mutants in higher plants, particularly crop plants, are required for understanding the elements of form and function in these plants, so that desired modifications are possible. One problem in raising mutants in higher plants is the complex nature of their form and function. To an extent, this complexity is reduced by using cells and protoplasts.

The need for generating new variability can be realized from the fact that the existing genetic potential of existing crop plants is diminishing fast as a result of specialized breeding, intensive cultivation, and vast dissemination. Some possible consequences are susceptibility to new pathogens, inability of the crops to adapt to changing environments, and an inadvertant spread of silent, deleterious genes. A good example is the U.S. corn blight of 1970 which resulted in a crop loss worth $1 billion. It was due to the widespread use of a cytoplasmic male sterile line which turned out to be susceptible to the blight.

II. CELL GENETICS — COMPLEMENTARY TO CONVENTIONAL GENETICS

Cell and protoplast cultures comprise an ideal system for the development of mutants. A plant cell geneticist can screen millions of cells in a small flask or dish using 10^4 to 10^5 cells or protoplasts per milliliter. Each cell or protoplast is a potential plant. On the other hand, a field test of an equal number of plants would require an enormous investment in terms of man, money, and space. In culture, it is easy to treat a large population of cells with modifying agents and subject them to selection pressure in controlled conditions. Also, in culture it is possible to establish a wide range of biochemical conditions which are not possible in the field.

A cell or protoplast has a limited metabolic pool and can be cultured on a defined medium. Complexities of form and function in higher plants are simplified at this level. Using haploid cells and protoplasts, it is even possible to select for recessive characters, which is not possible with a diploid plant. Also, a direct selection of recessive mutant plants is possible with anther or pollen-derived haploid plants. Therefore, the approach of a cell geneticist in increasing the pool of genetic variability is to complement the efforts of the plant breeder. This approach is believed to be more effective and is likely to yield rapid results.

III. TERMINOLOGY — MUTANT vs. VARIANT

Variants in tissue culture are selected on the basis of unusual phenotype. Any change, however, can be the result of a permanent, sexually transmissible, hereditary alteration in the primary structure of DNA, termed "mutation" or due to physiological basis, which is not sexually transmitted but is heritable, termed as "epigenetic change". The best examples of epigenetic changes are auxin and cytokinin habituation of plant cells.[185-187] This variation is stable and heritable at the cell level, but is not transmitted either through regeneration of plants or sexually. Therefore, the term mutation should be reserved for any change that is transmitted meiotically, according to the laws of inheritance. When the nature of heritable change is not known, the term variant should be applied.

A. Epigenetic Variation

As pointed out earlier, the best example of epigenetic variation is tissue habituation. Considered earlier to be spontaneous in origin, now, at least in some tissues, it can be induced.

The conversion of tobacco cells to cytokinin autonomy is a gradual, progressive process[187] which, unlike mutation, is strongly influenced by the physiological and developmental stage of cells.[188,189] The induction process is directed[189] and occurs at a high frequency — greater than 10^{-3}.

Once established, the habituated state is stable and heritable, passed on from cell to cell. However, reversion occurs at high frequency when the habituated tissue is made to regenerate into plants. The cells from regenerants, again, require cytokinin. Therefore, cytokinin habituation presumably involves the expression of a preexisting potentiality of the plant cell; it can be interpreted as the expression of a gene that is normally silent in pith cells of tobacco.[186]

To distinguish between an epigenetic change and a mutation, it has been proposed that there are two systems of inheritance:[213] a genetic system, concerned with transmission of developmental potentialities through sexual generation, and an epigenetic system, concerned with somatic transmission of the pattern of gene expression. Thus, epigenetic change is an heritable cellular alteration which does not result from a permanent change in the cell genome. An epigenetic change can be distinguished from mutation in a number of ways: (1) the change is directed and under inductive conditions the rate of variation is quite high, (2) the change, though stable, is potentially reversible again at high frequency, and (3) the epigenetic change is not transmitted meiotically.

B. Mutation

There exists some controversy in plant literature regarding the use of the term mutation; different criteria are applied to identify a cell line as mutant. In analogy with animal cell literature, some workers characterized variants as mutants on the basis of the following characters: (1) the phenotype is stable, (2) it occurs in a low frequency, and (3) it can be increased by mutagen. These three characters are sufficient for animal cell lines because regeneration is not possible from animal cells. In plants, where regeneration of whole plants is possible from cells, sexual transmission should be studied. Therefore, in addition to these three characters inheritance of variant phenotypes according to the rules of classical genetics (Mendelian segregation of nuclear gene mutants and uniparental transmission of cytoplasmic mutants) should be satisfied for any variant to qualify as a mutant.[170,171] However, plant cells in culture progressively lose the potential for regeneration. In such an eventuality (absence of regeneration), an altered gene product can be accepted as proof of mutation. It is necessary to understand why sexual transmission of a character or demonstration of altered gene product is an essential criterion to define a mutant because the first three characters described above can be ambiguous. For example, if we take the character of stability of phenotype in absence of selection pressure, then auxin- and cytokinin-habituated cells will also qualify as mutants. A low frequency of variant is also a poor criterion because epigenetic variants, such as cycloheximide-resistant cells, have been found to occur at a low frequency.[176]

Occasionally one comes across a paradoxical situation. The expression of a character is not always possible on regenerated plants, and character is expressed only at the level of a cell. For example, in a tobacco line resistant to 5-methyltryptophan, overproduction of tryptophan and an altered anthranilate synthetase were not expressed in plants regenerated but were expressed in cell cultures initiated from regenerates.[316]

IV. SPONTANEOUS MODIFICATION OR SOMACLONAL VARIATION

In earlier literature, one frequently finds instances of cell variation and the variant plants raised from them. These variations were, however, described in a rather cryptic way, because tissue or cell culture was considered to be basically a method to clone a particular genotype, and any variation was dismissed as an ''artifact of cultural conditions''. Nonetheless, tissue culture per se is an unexpectedly rich and novel source of genetic variability. In culture, there are changes in the cell environment, and new conditions may lead to a different cell behavior. Variation resulting in response to culture can be described as spontaneous modification or, in a precise term, somaclonal variation.[150] Regeneration of plants is possible from cell cultures originated from a somatic tissue as well as from a gametic tissue. Therefore, in order to distinguish between plants regenerating from two types of tissues, it is proposed that the use of the term somaclone should be restricted to plants regenerating from cell cultures originated from somatic tissue and a new term, gametoclone,[87] should be introduced to refer to the plants regenerated from cell cultures originated from gametic tissue. At this stage, it is worth noting that prior to this terminology, the term calliclone[274,275] was used to describe variant plants regenerating from callus cultures of geranium and the term protoclone[265] was used to describe the variant plants regenerated from protoplasts of potato. Further work will resolve the use of these different terms to describe the phenomenon because the importance of spontaneous modification is being realized.

One of the reasons for the late awareness about the importance of this phenomenon is the fact that the field of tissue and cell culture has been an exclusive reserve of plant physiologists, and only recently has it become a team effort[106] of physiologists, cytologists, geneticists, and plant breeders. This can be seen in the improvement of sugarcane,[220] tobacco,[69,228] pelargonium,[275] and potato.[265] This technology is applicable to any crop for which tissue culture-mediated plant regeneration protocol has been worked out.

A. Variant Plants with Desirable Characters

Although a large number of examples of variant plants with desired agronomic characters regenerated from tissue cultures are available, in most of them the nature of genetic variation has not been defined. One of the reasons for this may be the inability or impracticability of performing sexual crosses in vegetatively propagated plants. The work done in different crops is discussed below.

1. Sugarcane

Sugarcane is a vegetatively propagated cash crop and is a complex polyploid hybrid in which it is difficult to interpret the sexual crosses. The first attempt to utilize somaclonal variation began at the experimental station of the Hawaiian Sugarcane Planters Association. To begin with, variations were recorded in morphological, cytogenetic, and isozyme traits.[112,114] Some clones resistant to Fiji disease and eye spot disease[112] due to *Helminthosporium sacchari* were also obtained. A breakthrough was possible when some clones resistant to virus (Fiji disease) and downy mildew were obtained from susceptible pindar clones of sugarcane.[146] In tissue cultures, there was a predominant shift towards increased resistance to these diseases. These clones were tested in Fiji and Australia and were found to be resistant and showed a slight increase in yield of sugar.

The high frequency of variants resistant to eye spot disease was also possible from an Australian cultivar, Q101,[151] which is a highly desirable form but susceptible to eye spot disease. Callus cultures were initiated from the leaf base and maintained for 1 to 18 months prior to regeneration of the plants. The regenerated plants were tested for their resistance to disease, using a bioassay in which leaf discs were used to assess the leakage of electrolytes on exposure to the disease toxin. The resistant somaclones retained their resistance through subsequent generations. However, some underwent somatic regression.

Also, in Taiwan,[163] significant variations among sugarcane somaclones from eight varieties have been found in terms of sugar yield, stalk number, and cane length, etc. Some of these somaclones further improved over the performance of their parental clones. More interesting is the recovery of plants from tissue cultures resistant to the smut *Ustilago scitaminea*.[164]

As for the explanation, it has been proposed that commercial sugarcane cultivars are chromosomal mosaics. Hence, it is not surprising that plants regenerated from somatic tissues are highly variable.

2. Potato

A striking demonstration of somaclonal variation[265] was provided in potato cv. Russet Burbank, which has been cultivated for 70 years and represents 39% of the potato crop in the U.S. Starting with over 1000 clones raised from mesophyll protoplasts described as "protoclones", variant plants were found in terms of compactness of growth habit, date of maturity, tuber uniformity, tuber skin color, photoperiod requirement, etc. Some of these, such as tuber uniformity and early tuberization, are desirable characters over the parent. A more exciting part of the results relates to disease resistance;[264] 5 clones were resistant to toxin released by *Alternaria solani* (early blight) and 20 clones were resistant to race 0 of *Phytophthora infestans*. However, in somaclones of German potato, extensive variation could not be recorded,[309] but resistance to *P. infestans* toxin has been observed.[16] Variation in morphological characters is also on record in English potato cv. Maris Bard.[296] Unfortunately, these variations were not an improvement over the parental clone. Variation has also been reported for other potato cultivars.[5,279] Detailed cytological studies[133] have revealed the prevalence of aneuploidy in regenerated plants.

In addition to plants of protoplast origin, variation has also been recorded in plants regenerated from leaf, rachis, and petiole explants.[39] Recently, cytoplasmic DNA variation, in terms of mitochondrial DNA, has also been shown in potato plants regenerated from mesophyll protoplasts.[138] The data indicate that molecular diversity of mitochondrial genome is possible and can be of help in overcoming the cytoplasmic genetic uniformity prevalent in most major crops.

It is not clear why there is a high incidence of somaclonal variation in Russet Burbank potato.[100] This potato is a homozygous recessive for genes that cause spindle abnormalities leading to first division restitution (FDR), second division restitution (SDR), and double restitution (DR); as a result, no haploid pollen is produced. Protoclones of this potato were screened for mutation at this locus. Among 65 protoclones, 3 plants showed 20 to 30%, 1 plant showed 69%, and a mutant (M 248) showed 83% reduction in spindle abnormalities. The wide (20 to 69%) variation in reduction in four protoclones resulted from a reduced frequency of one or more types of spindle abnormalities, but only mutant M 248 showed a consistent and drastic reduction in all spindle abnormalities. Thus, M 248 produced 83% haploid pollen while Russet Burbank produced none. These results indicate a high mutation rate in cell cultures of this cultivar of potato.

Different genotypes of potato on culture under similar conditions differed in genetic stability[280] as assessed by rate and degree of polyploidization and aneuploidy. Monohaploids showed a more rapid rate of polyploidization than did dihaploid and tetraploid potatoes. It is concluded that differences in genetic stability are due to different ploidy levels and genetic makeup of genotypes.

3. Maize

From callus cultures of maize, plants resistant to corn leaf blight (*Drechslera maydis* Race T) were possible.[98] In confirmation[35,36] of this work it has been found that the frequency of resistant variants which are of spontaneous origin is very high.

4. Alfalfa

A high-yielding somaclone, over the parents, was possible in alfalfa.[247]

5. Geranium

New varieties in *Pelargonium*[274] released are somaclonal variants.

6. Brassica

Yellow-seeded variants of *Brassica juncea*[99] var. Rai 5 were isolated from a blackish-brown-seeded cultivar. The new variant has 2% more oil than the parent, and plants grown for three generations were found to breed true for yellow seed color.

B. Variant Plants with New Characters

Variant plants have been raised through tissue culture which show new characters which can be modified. Again, examples are taken from different crops.

1. Tobacco

The most extensively worked out plant in tissue culture is tobacco, but unfortunately, no desirable variant has been recovered.[53] However, one thing that has become clear in the study of somaclonal variation is that highly inbred cultivars of tobacco do, in fact, retain heterozygosity, and this becomes apparent in variants among dihaploid plants raised through anther culture. Some of these plants showed reduced vigor[44] and crumpled leaf phenotype.[68]

2. Rice

A number of variations are seen in somaclones from rice callus.[235] Many of the variants segregated in selfed progeny, but some were possible in a pure breeding state. Among the desired characters were increased grain number, panicle number, and grain weight.

Calrose 76, a commercial cultivar of rice[254] in the U.S., was developed for short stature and stiff straw. Doubled haploids derived from anther culture of this cultivar were 20 to 30% shorter and were 30% more tillering than the parent.[259]

3. Wheat

In tissue cultures of 4 cultivars of wheat,[2] of the 46 plants raised, only 28 were self-fertile. In Australian wheat Yaqui 50E, which is tip awned, through somaclones it has been possible to raise awnless variants.[151]

4. Tomato

Variant plants have been recovered in tomato[87] in respect to (1) male sterility, (2) jointless pedicel, (3) tangerine virescent leaf, (4) flower and fruit color, (5) lethal chlorophyll deficiency, and (6) virescence and mottled leaf appearance, and (7) dominant variations concerning fruit ripening and growth habit. On the basis of these characters it can be concluded that in tomato, (1) chromosome number variation can be recovered in regenerated plants and (2) the variants include dominant, semidominant, and recessive nuclear mutations. Of these, the more important factor is the development of new lines with higher pigment in the fruit. Also, a line has been obtained which shows resistance to *Fusarium oxysporum*.

C. Variants with Chromosome Modifications

Even before the recognition of the concept of somaclonal variation, there were a large number of reports showing variation in chromosome number in cell cultures and plants raised from them.[1] It is very rare to find the regeneration of normal diploid plants from an old culture. However, this is seen, despite the presence of abnormal cells[66] in the culture used for regeneration. This indicates that, at least in some species, a selective advantage exists for organogenesis by diploid cells. On the other hand, polyploid and aneuploid plants have been regenerated in vitro from a large number of plant species. Commercially important crop species where polyploid plants have been regenerated are *Pelargonium zonale*,[275] *Nicotiana tabacum*,[42] *N. alata*,[31] *Lycopersicon esculentum*,[86] and *Medicago sativa*.[247] Aneuploid plants have been recovered in *N. tabacum*[256] and *Saccharum*.[114] Polyploid as well as aneuploid plants were recovered in potato.[60] Most of these reports concern plants with a large number of chromosomes which are small in size and where it is difficult to carry out karyotypic analysis. In plants with large-sized chromosomes, such as *Allium cepa*,[223] plants regenerated from callus revealed acentric and centric fragments, ring chromosomes, bridges, and micronuclei. In another plant with large chromosomes, *Haworthia*,[225] translocations and deletions were seen in mitotic preparations. The study of anaphase I also revealed bridges and fragments, paracentric inversions, and subchromatid aberrations.

In a somaclone of wheat[210] a high frequency of univalents was seen. This suggested that several interchanges had occurred.

In somaclones of oats,[184] a high frequency of heteromorphic bivalents was seen which was interpreted as a result of chromosome deletions. The large frequency of deletions is likely to allow for selection and recovery of recessive mutants in diploid culture and, provided the locus of recessive mutation is matched by a deletion in the homologue, the mutation will express. This is referred to as culture-induced hemizygosity.[270] The occurrence of heteromorphic bivalents at low frequency is reported in maize,[106] wheat,[2] and rye grass hybrids.[1] Reciprocal translocation and inversion are also seen in wheat and rye grass hybrids; the same has been seen in *A. sativum*.[223]

Using Giemsa C-banding technique to examine large chromosomes of *Crepis capillaris*,[3] it has been shown that the chromosome rearrangements can be very extensive. The change was so apparent that it was difficult to relate chromosomes from cell cultures to chromosomes of root tip. An extension of this study is likely to yield valuable information.

D. Nature of Variation

The foregoing brief summary of somaclonal variation raises a number of questions.

1. Is Variation Preexisting?

At present, there is evidence which indicates that variation preexists as well as being induced in culture.

Studies on pineapple[302] and potato[301] indicate that the frequency of somaclonal variation is dependent on the explant source for initiating the culture. An application of fluctuation test analysis to tobacco plants[8] regenerated from callus culture during 30 to 55 days indicated that many of the variations preexisted or were generated very early in the process of culture. Also, in tobacco[134] a plant was discovered with a small sector of ruffled leaf and the raising of tissue from this region helped isolate plants with ruffled leaves.

However, other studies indicate that variations are induced and amplified in the process of culture. Variations among plants raised from a single protoplast of potato[295] provide good evidence for the occurrence of variation during the culture.

Other evidence in favor is the increase in frequency of variations during culture in garlic[223] and oat.[184] There is also evidence for early occurrence of genetic instability in protoplast cultures of potato.[281] Sectorial analysis of calli and variant plants regenerated from them support the culture phase origin of variation in rice[234,235] and maize.[80]

On chromosome doubling, rice plants[235] raised from microspores via anther culture proved homozygous as well as heterozygous. Similarly, in rapeseed[119] 9 out of 36 microspore-derived plants were heterozygous for some characters. Since donor plants were highly inbred and true breeding, the variation is interpreted to occur after diplodization.

2. Is Variation Simple or Complex?

Some of the variations are simple in inheritance, such as in tobacco[8] and maize,[80] whereas others show complex inheritance such as heading date and plant height in rice.[234] Although there is no information about the inheritance of disease resistance by somaclones of sugarcane,[113] in potato[264] complex inheritance is suggested.

3. Is Variation Heritable?

Somaclonal variations in many plants are heritable, indicating that they are solid. Potato somaclones derived from adventitious buds showed a high frequency of variation. Of the 360 lines examined through 3 tuber generations, 104 were true variants and 1 proved to be chimeric.[301]

E. Origin of Variation

Plant variants may be the consequence of heritably altered expression of genes. The altered expression may be due to (1) gene amplification, (2) gene deletion, (3) translocation of genes to position under different control signals, (4) modification of methylation patterns, and (5) movement of transposable elements or controlling elements to positions influencing expression of gene. For example, in somaclones of oats, chromosome deletion, translocation, and other rearrangements have been shown at the cytological level.[184] When these modifications are seen, the occurrence of changes at the fine structural level cannot be ruled out.

Of late, it has been possible to detect somaclonal variation at the molecular level. In potato[149] 2 out of 12 plants were shown to be variant by southern hybridization with one of the tester clones. As this clone represented 25S-rDNA both somaclonal variants may be regarded as mutants deficient in RNA genes. On the other hand, plants regenerated from tissue culture of maize,[34] when screened for alcohol dehydrogenase Adh1 and Adh2, were found to have a stable mutant of Adh1. The mutant gene (Adh1-usv) produces a functional enzyme with a slower electrophoretic mobility than that of progenitor Adh1-s allele and is stably transmitted to progeny. On cloning and sequencing gene Adh1-usv, a single base change in exon,[6] was the only alteration found.

F. Limitations and Prospects

The phenomenon of spontaneous modification has been shown in tissue cultures of many crop plants. However, in respect to variation, the investigator has no control over the change of characters, and it remains to be seen whether this can be of importance[132] in a plant improvement program. Recent discussions[29] and reviews on the importance and potential of plant breeding and biotechnology allude to the fact that the major portion of genetic variability needed for immediate crop improvement will come from existing gene pools. Cells, protoplasts, and recombinant DNA technologies offer attractive possibilities for the distant future.

Deletion is one of the most frequently reported cytological changes in tissue and cell cultures. Possibly, advantage can be taken of this. Commercial sugarcane is a hybrid between *Saccharum officinarum* × *S. spontaneum*. In the process of "nobilization", involving many years of backcrossing, the result is the loss of chromosomes of *S. spontaneum*. It is proposed that the process can be hastened through tissue cultures.[296,297] Similarly, tissue culture of *Triticale*[19] might result in improvement of grain quality since the shriveling of grain in triticale is due to telomeric heterochromatin on rye chromosomes. It is very likely that the deletion of telomeric heterochromatin on two rye chromosomes may reduce grain shriveling and result in an increase of yield.

An intervening callus phase and the resulting somaclones may help recover hybrid plants. Hybrid embryos and plants of *Hordeum distichum* × *Secale cereale* perished. However, a number of somaclones from an intervening callus phase of embryo or plantlets grew and tillered vigorously.[59]

V. INDUCED MODIFICATION

In contrast to the spontaneous modification occurring in culture (somaclonal variation), induced variation is also possible in cell cultures by exposing them to physical or chemical mutagenic agents which are capable of increasing the frequency of changes in the genetic material. Also, it is possible to subject the cells to selection pressure and recover variants directly from the untreated material. The two main types of possible variants are (1) auxotrophs, which require nutritional supplements for normal growth, and (2) resistant variants which are resistant to specific drugs, antimetabolites, or abnormal changes in environment, concerning nutrients and other conditions.

VI. SELECTION

The identification of variant cell lines can be through direct selection for resistance or indirect selection for auxotrophic and other variants.

A. Direct Selection for Resistance
In this method, a large number of cells are subjected to selection pressure (exposed to growth-inhibitory concentration) and resistant lines isolated. Resistance is defined as the ability to grow under selective condition.

Selection for resistance may be carried out at a borderline concentration that allows some growth,[173] at a high concentration which is inhibitory to growth,[312,313] or in the presence of an increasing concentration of a toxic substance.[182] In these approaches different types of variants, differing in their utility, are possible. However, the concentration of a drug required for selection is dependent on the inoculum size and physiological state of cells.

B. Indirect Selection for Conditional Mutants
The isolation of auxotrophs (nutritional mutants) and other variants, such as temperature-sensitive ones, involves retrieval from wild types under nonpermissive conditions. For example, on a minimal medium auxotrophs will not grow, whereas wild type will grow and multiply. Also, at restrictive temperature, temperature-sensitive cell lines will not grow against the growth of wild types. The methods for enrichment of auxotrophic cell lines will be taken up while describing the selection of auxotrophs.

It is easier to select for resistant variants than for conditional mutants.

C. Limits to Selection
The choice of selection is limited to traits that are expressed at the level of a cell, because many of the differentiated functions are not expressed at the level of a cell. For example, one cannot select for pigment-deficient variants from a cell culture that is not producing the pigment, whereas it is possible at the plant level.

VII. SYSTEM

For selection of variant cell lines different types of cultures successfully used are callus cultures,[98,173,177] cell suspension,[312,313] plated cell suspension culture,[47,198] and plated protoplasts.[30,48] However, callus cultures are least satisfactory, because in a callus culture only a small number of cells divide and growth is relatively much slower than in other systems. Cell suspension cultures are less than ideal, because they show cell aggregation, chromosome instability, and low regenerative capacity. By the time it is possible to obtain a fine cell suspension (after several passages), the cells lose regeneration potential. Therefore, protoplast culture is theoretically ideal[217] for the raising of variant cell lines because (1) protoplasts are individual cells, (2) it is possible to have them in large frequencies, and (3) it is possible to regenerate plants from protoplasts for analysis of meiotic inheritance of a trait. A significant drawback of cell culture is intercell feeding due to cell aggregation; it is also not found in protoplasts. However, protoplast culture is a laborious and tedius technique. Recent advances in the technology of isolation of protoplasts in large numbers and their culture at low density have made it possible to recognize a presumptive resistant variant within 20 to 25 days in *N. plumbaginifolia* and 30 to 40 days in *N. sylvestris*.[217]

A. Necessity of Plating Technique

Whether using protoplasts or cell suspension for isolation of variant lines, it is preferable to use the plating technique on agar or other jelling agent because the clones derived from single cells remain discrete and can be isolated easily. This is not possible in a liquid or callus culture.

A modification of the plating technique[307] using a membrane filter is quite suitable for transfer of cultures from one medium to another.

B. Mutagen Treatment

To begin with, callus cultures or cell suspensions were used to investigate the biological effect of mutagens on plant cells. The frequency of variants increases with treatments with mutagens (ethylmethylenesulfonate [EMS], *N*-methyl-*N'*-nitro-*N*-nitrosoguanidine [NG], *N*-ethyl-*N*-nitrosourea [NEU], ethyleneamine [EI], and ultraviolet [UV] and X-rays). The treatment is carried out in culture medium at 25 to 27°C, followed by a culture for a few generations to allow for the enrichment of variant phenotypes. Then, cultures are screened for expression of a desired phenotype. The efficiency of treatment[289] is assessed by a decrease in the frequency of dividing cells; the numbers are extrapolated from the growth curve. Survival has also been screened by staining.[222,314]

A study of radiation biology of plant cells in culture[124,311] has revealed that it is difficult to analyze the effect of ionizing radiation at the cellular, chromosomal, and molecular level because cell suspensions comprise clusters which dissociate at irregular rates, are rather difficult to synchronize, and have variable chromosome numbers and karyotype. In addition, they can have a previous mutation history. By contrast, protoplasts which comprise a free-cell system are suitable. It has been shown that (1) protoplasts are more sensitive to mutagen than cells, (2) chemical mutagens were more toxic to protoplasts than physical mutagens, (3) haploid protoplasts are more sensitive than diploid protoplasts, and (4) heterogeneity in protoplasts is dependent on age of leaf, culture substrate, species, and genotype.

While employing protoplasts for isolation of variant lines, freshly isolated protoplasts can be subjected to mutagenesis, as the removal of mutagen can be easily fitted in the washing step. Irradiation of haploid protoplasts 1 day after isolation has been used to induce auxotrophic mutants in *Hyoscyamus muticus*[94,266] and *N. plumbaginifolia*.[216]

As for physical mutagens, the cells are more sensitive during cell cycle events.[124] This is in agreement with animal systems where higher frequencies of mutations have been possible from cells undergoing proliferation.

Frequency of mutants increases with an increase in dose of mutagen.[289,307-310] Therefore, there is a tendency to use killing rates of 90 and even 99%. However, the choice of optimum mutagen dose should also take into account other possible effects. For example, in haploid cell suspension of tobacco,[79] regeneration of shoots from gamma-irradiated cells was impaired at a dose above LD_{50}, while the same occurred with UV-irradiated cells at a survival rate of 15%. Killing rates of 90% and more proportionately decrease the regeneration frequency of colonies formed.[45] Therefore, experiments at weak dosage or at different dosages are worth a trial, or permanent incubation at lower mutagen concentration can be attempted.[180] In a recent detailed study,[221] it was found that haploid protoplasts of *N. plumbaginifolia* were more sensitive than diploids to the lethal effect of gamma-irradiation, but the dose-response curves of gamma rays that induced mutagenesis were very similar. The irradiation dose capable of causing a tenfold increase in frequency of mutants resistant to valine was about 500 rads with both haploid and diploid protoplasts.[221]

C. Mutation Frequency

Mutation frequency in a microbial system is calculated by plating a population of cells on a selective medium and counting the number of mutant colonies that appear. Division of this number by the total number of viable cells in the plated population gives the mutation frequency. To determine the number of viable cells a series of dilutions of cell suspension is made and plated. The number of colonies appearing is calculated; then, from the size of aliquot plated and the dilution made, it is possible to calculate the frequency of viable cells in the original population. However, this method cannot be adopted for plant cells for obvious reasons.

In plant cell systems the mutation frequency[217] is expressed on the basis of "per colony forming unit". It is easy to do so in a protoplast system because the determination is based on discrete units. Calculations are made about total number of mutagenized protoplasts (Tm), survival rates in control (S), and survival rates in mutagenized plates (Sm). Survival rates of protoplast-derived cells is shown to average 50%.[46,219]

D. Ploidy of Plant Cells

Ploidy of the material to be employed for mutation will determine the nature of the mutant. When trying to recover all possible types of mutants, including recessive ones, haploid cells should be used. However, because of the prevalence of polyploidy in plants it is difficult to have true haploids — monohaploids. In crops such as potato, tobacco, and alfalfa from microspore, dihaploids (plants possessing two similar sets of chromosomes) are obtained.

Whenever it is anticipated that the desired mutant is dominant, recoveries may be equally effective from haploid as well as diploid tissues. Diploid cells are preferred when the desired mutant is recessive lethal or is rarer than unwanted recessive mutation that cannot be discriminated by the normal selection process.

E. Model System

Most of the cell variants concern species of *Daucus*, *Datura*, *Nicotiana*, and *Solanum*. These systems have their merits and demerits. However, an emerging model system is *N. plumbaginifolia* because of the following attributes:

1.   Easy protoplast culture
2.   High regeneration potential of cells
3.   Low chromosome number
4.   True haploid nature
5.   Availability of haploid plants

VIII. VARIANT CELL LINES

Induction of variant cell lines is dictated by their utility in fundamental or applied research. Among the basic utilities of cell lines are (1) markers in cell hybridization programs and

(2) unraveling the various metabolic pathways. As for application of this research, many of the modifications sought in plant cell cultures are the reflection of a constant endeavor to improve the nutritional and agronomic characters of crop plants and increase their productivity.

Every variant cell line raised raises a number of questions, the foremost of which are (1) how stable is the cell line, (2) can plants be obtained from it, (3) are the plants fertile, and (4) is the trait inherited?

A brief resume of various variant cell lines isolated so far is given below.

A. Pigment Variants

Because of their striking phenotype, cells altered in pigment biosynthesis can be identified easily. The first isolation concerned the orange-colored cell line of carrot,[81] containing a high level of carotene. The other lines[208] isolated in carrot are an orange line accumulating xanthophyll as well as α- and β-carotene and a line[288] containing large amounts of lycopene. However, plants[197] raised from these cell lines resembled the parent plant, discounting their genetic origin. Cell suspension of *Haplopappus gracilis*[84] on irradiation with X-rays and UV light resulted in strains with altered anthocyanin content.

Mutagenesis of *Datura innoxia* protoplasts with X-rays[262] and nitrosoguanidine[147] resulted in plants which were light green, yellow, albino, and one that did not contain anthocyanin in the stem. Also, mutagenesis of haploid callus cultures of *N. sylvestris*[169] with EI or MES led to the formation of chlorophyll-deficient shoots. An albino line was also obtained from ^{60}Co-irradiated protoplasts of *N. plumbaginifolia*.[178]

B. Alkaloid Variants

Selection for pigmented cell lines can also be used for selection of high alkaloid-containing variants, provided alkaloid synthesis is correlated with pigment synthesis. A search for yellow-colored cells of *Macleaya microcarpa*[143] resulted in selection for increased alkaloid content.

By UV irradiation of tobacco cells it was possible to select several lines that yielded high amounts of nicotine.[294] The higher biosynthetic potential of cells was transferred to plants.

C. Antibiotic Resistance

Since the mechanism of action of most antibiotics is known from work on microbes, antibiotics have been the natural choice for resistant variants; these cell lines have been more extensively analyzed than any other isolates. An advantage of having antibiotic-resistant cell lines is the prior knowledge about cellular components wherein the altered function can be identified. The mode of action of many antibiotics (streptomycin, kanamycin, and chloramphenicol) is based on differences in the physical structure of the translational machinery of prokaryotes and eukaryotes. Streptomycin and kanamycin are active on prokaryotic 70 S ribosome and are found in chloroplasts and mitochondria of eukaryotes. Hence, inheritance of resistance can be considered to be cytoplasmic.

The first cell lines resistant to streptomycin were isolated from haploid callus of *Petunia hybrida*[25,26] and then form *N. tabacum* and *N. sylvestris*.[173] To raise resistant cell lines, the callus tissue was subjected to a selection pressure of 0.5 mg/mℓ of streptomycin that inhibited cell division and chlorophyll synthesis. After 6 to 8 weeks, green tissue sectors were seen, from which the plants were regenerated. Resistance to streptomycin was constantly inherited maternally, suggesting an extranuclear change; it was a chloroplast mutation as revealed by analysis of ribosomal proteins from chloroplasts of a normal parent and resistant one (SRI). Furthermore, a difference in mobility of 1 out of 67 proteins was resolved.[324] These experiments demonstrated that similar to bacteria, streptomycin binds to a protein component, inhibits protein synthesis, and resistance is acquired by mutational change of protein. The cytoplasmic basis of streptomycin resistance was confirmed by the correlation of streptomycin with chloroplast transfer from *N. tabacum* SRI to *N. knightiana* in protoplast fusion.[190]

Also, in *N. tabacum* streptomycin resistance was demonstrated by another group.[300] Here, the resistance was tested by germinating the seeds from plants of the resistant cell line St-R 701. Yellow seedlings were obtained from normal plants, whereas green seedlings were produced from seeds resulting from crosses in which the streptomycin-resistant plant was the female parent.

Streptomycin causes degeneration of chloroplasts and mitochondria in sensitive tissue,[326] whereas mitochondria as well as chloroplasts remain normal in resistant tissue.[174] The drug causes a reduction in thylakoid formation and prevents normal development of grana. In mitochondria, it causes swelling of cristae and usually degeneration of organelles.

Inheritance of resistance to streptomycin is also possible as a Mendelian trait. Fertile plants raised from resistant cell lines of *N. sylvestris* produced resistant progeny when self-fertilized but not when crossed with a normal parent, indicating that resistance in the line SR 180 is determined by a recessive nuclear allele.[172] Another streptomycin-resistant cell line of *N. sylvestris* (SR 155) is sterile. Karyotypic analysis indicated possible translocation.[175]

Resistance to streptomycin is also on record in haploid cells of *Datura*.[299] It is possible to transfer the trait from cells to plant and back to cells.

Cross-resistance to antibiotics is also on record. A kanamycin-resistant cell line of *N. sylvestris*[72] was found to be resistant to streptomycin as well as neomycin. Also, two kanamycin-resistant cell lines of *N. tabacum* were found to be resistant to streptomycin.[236]

Resistance to the antibiotic cycloheximide (CH) which inhibits protein synthesis of eukaryotes is also on record. Haploid protoplasts of *N. tabacum*[176] became resistant to CH, but the resistance was lost on transfer to normal medium. Also, in carrot[108] the resistance to CH was lost in the absence of the drug. However, in other carrot cell lines, resistance to CH persisted in the absence of drug[289] and in one resistant cell line (WCH 105) plantlets were regenerated.[153] These plants were green but lacked normal dissected leaves. Inheritance of CH resistance was tested by somatic hybridization between resistant and sensitive cells. Hybrid cell clones were sensitive to the drug, indicating that resistance is a recessive character.[153]

Employing haploid cell lines of *N. sylvestris*, cell lines resistant to chloramphenicol (CAP)[73] were obtained. Plants from resistant cell lines gave rise to resistant calli. Chloramphenicol is an inhibitor of 70 S ribosomes in prokaryotes and eukaryotes. For this reason, in higher organisms it is used along with CH to differentiate between nuclear and extranuclear gene expression. Nondifferentiated callus tissue of tobacco was more sensitive to CAP than differentiated tissue shoots. The difference was especially evident in darkness and light-lacking UV and blue regions. Also, CAP solution photodegrades in light. Degradation of CAP occurs both in callus and shoot cultures. Therefore, CAP is not a good selection agent.[10,11]

Resistance to other antibiotics has also been possible. Cell lines of *N. plumbaginifolia* are resistant to lincomycin.[178] In one line, maternal inheritance of resistance was shown. Lincomycin-resistant mutants of *N. plumbaginifolia*[62] and *N. sylvestris*[65] were selected on the basis of greening. Resistance was inherited maternally. Transfer of resistance in cell fusion experiments[65] indicates that it is a chloroplast-controlled trait.

D. Resistance to Purine and Pyrimidine Analogues

Resistance to purine and pyrimidine analogues in animal cell lines is characterized by the lack of certain enzymes of purine and pyrimidine metabolism. These deficiencies are employed as genetic markers in cell fusion and in experiments on gene transfer. Hypoxanthine, aminopterin, and thymidine (HAT) medium is extensively used in animal cell system to select for somatic hybrids based on complementation of analogue-resistant cell lines. Only hybrids are able to incorporate exogenously supplied hypoxanthine and thymidine and survive. Aminopterin inhibits dihydrofolate reductase and hence inhibits synthesis of purine and pyrimidine in wild types and kills them. Cells resistant to purine and pyrimidine analogues are unable to incorporate exogenously supplied hypoxanthine and thymidine and die on HAT medium. These successes in animal cell lines prompted plant scientists to raise cell lines resistant to purine and pyrimidine analogues.

Cell lines resistant to the thymidine analogue 5-bromodeoxyuridine (BUDR) have been isolated in tobacco,[156,177] soybean,[226,227] sycamore,[37] and alfalfa.[166] Of these, tobacco,[177,181] soybean,[226,227] and sycamore[37] lines were not deficient in thymidine kinase. This is unlike mammalian cells[67] which are deficient in thymidine kinase, unable to phosphorylate, and consequently utilize external BUDR. In a soybean[226] cell line (BU-5), the resistance was not due to exclusion of the analogue because BUDR was freely incorporated into DNA. However, in the second cell line (BU-54), which was selected from BU-5 for resistance to higher concentration of the analogue, the incorporation of BUDR into DNA was possible only in the presence of FUDR. The cell line was also resistant to FUDR, an inhibitor to thymidylate synthetase, and aminopterin, an inhibitor of dihydrofolate reductase. A side effect of these types of resistances was an increased capacity to synthesize thymidylate. An overproduction of thymidylate[227] may have protected the cell against BUDR by competitively excluding the analogue from entering the DNA. This is consistent with the data that BUDR was incorporated into DNA of BU-54 cells only in the presence of FUDR when thymidylate synthesis was inhibited. In contrast, BUDR-resistant tobacco[181] cells incorporated the analogue into DNA to the same extent as sensitive cells. Therefore, here the resistance is by a different mechanism.

Due to lack of thymidine kinase deficiency, the BUDR-resistant cell lines of tobacco, soybean, and sycamore grow on HAT medium and cannot be used as markers in a cell hybridization program. However, five of the seven BUDR-resistant cell lines of alfalfa[166] failed to grow on HAT medium, indicating the possibility of their use in a HAT selection system.

Interestingly, cytokinin-habituated cells showed increased resistance to BUDR in the presence of cytokinin.[185] However, cytokinin habituation and sexually transmitted BUDR resistance were shown to be independent characters.[130]

Resistance to the purine analogues 8-azaguanine and 6-thioguanine results in loss of hypoxanthine guanine phosphoribosyl transferase (HGPRT) activity in mammalian cells.[41] Cell lines resistant to azaguanine are known in tobacco,[156] sycamore,[38] *Haplopappus gracilis*,[120] soybean,[307] and *Medicago sativa*.[166] Of these, sycamore cells show 50% reduction in HGPRT activity and it is reduced to 30% in *H. gracilis*. Therefore, these lines cannot be used as markers with HAT medium. However, two of four azaguanine-resistant cell lines of *M. sativa* failed to grow on HAT medium. Recently, a cell line of carrot[165] resistant to azaguanine was found to be deficient in the enzyme. This line could be counterselected on HAT medium. This line also carries α-amanitin resistance, which is a dominant trait. Due to these characters, this line is described to be a "universal hybridizer", because somatic hybrids of this line with any wild type could be selected by eliminating the parental strain on HAT medium and wild type with α-amanitin.

The cell line of *H. gracilis* resistant to 8-azaguanine was also resistant to 6-azauracil.[128] Resistance to another pyrimidine analogue, 5-fluorouracil, has been possible in carrot.[289] An increase in the endogenous level of pyrimidine can be ascribed to account for resistance.[291] Cell lines of *H. gracilis* and *Datura innoxia* resistant to 6-azauracil were characterized[127] as lacking activity of uracil phosphoribosyl transferase, a pyrimidine salvage enzyme that catalyzes the conversion of uracil and 6-azauracil to uridine-5'-monophosphate and 6-azauridine-5-monophosphate, respectively. The loss of this enzyme activity confers on the variants resistance to 6-azauracil. The resistant cells take up uracil from the medium, but do not convert it to a form that can be used for macromolecular synthesis.

Resistance to aminopterin, a folic acid antagonist that inhibits the production of tetra-hydrofolate required for thymidylate and purine biosynthesis, has been possible in *Datura*[182] and maize.[267] Another folate antagonist is methotrexate (MTX), a potent inhibitor of di-hydrofolate reductase (dHFR) in mammalian cells. Resistance to this drug is due to ampli-fication of gene coding for dHFR, the target protein of MTX. Cell lines of *Petunia*[12] showing resistance to the drug were possible on stepwise selection procedure. The resistant variants exhibited high values of MTX-binding proteins (60- to 400-fold higher than wild type), which declined to intermediate value on withdrawal of MTX. Also, cellular extracts from all variants showed a high staining of dHFR activity in gels. These results indicate that MTX resistance is mediated by elevation of the amounts of dHFR, probably a consequence of gene amplification. A cell line of carrot[50] resistant to MTX also showed three times higher activity of dHFR than wild-type cells. However, MTX-resistant cell lines of *Glycine*[306] did not show increased activity of dHFR.

Once again, borrowing the idea from animal cell literature[159] that resistance to hydroxyurea results in either an altered form of ribonucleotide reductase or an increase in the amount of this enzyme, tobacco[51] cell lines resistant to hydroxyurea were raised and plants regenerated. In experiments on crossing it was found that resistance[137] is conferred by a single dominant nuclear allele. Many of the previously isolated picloram-resistant cell lines were also resistant to hydroxyurea. This is the unique instance of unrelated genotypes arising simultaneously.

E. Resistance to Amino Acid Analogues

Amino acids are of central importance in food and feed crops and cell lines resistant to amino acids and their analogues are raised with the prospect that the knowledge gained will help in improving the nutritional value of crops.

The toxicity arising from inclusion of an individual amino acid into the nutrient medium is due either to interference with the synthesis of a related amino acid[30] or to the inhibition of nitrate or ammonium assimilation.[111] Amino acid analogues may be incorporated into the proteins resulting in nonfunctional synthesis. The inhibition can be overcome by feeding amino acids or normal metabolites.[30,316]

Resistance to amino acid analogues can arise through one or more of the following mechanisms. First, resistance to an analogue can be through overproduction of natural metabolites which may dilute the effect of an antimetabolite. Overproduction of an amino acid is due to deregulation of the enzyme normally subject to feedback inhibition. For example, resistance to the analogue 5-methyltryptophan results in the overproduction of tryptophan due to an altered anthranilate synthetase which becomes resistant to feedback inhibition. Second, resistance may also arise if the analogue is effectively excluded from the system (through mutation), affecting cell permeability. Cell lines altered in transport have been isolated, and these show resistance to more than one amino acid analogue.[22] Third, the compound may be degraded. For example, para-fluorophenylalanine (PFP) is degraded and this may result in overproduction of phenolic compounds which might be inhibitory to differentiation.[20] Fourth, resistance to an analogue may be due to overproduction of a secondary metabolite. Overproduction of a secondary metabolite may also result in suppression of regeneration. An increase in IAA synthesis in 5-methyltryptophan-resistant cell lines of *Daucus carota* inhibited embryogenesis.[290]

The first cell lines[312,313] resistant to 5-methyltryptophan were raised in carrot and tobacco. The analogue prevents growth by inhibiting the activity of anthranilate synthetase, the first enzyme in the biosynthesis of tryptophan. Extract from analogue-resistant cell lines revealed an enzyme that was less sensitive than the normal enzyme to inhibition by tryptophan and 5-methyltryptophan. Endogenous levels of free tryptophan in these resistant cell lines of tobacco and carrot were 2- and 27-fold higher than controls. Plants were regenerated from analogue-resistant cell lines. Cell cutures derived from these plants retained resistance to analogue like the original variant cell line and showed elevated levels of tryptophan (nine-fold). These characters qualify the resistant cell line to be a mutant.

The enzyme anthranilate synthetase has two isozymes - one sensitive and the other resistant to inhibition by tryptophan. Normal cell lines of tobacco[49] contain more of the feedback-sensitive form of enzyme, whereas tryptophan-resistant cell lines contain more of feedback-insensitive form of enzyme. Similarly, in cell lines of *Arabidopsis thaliana*[218] and *Catharanthus roseus*[263] resistant to 5-methyltryptophan, the latter possessed a feedback-insensitive form of enzyme (anthranilate synthetase) and excess level of tryptophan, but in the former, some of the lines showed an elevated level of tryptophan, but in the former, some of the lines showed an elevated level of tryptophan while others showed reduced capacity of the assimilation of the analogue.

Resistance to 5-methyltryptophan, altered anthranilate synthetase, and high, free levels of tryptophan was expressed in leaves of plants regenerated from variant cell lines of *Datura*.[245] These results demonstrate that amino acid overproduction phenotype can be selected at the cellular level and is expressed in plants regenerating from them.

Cell lines resistant to PFP, an analogue of phenylalanine, have been possible in tobacco, carrot,[237] sycamore,[92,93] and datura.[85] Carrot lines contain an elevated level of phenylalanine (sixfold), tyrosine (threefold), and an altered chorismate mutase which is not inhibited by phenylalanine or tyrosine. By contrast, in resistant cell lines of tobacco, only the tyrosine level is increased (threefold) and a normal level of phenylalanine was due to an increased level of phenylalanine ammonia lyase which prevents accumulation of phenylalanine by converting it into phenolic compounds.[22] One of the sycamore cell lines contains a normal level of phenylalanine and tyrosine but increased activity of phenylanaline ammonia lyase.[21] Resistance of the line is explained on the basis of either reduced uptake or increased catabolism of analogue.[92] In another cell line, a higher level of free phenylalanine and tyrosine was present.[93] The three resistant cell lines of tobacco[20] to PFP have been classified as permeation variant, overproducer, and variant exhibiting increased activity of phenylalanine ammonia lyase.

Of the individual amino acids, lysine is limiting in cereals and methionine limits the quality of protein in seed legumes. The amino acids lysine, threonine, isoleucine, and methionine are synthesized from aspartate. Enzymes of lysine and methionine synthesis, functioning as control points, are aspartokinase, dihydrodipicolinate synthetase (DHPS), and homoserine dehydrogenase. Because of allosteric control, the flow of carbon through the aspartate pathway can be interrupted by exposing cells either to a lysine analogue or an excess of lysine plus threonine. In some dicots, the aspartokinase activity is inhibited by lysine as well as threonine,[77,78] but in cereals the activity of this enzyme is not affected by threonine. It is therefore speculated that in cereals, lysine and threonine act singly to inhibit aspartokinase and homoserine dehydrogenase, respectively.[40] In either case, the inhibition of cell growth results from inhibition of methionine synthesis which can be reversed by addition of methionine.[77]

Cell lines of rice[52] resistant to lysine analogue S-2-aminoethylcysteine (SAEC) contained elevated levels of free lysine, isoleucine, methionine, leucine, valine, and several other amino acids. Unfortunately, plants could not be regenerated from these lines. Cell lines resistant to SAEC have also been possible in *N. tabacum*,[315] *Arabidopsis*,[218] and *N. plumbaginifolia*.[214] The amount of free lysine in tobacco cell line was tenfold higher than normal. In *N. plumbaginifolia*, cell lines accumulated 10 to 20 times more free lysine than the control. Also, lysine overproduction and SAEC resistance were expressed in plants regenerated. A feedback-insensitive form of DHPS, the pathway specific control enzyme for lysine synthesis, was detected in cell cultures and leaf extract of regenerated plants.

However, *Arabidopsis* lines contained a reduced level of lysine suggestive of a different mechanism of resistance. Also, in rice regenerants resistant to SAEC, no specific increase in lysine was noted.[260,261]

Inhibition of cell growth by lysine and threonine and its reversal by methionine was employed to raise a cell line overproducing methionine. A maize cell line resistant to lysine and threonine possessed an aspartokinase activity that was less sensitive than normal to inhibition by lysine. This cell line[117] contained increased levels of free lysine, isoleucine, threonine, and methionine. A plant raised from the resistant cell line produced resistant progeny when selfed and both resistant and sensitive progeny when crossed with normal parent, indicating that resistance[116] is due to a single semidominant nuclear allele. These results raise the hope that cell cultures can be used for improvement of crop plants.

Resistance to ethionine, a methionine analogue, results in a tenfold increase in free methionine in a cell line of carrot.[315] Similarly, tobacco[104] cell lines resistant to ethionine contained substantial greater accumulation of methionine ($110\times$), threonine ($18\times$), and lysine ($5\times$) than sensitive lines. Also, the resistant line showed 16 times more activity of lysine-sensitive aspartate kinase activity than the control.

Similarly, resistance to hydroxyproline, an analogue of proline, results in a tenfold greater accumulation of free proline in a cell line of carrot.[315] A practical utility of hydroxyproline resistance can be in raising plants resistant to drought, such as *Hordeum vulgare*.[148]

Biosynthesis of valine, leucine, and isoleucine proceeds through two series of intermediates, but a common set of enzymes catalyze the first three reactions. Therefore, an excess of one end product of branched pathway is likely to be inhibitory to synthesis of amino acids produced by other branches. This was shown in tobacco[30] cells where valine inhibits cell division and this was reversed by isoleucine. Two cell lines of *Valeriana wallichii*[14] resistant to the leucine analogue $5',5',5'$-trifluoro-DL-leucine accumulated higher amounts of aspartic acid and glutamic acid. One of these lines also accumulated higher amounts of valine and leucine. Resistance to an isoleucine analogue O-methylthreonine (OMT) is accompanied by a two- to threefold increase of the free isoleucine pool in cells of *Rosa*.[286] The cells are also cross-resistant to another isoleucine analogue, DL-4-thioisoleucine. These analogues inhibit the growth of wild-type *rosa* cells and isoleucine is able to reverse efficiently and specifically OMT toxicity. The OMT-resistant variants showed a reduction in sensitivity of the enzyme L-threonine deaminase to feedback inhibition by isoleucine.

F. Resistance to Toxins

One of the applications of genetic modification of plant cells can be cell lines and plants resistant to toxins produced by pathogens. The first such attempt was in tobacco,[48] when cell lines and plants resistant to methionine sulfoximine (MSO) were raised which were in turn resistant to wildfire disease of tobacco caused by *Pseudomonas tabacci*. This pathogen produces a toxin which is structurally similar to methionine. An analogue of methionine, MSO, causes the characteristic chlorotic halos on leaves of tobacco similar to that produced by natural toxin. The cause of resistance was traced to an increased concentration of endogenous methionine which was higher than that of the susceptible normal plants, thereby suggesting that resistance to an analogue is by way of overproduction of a natural metabolite. Self-fertilization of heterozygous plants yielded resistant, intermediate, and sensitive seedlings to drug in the ratio of 1:2:1, indicating that resistance was due to single, semidominant nuclear mutation.

Similarly, maize plants[97] resistant to toxin produced by *Helminthosporium maydis* were raised by subjecting cell lines of maize with Texas male sterile cytoplasm (cms T). The resistant plants were also male fertile. Therefore, no practical benefit could be derived from these plants. It is very likely that male sterility and susceptibility to toxins are determined by a single extranuclear locus, and any change in this confers male fertility as well as resistance to disease.[98]

Another instance of toxin resistance is seen in potato[15] cell lines resistant to toxin produced by *Phytophthora infestans*. Regenerated plants, as well as cell lines raised from it, were toxin-resistant. These plants[16] were 25% more resistant to toxins than the parents. Subsequently, potato plants resistant to wilt disease[17] were obtained by exposing cell culture to culture filtrates of *Fusarium oxysporum*. Similarly, plants of *Brassica napus*,[255] *Oryza sativa*,[161] and *Avena sativa*[249] resistant to *Phoma lingam, H. oryzae*, and *H. victorae*, respectively, were raised by subjecting cells to culture filtrates of these fungi. In *A. sativa*, the cell line used was resistant to crown rust (*Puccinia coronata*). These cell lines and the plants obtained from them became resistant to the toxin victorin produced by *H. victorae*, but became susceptible to rust. This is an interesting case of losing resistance to one pathogen in the process of acquiring resistance to another.

G. Resistance to Herbicides

The use of selective herbicides has been recommended to reduce weed populations that compete with crop plants for food and space. However, unfortunately, herbicides in current use are not selective. At a concentration required to kill the weeds, the herbicides are also toxic to crop species. Hence, the development of resistant variants of crop species to herbicides, employing cell culture, is a fascinating and promising aspect of plant cell culture.

This type of work, however, has its limitation. Resistance to herbicides can only be seen at the whole plant level and not at the cell level. For example, matribuzin, a highly effective herbicide for photosynthetic seedlings, acting on a photosynthetic electron transport system will have no effect on callus cultures.[82] An approach that can be employed is to raise plants resistant to herbicides. Resistance to the herbicides bentazone and phenmedipharm was not possible at the level of callus cultures, but herbicide treatment of mutagenized plants revealed localized green areas of resistant cells. When these areas were isolated, cultures raised, and plants regenerated, these plants were resistant to herbicides.[243] More recently, an alternate approach has been to raise herbicide-resistant green cells from photomixotrophic cultures.[64] In this way, terbutryn-resistant cell lines were isolated from photomixotrophic cell cultures of *N. plumbaginifolia*. Terbutryn is a triazine herbicide inhibiting photosynthesis by interruption of electron flow at the acceptor side of photosystem II. The primary symptom is extensive bleaching of plants/cells. Selection was based on the greening ability of resistant cells in the presence of an inhibitory concentration of triazine. The plants regenerated showed maternal inheritance of resistance.

Resistant cell lines have been possible against a number of herbicides — phenoxy herbicides, the triazoles, the carbamates, dipyridylium, and the herbicides picloram and glyphosphate. However, only in a few cases has regeneration of plants from these lines been possible, and mostly in *Nicotiana*. Tobacco[7] plants resistant to different herbicides raised are resistant to isopropyl-*N*-carbamate, aminotriazole,[9] picloram,[54] paraquat[195] and amitrole,[271,293] chlorsulfuron, and sulfometuron.[55] Sexual transmission of amitrole resistance[293] to F_1 progeny does not seem to obey Mendelian rules. The genetic data also do not support the possibility of cytoplasmic maternal inheritance. Plants have also been possible in tomato hybrid (*Lycopersicon peruvianum* × *L. esculentum*) cells resistant to paraquat[295] and celery cells resistant to asulam.[194]

The commercial thiocarbamate herbicides, because of their volatile nature and low water solubility, are not suitable selection agents. Instead, structurally related compounds with more attractive physical properties can be used. Using the nonvolatile thiocarbamate R-14705 (*S*-3 methylpyridyl *N,N*-dibutylthiocarbamate), tobacco cells resistant to it were found to be cross-resistant to commercial thiocarbamates.[88] Similarly, tobacco cells resistant to *S*-propyl *N,N* dipropylthiocarbamate were found to be cross-resistant to other thiocarbamate herbicides.[90]

Glyphosate [(*N*-phosphonomethyl) glycine] is a nonselective postemergence herbicide widely used in sugarcane plantations for the control of perennial weeds. A sugarcane[142] cell line resistant to glyphosate has been possible. Glyphosate is a broad-spectrum herbicide that leaves no active soil residue and has low toxicity in mammals, fish, and aquatic invertebrates. Crop plants resistant to glyphosate would greatly enhance the potential range of herbicide application, while minimizing potential damage to the environment. Glyphosate ultimately perturbs a variety of biochemical processes including protein synthesis, nucleic acid synthesis, photosynthesis, and respiration. The shikimic acid pathway enzyme 5-enolpyruvyl-shikimic acid-3-phosphate synthase (EPSP) is strongly inhibited in *N. sylvestris*.[253] This enzyme was 12-fold higher in *Daucus carota*[209] cells adapted to glyphosate. The adapted cell also had a higher level of amino acids, especially threonine, methionine, tyrosine, phenylalanine, tryptophan, histidine, and arginine, than nonadapted cells. Tobacco cell lines tolerant to glyphosate were also tolerant to amitrole[272] (3-amino-1,2,4-triazole). Similarly, amitrole-tolerant lines were also tolerant to glyphosate.

Cross-resistance of lines to similar or dissimilar herbicides is known. Paraquat-resistant cell lines of soybean and tobacco are also resistant to a related dipyridylium herbicide, diquat.[125] Some of the picloram-resistant plants of tobacco[53,159] were also resistant to hydroxyurea.

One of the most widely used herbicides is 2,4-D. Attempts have been made to raise crops resistant to 2,4-D. A tobacco[232] cell line resistant to 2,4-D stably maintained resistance under nonselective conditions. However, plants regenerated from this cell line were not resistant to 2,4-D, but callus reinduced from these plants showed 2,4-D resistance,[233] indicating stable genetic modification. The resistant cells took up 2,4-D more rapidly than wild-type cells. The resistance was possibly due to immobilization of 2,4-D because there was a significantly lower rate of efflux out of cells. These cells were also cross-resistant[212] to other auxins, IAA, NAA, and picloram.

H. Resistance to Stress

Stress can be defined as any environmental factor[158] that brings about a change, physical or chemical, in cells or plants which is reverisble or permanent. According to agronomic definition,[28] the stress can be any environmental factor capable of reducing yield. A study of the stress phenomenon and related resistance mechanisms is essential because a large areas of land are increasingly becoming unfit for cultivation because of limitations imposed by natural or man-made environmental stress. Recently, physiological aspects of stress and its resistance have been studied with the technique of cell culture, and attempts are being made to develop plants that are better adapted to stress.

1. Chilling Stress

Plants of tropical and subtropical regions are sensitive to chilling stress[158] which can be induced by low temperature (10 to 15°C) or freezing temperature (0°C). Direct effects of chilling apparent within hours are changes in membrane permeability, consequent leakage of solutes, and alteration in respiratory activity.[323] Indirect effects which follow, apparent in a few days, are different metabolic disturbances. Primary effects[158] of chilling injury are temperature-dependent transition of membrane lipids from liquid crystalline to solid state and also temperature-dependent alteration in the hydrophobic nature of protein.

Attempts were made to raise chilling-resistant cell lines of *N. sylvestris* and *Capsicum annuum*[76] by subjecting the cells for 21 days at 3 to 5°C with or without mutagen (EMS) treatment. In the cell line showing resistance, mutagen helped isolation and retained it after a period of growth at normal temperature (24°C). Some of the cell lines showed high resistance and some low resistance to chilling. Regeneration of plants was possible from the latter. However, tissue cultures raised from regenerated plants did not show resistance to chilling.[70] Therefore, resistance of cells to chilling was considered to be an epigenetic change.

2. Heat Stress

The transfer of living organisms or their cells to an elevated temperature produces stress. Concurrent with the development of transient thermotolerance is the accumulation of novel proteins — heat shock proteins (HSP). The extent of stress is primarily determined by temperature differential, duration of exposure, and previous growth conditions. The ability to make 70-kdalton HSP under condition of heat shock is higher in a fruit fly population that has been maintained for 7 years at 25°C than in a replicate population grown at 14°C. The second population was shown to be more sensitive to heat shock at 40°C than the first. It has been theorized[196] that HSP protect the cells from heat injury by stabilizing other proteins in a nonspecific manner.

Tobacco cells[129] grown as suspension at 26°C produce HSP when exposed to elevated temperatures of 34 to 42°C. At 34 and 38°C, synthesis of normal proteins is maintained while HSP are expressed within 30 min after the initiation of shock. At 42°C, normal proteins are made either at a reduced rate or not at all, but HSP are expressed. Cells growing in the log phase produce HSP at a higher rate than those in a stationary phase.

It is not clear how plant cells acquire heat tolerance. Heat tolerance of cultured pear cells[321] induced by heat shock is different than heat tolerance acquired during growth at elevated temperature. Cells grown at 22°C on exposure to 38°C for 20 min (heat shock) showed maximum tolerance within 6 hr, whereas cells grown at 30°C developed maximum tolerance after 5 to 6 days. This maximum was well below that induced by heat shock. Heat tolerance induced by shock treatment was retained at 22°C for 2 days and was partly lost after 4 days. However, cells acclimated to 30°C lost all tolerance 1 to 2 days after transfer to 22°C.

3. Salt Stress

A widespread chemical condition that inhibits plant growth in nature is the salinity of soil, primarily due to NaCl. To overcome salinity, soil reclamation and desalination of water for irrigation are expensive practices. Therefore, raising resistant plants is a better proposition, and in this context, tissue and cell cultures appear to be promising.

Cell lines of *N. sylvestris*[328] and *N. tabacum*[205] were obtained which were able to grow on 0.17 and 0.09 *M* NaCl, respectively. Similarly, cell lines of *N. sylvestris* and *C. annuum*[75] were obtained which were able to grow at 0.34 *M* NaCl. Some of the cell lines retained resistance to salt after several passages through medium lacking NaCl. Plants were possible from these cell lines.[71] However, tissue raised from *N. sylvestris* plants showed limited capacity to grow on salt medium, but tissue from *Capsicum* grew well on NaCl medium.

Salt-resistant cell lines of alfalfa[61] and rice[244] able to grow on 0.17 and 0.25 *M* NaCl, respectively, were possible. Unfortunately, plants could not be regenerated from these resistant cell lines.

Inheritance of salt resistance was demonstrated when plants regenerated from salt-resistant cell lines of tobacco[206] were shown to grow on salt medium up to two generations. Similarly, plants have been possible from salt-resistant cell lines of *Kickxia ramosissima*,[183] *Datura innoxia*[293] *Medicago sativa*[277,278] *Citrus sinensis* and *C. aurantium*,[144] *Avena*,[207] *Sorghum*,[24] *Colocasia*,[224] *Ipomea batata*,[257] and *Vitis rupestris*.[154] Salt-resistant cell lines have also been obtained in *Pennisetum*[56,246] *Cicer arietinum*,[238] *Pisum sativum*, and *Vigna radiata*.[105]

In studies on adaptation of sweet potato[257] and tobacco[27] cells to salt, the adapted cells showed a reduced cell expansion. This adaptation in cell expansion of tobacco cells was not due to the failure of cells to maintain turgor, since tolerant cells underwent osmotic adjustment in excess of the change in water potential caused by addition of NaCl to the medium. Tolerance of the adapted cells, as indicated by fresh or dry weight gain, did not increase proportionately with an increase in turgor. Adaptation of cells appears to be due to mechanisms involving an altered relationship between turgor and cell expansion.

An understanding of osmotic adjustment of cells to salt is likely to be helpful in isolating salt-resistant cell lines and plants. In a study of effects of seawater solution of different inorganic salts and mannitol on growth and discoloration of carrot[103] cells, growth inhibition was attributed to an increase in osmotic pressure and discoloration and necrosis due to toxicity of salts. Further, it was found that single salt (NaCl) was more toxic than seawater.[57] Therefore, it was recommended that salt-resistant cell lines should be isolated with seawater instead of NaCl.

One of the most remarkable metabolic consequences of salt stress in plants is the rapid accumulation of proline.[283,284] Also, at the level of a cell in the callus of *Lycopersicon esculentum*[136] at increasing salt concentration there was an increase in endogenous levels of proline. The plating efficiency of the protoplasts of this plant was reduced by inclusion of NaCl, but it could be improved by simultaneously including proline.[250] Accumulation of proline in response to salinity is also seen in saline-resistant cell lines of *N. sylvestris*[74] and *N. tabacum*.[304,305] After growth for 24 generations in the absence of salt, on transfer to salt, proline reaccumulated rapidly, indicating the stability of this character. Proline accumulation was fully reversible. Also, salt-tolerant cells of *Cicer arietinum*[240] accumulated free proline in response to increasing NaCl concentration while sensitive lines did not. The tolerant cells grown in the absence of NaCl accumulated proline on reculture on NaCl medium. The addition of proline to medium containing NaCl increased fresh and dry weights and the free proline content of both cell types.

It is interesting to note that specific proteins in quality and quantity are produced in salt-resistant tobacco cells.[83] On transfer to medium without NaCl, these proteins do not disappear until second passage. Amino acid analysis of these proteins revealed that they contain some hydroxyproline. Contrarily, in carrot[248] cells, overproduction of proline is not directly related to salt tolerance. (Overproduction of proline in this cell line is due to resistance to azetidine 2-carboxylic acid.) Cells characterized by differences in overproduction of proline showed similar osmotolerance, suggesting that factors other than proline may be involved.

Plants exposed to a saline environment may suffer from ion excess which may result in nutritional imbalance and specific ion deficiencies.[107,322] In salt-sensitive and salt-resistant cells of tobacco,[305] internal Na^+ concentration rose steadily as a result of external NaCl level. Sensitive cells lost a part of their K^+ content in response to NaCl, but resistant cells did not reveal this loss. Similarly, cells of *C. arietinum*[239] showed an increased accumulation of Na^+ and Cl^- with an increase in salinity. This accumulation occurred during the first week and growth inhibition was noticed during the third week of culture. In this cell line, contrary to that of tobacco, tissue K^+ decreased with increased salinity. Also, in *Citrus aurantium*[18] an exposure of cells of NaCl-sensitive and NaCl-tolerant lines to an equal external concentration of NaCl resulted in a greater loss of K^+ from the NaCl-sensitive line. This observation lends credence to the conclusion that growth and ability to retain a high level of internal K^+ are correlated. However, analyses of internal Na^+, K^+, and Cl^- concentrations failed to identify any particular internal ion concentration which could serve as a reliable marker for salt tolerance.

As a result of stress, in some plants there is accumulation of quaternary ammonium compounds such as glycine betaine.[118] In salt-stressed plants of sorghum,[109] accumulation of betaine has been shown. Proline and glycine betaine have also been found to be osmoprotective compounds against osmotic stress in bacteria.[287] A proline-overproducing strain of *Escherichia coli* was found to be salt tolerant.

Cross-resistance between drought and salt stresses is also known. Tomato[32] cells subjected to polyethylene glycol (PEG) for drought resistance were found to be resistant to NaCl. Tobacco[273] cells tolerant to NaCl and PEG showed quantitative and qualitative changes in proteins on gels. The intensity of some proteins increased with increasing concentration of NaCl while intensity of others decreased. Interestingly, abscisic acid (ABA) is described as accelerating adaptation of tobacco[152] cells to salt.

4. Drought Stress

It is estimated that the worldwide shortage of food[145] is due more to scarcity of water than all other factors combined. Therefore, selection of drought-tolerant crops is desirable.

Cell lines of tobacco[115] and tomato[32] were grown in an increasing concentration of polyethylene glycol (PEG) in the hopes of raising drought-resistant plants. The cells became tolerant to PEG, but this was lost in the absence of selection pressure; therefore the change was described as an adaptation. However, clonal variation was found in cells tolerant to PEG.[110] Some of the cell clones retained resistance for many generations in the absence of selection pressure, raising the hope that drought-tolerant plants are possible.

In order to recover cells resistant to water stress, an attempt was made to raise cell lines resistant to the growth inhibitor ABA. In cell lines of *N. tabacum* and *N. sylvestris*[318,319] resistance to ABA was retained in the absence of selection pressure, but this did not alter the response of cells to water stress.

5. Metal Stress

Industrial, mining, and sewage disposal operations result in contamination of soils with metals and other unwanted changes.

Tomato cells tolerant to aluminum[193] and cell lines of carrot tolerant to aluminum and manganese[229,230] were obtained by subjecting cells to $AlCl_3$ and $MnCl_2$. The Al-tolerant cells were not tolerant to Mn, and the Mn-tolerant were not tolerant to Al. The Al tolerance was stable because selected cells retained tolerance to aluminum in the absence of selection pressure. Al-tolerant cells were also tolerant to gallium, an allied metal. Al-tolerant cells[230] abundantly released citric acid into the medium. The conditioned medium from Al-tolerant cells reduced the Al toxicity of wild-type cells.[231] Plants could not be obtained from these cell lines. However, plants were possible from Al-tolerant cells line of *Sorghum*.[276] The Al-selected callus grew better in the presence of Al. Also, fertile plants were recovered from Al-resistant callus cultures of tobacco.[58] All these transmitted Al resistance to their seedling progeny (selfed and back-crossed) in segregation ratios expected for a single dominant mutation.

Of the heavy metals, a cell line of petunia[222] was raised which was resistant to $HgCl_2$ and the resistance was retained in regenerated plants. Cadmium is another heavy metal which is a common environmental contaminant. *Datura*[126] cell lines resistant to cadmium were raised which could grow at a lethal concentration of the metal. Resistance to cadmium was correlated with the synthesis of low-molecular-weight, cystein-rich, cadmium-binding proteins. These proteins were absent in controls. Synthesis of these proteins was induced rapidly, within 1 hr, in response to cadmium and reached a maximum within 10 to 12 hr of exposure.

I. Developmental Variants

Developmental variants of higher plants have been isolated in the search for a better understanding of the mechanisms involved in the control of morphogenetic processes. Recovery of cell mutants impaired in early stages of embryogenesis and temperature sensitiveness has been possible in carrot.[33] One cell type is blocked to grow on an embryo-forming medium. In the second, the cells can grow, but fail to form embryos. The third is blocked at the globular stage of embryo development. More recently, similar variants have also been recovered by another group.[101]

Polyamines are considered to regulate development and differentiation. Tobacco[168] mutants resistant to methyl-glyoxal-bis-guanylhydrazone (MGBG), a spermidine synthesis inhibitor, had abnormal polyamine levels and abnormal floral development.

1. Growth Regulator Variants

On mutagenesis of sycamore[157] cells, an auxin autotrophic cell line was possible which was stable and had an altered auxin oxidase.[155] Auxin autotrophy is also seen in 5-methyl-tryptophan-resistant cell lines[316] of *Daucus carota*, *N. tabacum*, and *S. tuberosum*. Auxin autotrophy of *Daucus*[292] cell line resistant to 5-methyltryptophan has been confirmed and it has been shown that it is due to an overproduction of IAA.

Crown gall cells are auxin autotrophic, but on mutagenesis with gamma rays it has been possible to recover auxin heterotrophic cells.[4]

Cells resistant to ABA were mentioned in the preceding section on drought stress.

A cell line of tobacco[232] resistant to 2,4-D was found to retain resistance in the absence of selection pressure. It was also cross-resistant[212] to other auxins (IAA, NAA, and picloram). To understand the resistance mechanism, a comparison of uptake and metabolism of 2,4-D between the variant and its wild type indicated that neither uptake nor metabolism of 2,4-D is related to resistance. Later, in a study of macromolecular synthesis,[211] it was found that wild-type cells differed from resistant cells in their sensitivity of DNA synthesis, whereas no significant differences were found in RNA and protein synthesis.

Mutants of tobacco resistant to the toxic effect of NAA were isolated from a mutagenized population of haploid mesophyll protoplasts of tobacco.[201] Among the plants regenerated from these cell lines, two clones impaired in root morphogenesis were recovered. The trait was transmissible to progeny as a single nuclear dominant trait and with resistance to NAA at the cellular level.

2. Transcriptional Variants

The availability of transcriptional mutants in plants would provide systems for probing transcriptional processes. Such mutants can be obtained through the use of selective inhibitors of transcription, such as α-amanitin, and inhibition can be probed through alteration in nuclear RNA polymerase which is required for DNA-dependent transcription.

Carrot[162] lines resistant to α-amanitin were ascribed through an altered sensitivity of RNA polymerase II. In a detailed study, an evaluation of protoplast and cell suspension of carrot indicated that protoplasts were tenfold more refractory than cells to three ametoxin derivatives (α-amanitin, 6'-deoxy-α-amanitin, and 6'-*o*-methyl-α-amanitin). Of these α-amanitin was selectively degraded. Derivatives in which the 6'-hydroxy moiety is either absent or converted to the corresponding methyl ether are more potent inhibitors of cells and are resistant to degradation in vitro. Plant oxidases may be involved in the activation of hydroxylindole moiety of α-amanitin.

J. Metabolic Variants

A spectrum of variants is possible, which are altered in regulatory and structural functions.

1. Alternate Respiratory Chain

The efficiency of oxidation is lower in a cyanide-resistant respiratory chain than in a cyanide-sensitive chain. Elimination of a cyanide-resistant respiratory chain was attempted by selection for resistance to carboxin. Selection of resistant cell lines and plants has been possible in tobacco.[242] The resistance is inherited through seeds.

2. Carbon Metabolism

Higher plant cells are unable to utilize lactose and galactose. However, a tobacco[132] cell line able to utilize lactose was isolated by growing cells on lactose and sucrose. This cell line had an increased level of β-galactosidase activity. Similarly, a sugarcane[179] cell line was isolated that was capable of growing on galactose as the sole carbon source. These cells contained tenfold more uridine diphosphate (UDP) galactose 4-epimerase than the wild-type cells on sucrose medium. Increased activity of this enzyme which catalyzes the conversion of UDP galactose to UDP glucose prevented the accumulation of UDP galactose, occurring in normal cells on transfer to galactose medium, and enhanced the capacity of these variant cells to channel this metabolite into glycolysis.

Soybean[160] suspension cells are able to grow very slowly on maltose medium. Variants were recovered that grew rapidly on this medium.

A tobacco[51] cell line isolated is able to grow on glycerol medium. The plants regenerated from this cell line, when backcrossed reciprocally with normal plants, produced an approximately equal number of resistant and normal progeny. This was expected because the original cell line was heterozygous.

3. Sulfur Metabolism

The selection of regulatory variants derepressed for sulfate[89] assimilation was attempted on the basis of resistance to solenocysteine and solenomethionine. This scheme favors the isolation of variants accumulating cysteine as well as methionine. Overproduction of these amino acids is expected as a result of derepression of sulfate assimilation. One double-resistant line, SCM 7 of tobacco, is more sensitive to solenate, the toxic analogue of sulfate, when grown on medium containing cysteine as a sulfur source, that represses sulfate assimilation.

4. Carbon Assimilation

Resistance towards inhibitors of photophosphorylation is sought with a view to raise plants which will be photosynthetically efficient. In photophosphorylation, ribulose 1,5-bisphosphate is converted into glycine from which serine and CO_2 are formed. The last reaction is inhibited by isonicotinic acid hydrazide (INH) and glycine hydroxamate (GH). Cell lines of tobacco resistant to INH[23] have been obtained; these are expected to have a change in photorespiratory pathway. Plants have been possible from this cell line and resistance to INH is inherited through seeds.[327]

5. Nitrate Assimilation

In higher plants, nitrate is assimilated in a series of steps:

1. Nitrate enters the cell with the help of permease.
2. Nitrate is reduced to nitrite by nitrate reductase (NR).
3. Nitrite is reduced to ammonium by nitrate reductase.
4. Ammonium is combined with glutamate to yield glutamine via glutamic synthetase.

Cell variants have been found in which the first two steps are impaired. An understanding of control of the nitrate pathway will form a basis for constructing mutants which will be more effective in the utilization of added fertilizers.

Tobacco cells in culture on a medium containing nitrate as a sole source of nitrogen were inhibited by single amino acid threonine and a mixture of amino acids, casein hydrolysate. A variant[111] cell line was isolated which was resistant to threonine inhibition when grown on nitrate medium. In normal cells, casein hydrolysate inhibited both uptake of nitrate and synthesis of NR, whereas in variant cell lines it inhibited only NR, thereby indicating that uptake of nitrate was altered in the variant cell line. Unfortunately, no genetic analysis was carried out.

Mutants impaired in the activity of NR have been isolated and characterized. NR is a complex multisubunit enzyme that catalyzes NADPH-dependent reduction of nitrate to nitrite. Reduced methyl viologen $FMNH_2$ or $FADH_2$ can also act as an electron donor. These activities depend on the presence of molybdenum as a cofactor. In the absence of a cofactor the enzyme has only cytochrome C reductase activity.

In the absence of NR activity, chlorate is not converted to toxic chlorite and hence NR-deficient cell lines are resistant to chlorate. In microbes, mutants defective in nitrate assimilation were isolated employing chlorate. In a similar way, mutants defective in NR have been isolated in *N. tabacum*[191,199,200] by subjecting cells to chlorate. Variants lacking NR activity could not utilize nitrate but could grow on medium containing a reduced form of nitrate such as amino acids. The class of variants designated as *CnX* was deficient in xanthine dehydrogenase. Since xanthine dehydrogenase and NR use the same molybdenum cofactor, the *CnX* allele is believed to affect the molybdenum cofactor and not the NR apoprotein. Another class of variants designated as *nia* lack nitrate dehydrogenase activity, indicating that mutation affects apoprotein. That these independent functions were impaired in two classes of mutants was confirmed by in vitro complementation studies; a low level of NR activity was restored by mixing extracts of two types of variants. Partial reconstitution of NR activity was also possible by fusing protoplasts[102] obtained from two variant cell lines. These hybrid cells were able to utilize nitrate and regenerate to form plants, the activities not depicted by original cell lines. NR-deficient cell lines of tobacco,[192] *CnX* type, could be repaired in vitro by (1) molybdate and (2) preparation of active molybdenum cofactor of homologous or heterologous origin. The repair was 20 and 80%, respectively.

Cell lines lacking NR have been isolated in *Datura innoxia*,[141] *Rosa damascena*,[202] and *Hyoscyamus muticus*.[285] *Datura* and *Rosa* cells were subjected to chlorate. Chlorate-resistant cells of *Datura* were able to take nitrate but could not utilize it due to deficiency of NR.

NR-deficient cell lines were also obtained from mutagenized haploid protoplasts of *N. plumbaginifolia*[178] that were resistant to chlorate. Plant regeneration was possible in only two lines which had residual NR activity. Therefore, regeneration seems to be dependent on residual NR activity. In the rest of the lines, xanthine dehydrogenase was detected in some, but not in others. The latter are probably defective in molybdenum cofactor biosynthesis and the former are impaired in NR apoprotein. NR activity in the two lines could be partially restored by increasing the molybdenum concentration of the culture medium. These lines are therefore similar to *CnX* lines of *N. tabacum*.[203] Other lines do not respond to molybdenum cofactor and are probably new types.

Variant cells of *Hyoscyamus*[91] required casein hydrolysate for growth and were found to be chlorate resistant and deficient in NR. These mutants had the characteristics of *CnX*-type mutants in terms of (1) lack of NR and xanthine dehydrogenase activities, but presence of cytochrome C reductase and nitrite reductase activities and (2) in vitro complementation with a molybdenum cofactor source. Also, two cell lines of rice,[303] developing from microspores on anther culture, which were resistant to chlorate and could not utilize nitrate were lacking in NR. In these lines xanthine dehydrogenase was present, but not cytochrome C reductase. Of the NR-deficient cell lines of petunia[282] isolated showing resistance to chlorate, two were *CnX* type and one *nia* type.

Contrary to the results described above, it has also been shown that chlorate-resistant mutants of higher plants[43,202] possess NR. The basis of resistance of these mutants is unknown. The explanations advanced are (1) chlorate does not enter the cell, (2) chlorate enters the cell but it is sequestered in a place where it cannot cause damage, (3) the cells lack chlorate reductase, even though they possess NR (this implies that the two activities are not related), (4) the cells convert chlorate to a harmless form, or (5) cells lack sensitivity to chlorate. Further, the cells of *R. damascena* resistant to chlorate and having NR[204] were found to take up chlorate, and reduce it to chlorite, but this was much less than wild-type cells. The slower production of chlorite apparently accounted for the resistance of the cells to chlorate.

K. Auxotrophs

More desirable are auxotrophic mutants of higher plants. Auxotrophic mutants of microbes have been valuable in defining metabolic sequences and understanding gene regulation. They have also served as markers in experiments on cell fusion and transfer of genes. However, in higher plants there is paucity of auxotrophic mutants as compared to resistant mutants. This is partly due to the nature of higher plant cells, where growth is density dependent and cross-feeding is unavoidable. Also, the diploid nature of higher plant cells precludes the recovery of recessive mutants. Further, unlike microbial cells, replica plating is not possible with higher plant cells, and the testing of thousands of individual cells for their nutritional deficiency is a formidable task.

The first attempt to obtain auxotrophic mutants was in haploid cells of *N. tabacum*,[47] following the technique employed for enrichment of auxotrophic cells in mammalian cells.[131] This procedure is in fact based on selective killing of prototrophic cells from a cell population and recovery of potential auxotrophs. The killing of prototrophic cells is based on photolability of DNA due to incorporation of BUDR. When a population of cells comprising prototrophs and auxotrophs is incubated on a minimal medium containing BUDR, the growing population of prototrophs on this medium will incorporate BUDR into their DNA and will be killed on exposure to bright light, whereas nongrowing auxotrophs will not incorporate BUDR and will survive. The auxotrophs are then rescued on transfer to complete medium and then tested individually for specific requirements. It is clear from mammalian cell literature that two conditions specifically affect the isolation process. First, it is essential that only prototrophic cells should incorporate BUDR. The auxotrophic cells should be starved before BUDR treatment and be depleted of their metabolic pools. Second, the duration of BUDR treatment should be enough to kill the prototrophs. This can be determined by incubating cells for different periods in BUDR in the dark, then plating aliquots from these cells and exposing the plates to light. When no cells survive this can be taken as the optimal period of treatment.

A haploid cell suspension of *N. tabacum* was mutagenized and after washing the mutagen, the cell suspension was incubated in minimal medium for 96 hr to starve the auxotrophic cells. Following this, the culture was fed with BUDR for 36 hr in the dark, then transferred to supplemented medium and exposed to light. Out of a total population of 1.75×10^6 haploid cells treated in this way, 119 colonies appeared on supplemented medium. When tested for growth on minimal medium, only six of these colonies failed to grow and were potential auxotrophs. Of these, three were amino acid requiring (arginine, lysine, and proline), two were vitamin requiring (biotin and *p*-aminobenzoic acid), and one was purine requiring (hypoxanthine). However, all six auxotrophs were leaky; they grew slowly without these substances on a minimal medium. Therefore, these colonies were not true auxotrophs.

Unfortunately, subsequent workers[6] could not confirm the BUDR-induced killing of plant cells and hence the selection of auxotrophs in higher plants suffered a setback. Nonetheless, BUDR was employed for selection of temperature-sensitive mutants[167] in *N. tabacum*. In the selection of auxotrophs, BUDR helps rescue these variants due to their inability to grow on minimal medium, and in this case BUDR helped to rescue temperature-sensitive mutants by their inability to grow at elevated temperature (33°C). Of the 84 colonies recovered, 2 temperature-sensitive lines were characterized. One cell line grew well at 26°C and slowly at 33°C. The second cell line had a reddish brown color and grew slowly at 26°C, but not at 33°C.

In efforts to find an alternate selection scheme[241] for auxotrophs, an arsenate selection scheme was proposed. Arsenate is a respiratory poison and can be employed to kill growing cells; the survivors are screened for auxotrophy. A population of mutagenized soybean cells was incubated for 1 day in medium containing 2 m*M* arsenate. The cells were washed and plated on enriched medium; 12 colonies appearing on this medium did not grow on minimal medium and could be stimulated to grow on medium enriched with a mixture of amino acids.

For want of a proper enrichment and selection method for the isolation of auxotrophs, a nonselective total isolation method described for microbes[13] was adopted for plant cells.[94,140,258] To avoid the problem of intercell feeding, a haploid cell suspension of *D. innoxia* was filtered to obtain a suspension comprising 60% free cells. On plating this suspension on supplemented medium at a low density, it was possible to isolate discrete clones. A screening of 2370 such clones resulted in recovery of 1 auxotroph, requiring pantothenate,[258] and of the many presumptive "leaky" mutants 1 was adenine requiring. In a characterization[140] of these mutants, adenine-requiring cells also grew well on a medium containing 5-aminoimidazole-4-carboxamide ribotide or inosine instead of adenine. The pantothenate-requiring cells grew well in the presence of pantoic acid alone. These mutants on starvation of pantothenate and adenine for 4 days resumed growth when transferred to appropriate media. Inclusion of wild-type cells could not cross-feed these mutants.

Similarly, using a total isolation and nonselective method, screening of 29,000 cell colonies resulting from haploid protoplasts of *H. muticus*[94] yielded two auxotrophs — one for histidine and one for tryptophan — and three clones for nicotinamide. Two temperature-sensitive mutants were also found; one stops growing at 32°C and the other undergoes chlorosis and accumulates an insoluble brown pigment. A characterization[95,96] of histidine- and tryptophan-requiring mutants confirmed their auxotrophy. No cross-feeding was detected either between wild types or auxotrophic cells. The auxotrophic phenotypes behaved as recessive traits in experiments on protoplast fusion.

Also, in *N. plumbaginifolia* screening of 14,229 colonies resulting from haploid protoplasts, 3 lines requiring uracil, leucine, and isoleucine were obtained. In the isoleucine-requiring line auxotrophy was due to deficiency of threonine diaminase.[268]

The total isolation method is time and resource consuming. Hence after characterization of adenine- and pantothenate-requiring auxotrophs an attempt was made to reisolate[121] them using the arsenate counterselection method. It was possible to practice this reisolation method because (1) both auxotrophs could be starved for several days on minimal medium and then rescued on transfer to required medium and (2) no cross-feeding was detected when wild-type and auxotrophs were cultured together. Using arsenate counterselection it was shown that 2 m*M* of arsenate killed growing cells on minimal medium, and nongrowing auxotrophs were spared due to starvation. Later, these were recovered on enriched medium. However, arsenate selection gave erratic results; in some experiments many colonies survived and in some there were no survivors. These results confirmed the basic observation that arsenate counterselection can be used for selection of auxotrophs, but requires refinement. Later[122] it was found that plating of cells at high density after arsenate treatment for recovery of auxotrophs (Ad-requiring) was the major source of variability of results. It was suspected that along with auxotroph, wild-type cells were recovered. This could be rectified in low-density plating of cells on feeder plates. In this way, a greater frequency of auxotroph could be recovered.

In a test of several enrichment agents — 1-β-4-arabinofuranosylcytosine (arac-c), 5-FU, FUDR, nystatin, and arsenate — the first three did not discriminate between nongrowing and growing Ad-requiring cells; only nystatin and arsenate could discriminate and could be used for counterselection. However, neither of these chemicals could be used[139] for discrimination of growing and nongrowing cells of two other auxotrophs — isoleucine + valine, pantothenate-requiring. From these results it becomes clear[123] that a particular counterselecting agent may be effective for selection of a narrow spectrum of auxotrophs and under certain conditions of growth.

That an enrichment agent is limited in its specificity towards certain auxotrophs is also seen in another recent investigation[215] in which BUDR was the selection agent. This work reemphasizes the use of BUDR as a selection agent. Employing BUDR auxotrophs recovered required, his, ile leu, ile + val, met, and tyr, but none lacking in NR could be obtained, whereas many NR-deficient clones were recovered either by positive selection or total isolation procedure. Effectiveness of BUDR depended on the time of its presence. The lowest escape rate (0.1%) was possible when BUDR was present during first division, at day 3 or 4 of protoplast culture.

IX. CHROMOSOME ELIMINATION: A NEW METHOD FOR CELL GENETICS

Controlled chromosome elimination has proved to be a valuable technique for genetic analysis of mammalian cells,[135] slime mold,[317] and yeast.[320] As a substitute for segregation, chromosome loss can be used to reveal genetic attributes otherwise marked by dominant alleles.

As for plant cells, induction of chromosomal loss was demonstrated, for the first time, in soybean[252] cell suspension using a herbicide — isopropyl-*N*-(3-chlorophenyl) carbamate (CIPC). Partial haploid[251] cell lines of soybean were prepared by CIPC treatment of a heterozygote cell line originating from the crossing of two homozygous lines. Cell lines which lacked chromosomes were characterized physiologically and in respect to variety of isozymes. The loss of chromosomes revealed a phenotype corresponding to recessive parental genotype. New phenotypes were also observed, indicating a complex genotype and the interaction of several genes.

CIPC treatment could also be used for genetic analysis[251] of a mutant cell line which exhibited more than one phenotype. The mutant cell line required asparagine for growth, but also had acquired the ability to grow on allantoin as the sole source of nitrogen. When treated with CIPC to remove chromosomes, the requirement for asparagine could be separated from the ability to use allantoin, indicating that these phenotypes were the result of separate mutations.

These results will serve as a new milestone in the cell genetics of higher plants.

X. PROSPECTS

From this account, it can be concluded that in vitro systems are proving to be an excellent source of raising variant cell lines and plants for use in fundamental as well as applied research. However, the present state of affairs can be described as the beginning of cell genetics. Production of plants with improved amino acid composition, tolerance to toxins, and improved photosynthetic efficiency is an indication of the possible utility of cell cultures for crop improvement. However, many of the variant cell lines have yet to be characterized, and regeneration of plants from them and transmission of their characters have yet to be demonstrated. Also, many of the variant cell lines concern Carrot, *Datura*, *Nicotiana*, and *Solanum*. This range must be extended to important crop plants, such as cereals and legumes.

REFERENCES

1. **Ahloowalia, B. S.**, Chromosome changes in parasexually produced rye grass, in *Current Chromosome Research*, Jones, K. and Brandham, P., Eds., Elsevier, Amsterdam, 1976, 115.
2. **Ahloowalia, B. S.**, Plant regeneration from callus culture in wheat, *Crop Sci.*, 22, 405, 1982.
3. **Ashmore, S. E. and Gould, A. R.**, Karyotype evolution in a tumour-derived plant tissue culture analysed by Giemsa C-banding, *Protoplasma*, 106, 297, 1981.
4. **Atsumi, S.**, Induction, selection and isolation of auxin heterotrophic and auxin-resistant mutants from cultured crown gall cells irradiated with gamma rays, *Plant Cell Physiol.*, 21, 1041, 1980.
5. **Austin, S. and Cassells, A. C.**, Variation between plants regenerated from individual calli produced from separated potato stem callus, *Plant Sci. Lett.*, 31, 107, 1983.

6. **Aviv, D. and Galun, E.**, An attempt at isolation of nutritional mutants from cultured tobacco protoplasts, *Plant Sci. Lett.*, 8, 299, 1976.

7. **Aviv, D. and Galun, E.**, Isolation of tobacco protoplasts in the presence of isopropyl-N-phenylcarbamate and their culture and regeneration into plants, *Z. Pflanzenphysiol.*, 83, 267, 1977.

8. **Barbier, M. and Dulieu, H. L.**, Effets genetiques observes sur des plantes de Tabac regenerees a partir de cotyledons par culture in vitro, *Ann. Amelior Plantes*, 30, 321, 1980.

9. **Barg, R. and Umiel, N.**, Development of tobacco seedlings and callus cultures in the presence of amitrol, *Z. Pflanzenphysiol.*, 83, 437, 1977.

10. **Barg, R., Pealeg, N., and Nitzan, Y.**, Sensitivity of tobacco tissue culture to chloramphenicol and its photodegradation products, *Plant Cell Environ.*, 6, 77, 1983.

11. **Barg, R., Pealeg, N., and Nitzan, Y.**, Fate of chloramphenicol in tobacco tissue cultures under various light regimes, *Plant Cell Environ.*, 6, 83, 1983.

12. **Barg, R., Pealeg, N., Perl, M., and Beckmann, J. S.**, Isolation of methotrexate resistant cell lines of *Petunia hybrida* upon stepwise selection procedure, *Plant Mol. Biol.*, 3, 303, 1984.

13. **Beadle, G. W. and Tatum, E. L.**, Neurospora. II. Methods of producing and detecting mutations concerned with nutritional requirements, *Am. J. Bot.*, 32, 678, 1945.

14. **Becker, H. and Baumer, J. J.**, Isolation and characterization of cell lines of *Valeriana wallichi* resistant to trifluoroleucine, *Z. Pflanzenphysiol.*, 112, 43, 1983.

15. **Behnke, M.**, Selection of potato callus for resistance to culture filtrates of *Phytopthora infestans* and regeneration of resistant plants, *Theor. Appl. Genet.*, 55, 69, 1979.

16. **Behnke, M.**, General resistance to late blight of *Solanum tuberosum* plants regenerated from callus resistant to callus filtrates of *Phytopthora infestans, Theor. Appl. Genet.*, 56, 151, 1980.

17. **Behnke, M.**, Selection of dihaploid potato callus for resistance to the culture filtrate of *Fusarium oxysporum*, *Z. Pflanzenzuecht.*, 85, 254, 1980.

18. **Ben-Hayyim, G., Spiegel-Roy, P., and Neumann, H.**, Relationship between ion accumulation of salt-sensitive and isolated stable salt-tolerant cell lines of *Citrus aurantium*, *Plant Physiol.*, 78, 144, 1985.

19. **Bennett, M. D.**, Nuclear instability and its manipulation in plant breeding, *Philos. Trans. R. Soc. London B*, 292, 475, 1981.

20. **Berlin, J.**, Parafluorophenylalanine resistant cell lines of tobacco, *Z. Pflanzenphysiol.*, 97, 317, 1980.

21. **Berlin, J. and Widholm, J. M.**, Correlation between phenylalalanine ammonia lyase activity and phenolic biosynthesis in p-fluorophenylalanine-sensitive and -resistant tobacco and carrot tissue cultures, *Plant Physiol.*, 59, 550, 1977.

22. **Berlin, J. and Widholm, J. M.**, Amino acid uptake by amino acid analog resistant tobacco cell lines, *Z. Naturforsch.*, 33, 634, 1978.

23. **Berlyn, M. B.**, Isolation and characterization of isonicotinic acid hydrazide resistant mutants of *Nicotiana tabacum, Theror. Appl. Genet.*, 58, 19, 1980.

24. **Bhaskaran, S., Smith, R. H., and Schertz, K.**, Sodium chloride tolerant callus of *Sorghum bicolor*, *Z. Pflanzenphysiol.*, 112, 459, 1983.

25. **Binding, H.**, Selection im Kallus Kulturen mit haploid Zellen, *Z. Pflanzenzuecht.*, 67, 33, 1972.

26. **Binding, H., Binding, K., and Straub, I.**, Selection in Gewebekulturen mit haploiden Zellen, *Naturwissenschaften*, 57, 138, 1970.

27. **Binzel, M. L., Hasegawa, P. M., Handa, A. K., and Bressan, R. A.**, Adaptation of tobacco cells to NaCl, *Plant Physiol.*, 79, 118, 1985.

28. **Blum, A.**, Genetic improvement of drought adaptation, in *Adaptation of Plants to Water and High Temperature Stress*, Turner, N. C. and Kramer, P. J., Eds., John Wiley & Sons, New York, 1980, 450.

29. **Borlaug, N. E.**, Contributions of conventional plant breeding to food production, *Science*, 219, 689, 1983.

30. **Bourgin, J. P.**, Valine resistant plants from in vitro selected tobacco cells, *Mol. Gen. Genet.*, 161, 225, 1978.

31. **Bravo, J. E. and Evans, D. A.**, Somaclonal variation in the ornamental, *Nicotiana alata*, 1984, unpublished.

32. **Bressan, R. A., Hasegawa, P. M., and Handa, A. K.**, Resistance of cultured higher plant cells to polyethylene glycol-induced water stress, *Plant Sci. Lett.*, 21, 23, 1981.

33. **Breton, A. M. and Sung, Z. R.**, Temperature-sensitive carrot variants impaired in somatic embryogenesis, *Dev. Biol.*, 90, 58, 1982.

34. **Brettell, R. I. S., Dennis, E. S., Scowcroft, W. R., and Peacock, W. J.**, Molecular analysis of a somaclonal mutant of maize alcohol dehydrogenase, *Mol. Gen. Genet.*, 202, 235, 1986.

35. **Brettell, R. I. S. and Ingram, D. S.**, Tissue culture in the production of novel disease resistant crop plants, *Biol. Rev.*, 54, 329, 1979.

36. **Brettell, R. I. S., Thomas, E., and Ingram, D. S.**, Reversion of Texas male sterile cytoplasm in culture to give fertile T-toxin resistant plants, *Theor. Appl. Genet.*, 58, 55, 1980.

37. **Bright, S. W. J. and Northcote, D. H.**, Protoplast regeneration from normal and bromodeoxyuridine-resistant sycamore callus, *J. Cell Sci.*, 16, 445, 1974.

38. **Bright, S. W. J. and Northcote, D. H.,** A deficiency of hypoxanthine phosphoribosyltransferase in a sycamore callus resistant to azaguanine, *Planta*, 123, 79, 1975.

39. **Bright, S., Jarrett, Y., Nelson, R., Greisen, G., Karp, A., Franklin, J., Norbury, P., Kueh, J., Rognes, S., and Miflin, B.,** Modifications of agronomic traits using in vitro technology, in *Plant Biotechnology*, Mantell, S. H. and Smith, H., Eds., Cambridge University Press, Cambridge, 1983, 251.

40. **Bright, S. W. J., Wood, E. A., and Miflin, B. J.,** The effect of aspartate derived amino acids (lysine, threonine, methionine) on the growth of excised embryos of wheat and barley, *Planta*, 139, 113, 1978.

41. **Brockman, R. W. and Anderson, E. P.,** Biochemistry of cancer (metabolic aspects), *Annu. Rev. Biochem.*, 32, 463, 1963.

42. **Brossard, D.,** The influence of kinetin on formation and ploidy levels of buds arising from *Nicotiana tabacum* pith tissue grown in vitro, *Z. Pflanzenphysiol.*, 78, 323, 1976.

43. **Buchanan, R. J. and Wray, J. L.,** Isolation of molybdenum cofactor defective cell lines of *Nicotiana tabacum, Mol. Gen. Genet.*, 188, 228, 1982.

44. **Burk, L. G. and Metzinger, D. F.,** Variation among anther-derived doubled haploids from an inbred line of tobacco, *J. Hered.*, 67, 381, 1976.

45. **Caboche, M. and Muller, J. F.,** Use of a medium allowing low cell density growth for in vitro selection experiments, isolation of valine resistant clones from nitrosoguanidine mutaginised cells and gamma-irradiated tobacco plants, in *Plant Cell Culture: Results and Perspectives*, Sala, F., Parisi, B., Cella, R., and Cifferi, O., Eds., Elsevier, Amsterdam, 1980, 133.

46. **Caboche, M.,** Nutritional requirements of protoplast-derived haploid tobacco cells grown at low densities in liquid medium, *Planta*, 149, 7, 1980.

47. **Carlson, P. S.,** Induction and isolation of auxotrophic mutants in somatic cell cultures of *Nicotiana tabacum, Science*, 168, 487, 1970.

48. **Carlson, P. S.,** Methionine sulfoxamine resistant mutants of tobacco, *Science*, 180, 1366, 1973.

49. **Carlson, P. S. and Widholm, J. M.,** Separation of two forms of anthranilate synthetase from 5-methyltryptophan susceptible and resistant cultured *Solanum tuberosum* cells, *Physiol. Plant.*, 44, 251, 1978.

50. **Cella, R., Albani, D., Biasini, M. G., Carbonera, D., and Parisi, B.,** Isolation and characterization of a carrot cell line resistant to methotrexate, *J. Exp. Bot.*, 35, 1390, 1984.

51. **Chaleef, R. S.,** *Genetics of Higher Plants: Applications of Cell Culture*, Cambridge University Press, New York, 1981.

52. **Chaleef, R. S. and Carlson, P. S.,** In vitro selection for mutants of higher plants, in *Genetic Manipulation with Plant Material*, Ledoux, L., Ed., Plenum Press, New York, 1975, 351.

53. **Chaleef, R. S. and Keil, R. L.,** Genetic and physiological variability among cultured cells and regenerated plants of *Nicotiana tabacum, Mol. Gen. Genet.*, 131, 254, 1981.

54. **Chaleef, R. S. and Parsons, M. F.,** Direct selection in vitro for herbicide resistant mutants of *Nicotiana tabacum, Proc. Natl. Acad. Sci. U.S.A.*, 75, 5104, 1978.

55. **Chaleef, R. S. and Ray, T. B.,** Herbicide-resistant mutants from tobacco cell cultures, *Science*, 223, 1148, 1984.

56. **Chandler, S. F. and Vasil, I. K.,** Selection and characterization of NaCl tolerant cells from embryogenic cultures of *Pennisetum purpureum* (Napier grass), *Plant Sci. Lett.*, 37, 157, 1984.

57. **Chen, H. H., Gavinlertvatana, P., and Li, P. H.,** Cold acclimation of stem-cultured plants and leaf callus of *Solanum* species, *Bot. Gaz.*, 140, 142, 1979.

58. **Conner, A. J. and Meredith, C. P.,** Large scale selection of aluminium-resistant mutants from plant cell culture: expression and inheritance in seedlings, *Theor. Appl. Genet.*, 71, 159, 1985.

59. **Cooper, K. V., Dale, J. E., Dyer, A. F., Lyne, R. L., and Walker, J. T.,** Hybrid plants from barley and rye cross, *Plant Sci. Lett.*, 12, 293, 1978.

60. **Creissen, G. P. and Karp, A.,** Karyotypic change in potato plants regenerated from protoplasts, *Plant Cell Tissue Organ Culture*, 4, 171, 1985.

61. **Croughan, T. P., Stavarek, S. J., and Rains, D. W.,** Selection of NaCl tolerant line of cultured alfalfa cells, *Crop Sci.*, 18, 959, 1978.

62. **Cseplo, A. and Maliga, P.,** Lincomycin resistance, a new type of maternally inherited mutation in *Nicotiana plumbaginifolia, Curr. Genet.*, 6, 105, 1982.

63. **Cseplo, A. and Medgysey, P.,** Characteristic symptoms of photosynthesis inhibition by herbicides are expressed in photomixotrophic tissue cultures of *Nicotiana, Planta*, 168, 24, 1986.

64. **Cseplo, A., Medgyesy, P., Hideg, E., Demetor, S., Marton, L., and Maliga, P.,** Triazine resistant *Nicotiana* mutants from photomixotrophic cell cultures, *Mol. Gen. Genet.*, 200, 508, 1985.

65. **Cseplo, A., Nagy, F., and Maliga, P.,** Rescue of a cytoplasmic lincomycin resistance factor from *Nicotiana sylvestris* into *N. plumbaginifolia* by protoplast fusion, in *Int. Protoplast Symp.*, Potrykus, I., Harnes, C. T., Hinnen, A., Hutter, R., King, P. J., and Shillitto, R. D., Eds., Birkhauser-Verlag, Basel, 126, 1983.

66. **D'Amato, F.**, Cytogenetics of differentiation in tissue and cell culture, in *Applied and Fundamental Aspects of Plant Cell Tissue and Organ Culture*, Reinert, J. and Bajaj, Y. P. S., Eds., Springer-Verlag, Berlin, 1977, 343.

67. **Deng, Q. I. and Ives, D. H.**, Modes of nucleoside phosphorylation in plants: studies on the apparent thymidine kinase and true uridine kinase of seedlings, *Biochim. Biophys. Acta*, 277, 235, 1972.

68. **De Paepe, R., Belton, R., and Gnangbe, F.**, Basis and extent of genetic variability among doubled haploid plants obtained by pollen culture in *Nicotiana sylvestris*, *Theor. Appl. Genet.*, 59, 177, 1981.

69. **Devreux, M. and Laneri, V.**, Anther culture, haploid plants, isogenic lines and breeding research in *Nicotiana tabacum*, in *Polyploidy and Induced Mutations in Plant Breeding*, International Atomic Energy Agency, Vienna, 1974, 101.

70. **Dix, P. J.**, Chilling resistance is not transmitted sexually in plants regenerated in *Nicotiana sylvestris* cell lines, *Z. Pflanzenphysiol.*, 84, 223, 1977.

71. **Dix, P. J.**, Environmental stress resistance, selection and plant cell cultures, in *Plant Cell Cultures: Results and Perspectives*, Sala, F., Parisi, B., Cell, R., and Ciferi, O., Eds., Elsevier, Amsterdam, 1980, 183.

72. **Dix, P. J.**, Cross-resistance in cell lines of *Nicotiana sylvestris* selected for resistance to individual antibiotics, *Ann. Bot.*, 48, 321, 1981.

73. **Dix, P. J.**, Inheritance of chloramphenicol resistance, a trait selected in cell cultures of *Nicotiana sylvestris*, *Ann. Bot.*, 48, 315, 1981.

74. **Dix, P. J. and Pearce, R. S.**, Proline accumulation in NaCl-resistant and sensitive cell lines of *Nicotiana sylvestris*, *Z. Pflanzenphysiol.*, 102, 243, 1981.

75. **Dix, P. J. and Street, H. E.**, Sodium chloride resistant cultured cell lines from *Nicotiana sylvestris* and *Capsicum annuum*, *Plant Sci. Lett.*, 5, 231, 1975.

76. **Dix, P. J. and Street, H. E.**, Selection of plant cell lines with enhanced chilling resistance, *Ann. Bot.*, 40, 903, 1976.

77. **Dunham, V. L. and Bryan, J. K.**, Synergistic effects of metabolically related amino acids on the growth of a multicellular plant, *Plant Physiol.*, 44, 1601, 1969.

78. **Dunham, V. L. and Bryan, J. K.**, Synergistic effects of metabolically related amino acids on the growth of a multicellular plant. II. Studies of ^{14}C amino acid incorporation, *Plant Physiol.*, 47, 91, 1971.

79. **Eapen, S.**, Effect of gamma and UV-irradiation on survival and totipotency of haploid tobacco cells in culture, *Protoplasma*, 89, 149, 1976.

80. **Edallo, S., Zucchinali, C., Perenzin, M., and Salamini, F.**, Chromosome variation and frequency of spontaneous mutation associatd with in vitro culture and plant regeneration in maize, *Maydica*, 26, 39, 1981.

81. **Eichenberger, M. E.**, Sur une mutation survenue dans une culture de tissus de carotte, *C.R. Soc. Biol.*, 145, 239, 1951.

82. **Ellis, B. E.**, Non-differential sensitivity to the herbicide, matribuzin in tomato cell suspension cultures, *Can. J. Plant Sci.*, 58, 775, 1978.

83. **Ericson, M. C. and Alfinito, S. H.**, Proteins produced during salt stress in tobacco cell cultures, *Plant Physiol.*, 74, 506, 1984.

84. **Eriksson, T.**, Effects of ultraviolet and X-ray radiation on in vitro cultivated cells of *Haplopappus gracilis*, *Physiol. Plant.*, 20, 507, 1967.

85. **Evans, D. A. and Gamborg, O. L.**, Effects of para-fluorophenylalanine on ploidy levels of cell suspension cultures of *Datura innoxia*, *Environ. Exp. Bot.*, 19, 269, 1979.

86. **Evans, D. A. and Sharp, W. R.**, Single gene mutations in tomato plants regenerated from tissue culture, *Science*, 221, 949, 1983.

87. **Evans, D. A., Sharp, W. R., and Medina-Filho, H. P.**, Somaclonal and gametoclonal variation, *Am. J. Bot.*, 71, 759, 1984.

88. **Flashman, S. M.**, Use of a non-volatile thiocarbamate to select for herbicide tolerant tobacco cell lines, *Plant Sci.*, 38, 149, 1985.

89. **Flashman, S. M. and Filner, P.**, Selection of tobacco cell lines resistant to solenoamino acids, *Plant Sci. Lett.*, 13, 219, 1978.

90. **Flashman, S. M., Meredith, C. P., and Howard, J. A.**, Selection for increased vernolate tolerance in tobacco cell cultures, *Plant Sci.*, 38, 141, 1985.

91. **Frankhauser, H., Bucher, F., and King, P. J.**, Isolation of biochemical mutants using haploid mesophyll protoplasts of *Hyoscyamus muticus*, *Planta*, 60, 415, 1984.

92. **Gathercole, R. W. E. and Street, H. E.**, Isolation, stability and biochemistry of a p-fluorophenylalanine-resistant cell line of *Acer pseudoplatanus*, *New Phytol.*, 77, 29, 1976.

93. **Gathercole, R. W. E. and Street, H. E.**, A p-fluorophenylalanine resistant cell line of sycamore with increased contents of phenylalanine, tyrosine and phenolics, *Z. Pflanzenphysiol.*, 89, 283, 1978.

94. **Gebhardt, C., Schnebli, V., and King, P. J.**, Isolation of biochemical mutants using haploid mesophyll protoplasts of *Hyoscyamus muticus*, *Planta*, 153, 81, 1981.

95. **Gebhardt, C., Shimamoto, K., Lazar, G., Schnebli, V., and King, P. J.**, Isolation of biochemical mutants using haploid mesophyll protoplasts of *Hyoscyamus muticus*. III. General characterization of histidine and tryptophan auxotrophs, *Planta*, 159, 18, 1983.

96. **Gebhardt, C., Shimamoto, K., Lazar, G., Schnebli, V., and King, P. J.**, Isolation of biochemical mutants using haploid mesophyll protoplasts of *Hyoscyamus muticus*, *Planta*, 159, 18, 1985.

97. **Gengenbach, B. G. and Green, C. E.**, Selection of T-cytoplasm maize callus cultures resistant to *Helminthosporium maydis* race T pathotoxin, *Crop. Sci.*, 15, 645, 1975.

98. **Gengenbach, B. G., Green, C. E., and Donovan, C. M.**, Inheritance of selected pathotoxin resistance in maize plants regenerated from cell cultures, *Proc. Natl. Acad. Sci. U.S.A.*, 74, 5113, 1977.

99. **George, L. and Rao, P. S.**, Yellow-seeded variants in in vitro regenerants of mustard (*Brassica juncea* var. Rai-5), *Plant Sci. Lett.*, 30, 327, 1983.

100. **Gill, B. S., Kam-Morgan, L. N. W., and Shepard, J. F.**, An apparent meiotic mutation in a mesophyll cell protoclone of the "Russet Burbank" potato, *J. Hered.*, 76, 17, 1985.

101. **Giuliano, G., LoSchiavo, F., and Terzi, M.**, Isolation and characterisation of temperature sensitive carrot cell variants, *Theor. Appl. Genet.*, 67, 179, 1984.

102. **Glimelius, K., Eriksson, T., Grafe, R., and Muller, A. J.**, Somatic hybridization of nitrate reductase deficient mutants of *Nicotiana tabacum* by protoplast fusion, *Physiol. Plant.*, 44, 273, 1978.

103. **Goldner, R., Umiel, N., and Chen, Y.**, The growth of carrot callus cultures at various concentrations and composition of saline water, *Z. Pflanzenphysiol.*, 85, 307, 1977.

104. **Gonzales, R. A., Das, P. K., and Widholm, J. M.**, Characterization of cultured tobacco cell lines resistant to ethionine, a methionine analog, *Plant Physiol.*, 74, 640, 1984.

105. **Gosal, S. S. and Bajaj, Y. P. S.**, Isolation of sodium chloride resistant cell lines in some grain legumes, *Indian J. Exp. Biol.*, 22, 209, 1984.

106. **Green, C. E., Phillips, R. L., and Wang, A. S.**, Cytological analysis of plants regenerated from maize tissue cultures, *Maize Genet. Coop. Newsl.*, 51, 53, 1977.

107. **Greenway, H. and Munns, R.**, Mechanisms of salt tolerance in nonhalophytes, *Annu. Rev. Plant Physiol.*, 31, 149, 1980.

108. **Gresshoff, P. M.**, Cycloheximide resistance in *Daucus carota* cell cultures, *Theor. Appl. Genet.*, 54, 141, 1979.

109. **Grieve, C. M. and Maas, E. V.**, Betaine accumulation in salt-stressed sorghum, *Physiol. Plant.*, 61, 167, 1984.

110. **Handa, A. K., Bressan, R. A., Handa, S., and Hasegawa, P. M.**, Clonal variation for tolerance to polyethylene glycol-induced water stress in cultured tomato cells, *Plant Physiol.*, 72, 645, 1983.

111. **Heimer, Y. M. and Filner, P.**, Regulation of the nitrate assimilation pathways of cultured tobacco cells. II. Properties of a variant cell line, *Biochim. Biophys. Acta*, 215, 152, 1978.

112. **Heinz, D. J.**, in *Induced Mutations in Vegetatively Propagated Plants*, International Atomic Energy Agency, Vienna, 1973.

113. **Heinz, D. J., Krishnamurthi, M., Nickell, L. G., and Maretzki, A.**, Cell tissue and organ culture in sugarcane improvement, in *Applied and Fundamental Aspects of Plant Cell Tissue and Organ Culture*, Reinert, J. and Bajaj, Y. P. S., Eds., Springer-Verlag, Berlin, 1977, 3.

114. **Heinz, D. J. and Mee, G. W. P.**, Morphologic, cytogenetic and enzymatic variation in *Saccharum* species hybrid clones derived from callus tissue, *Am. J. Bot.*, 58, 237, 1971.

115. **Heyser, J. W. and Nabors, M. W.**, Osmotic adjustment of tobacco cells and plants to penetrating and non-penetrating solutes, *Plant Physiol.*, 63 (Suppl.), 720, 1979.

116. **Hibberd, K. A. and Green, C. E.**, Inheritance and expression of lysine plus threonine resistance selected in maize tissue culture, *Proc. Natl. Acad. Sci. U.S.A.*, 79, 559, 1982.

117. **Hibberd, K. A., Walter, T., Green, C. E., and Gengenbach, B. G.**, Selection and characterisation of a feedback insensitive tissue culture of maize, *Planta*, 148, 183, 1980.

118. **Hitz, W. D. and Hanson, A. D.**, Determination of *Glycine betaine* by pyrolysis gas chromatography in cereals and grasses, *Phytochemistry*, 19, 2371, 1980.

119. **Hoffmann, F., Thomas, E., and Wenzel, G.**, Anther culture as a breeding tool in rape. II. Progeny analysis of androgenetic lines and induced mutants from haploid cultures, *Theor. Appl. Genet.*, 61, 225, 1982.

120. **Horsch, R. B. and Jones, G. E.**, 8-Azaguanine-resistant variants of cultured cells of *Haplopappus gracilis*, *Can. J. Bot.*, 56, 2660, 1978.

121. **Horsch, R. B. and King, J.**, Isolation of an isolecuine-valine-requiring auxotroph from *Datura innoxia* cell cultures by arsenate counter selection, *Planta*, 159, 12, 1983.

122. **Horsch, R. B. and King, J.**, The isolation of auxotrophs from *Datura innoxia* cell cultures following recovery of arsenate-treated cells on feeder plates, *Planta*, 160, 168, 1984.

123. **Horsch, R. B. and King, J.**, Arsenate counter selective enrichment for auxotrophic plant cells works well in theory but not in practice, *Can. J. Bot.*, 63, 2115, 1985.

124. **Howland, G. P. and Hart, R. W.,** Radiation biology of cultured plant cells, in *Applied and Fundamental Aspects of Plant Cell, Tissue and Organ Culture,* Reinert, J. and Bajaj, Y. P. S., Eds., Springer-Verlag, Berlin, 1977, 731.

125. **Hughes, K. W.,** Diquat resistance in a paraquat resistant soybean cell line, in *Abstr. 4th Int. Congr. Plant Tissue and Cell Culture,* IAPTC, Calgary, Canada, 1978, 170.

126. **Jackson, P. J., Roth, E. J., McClure, P. R., and Naranjo, C. M.,** Selection, isolation and characterization of cadmium resistant *Datura innoxia* suspension cultures, *Plant Physiol.,* 75, 914, 1984.

127. **Jones, G. E.,** 6-Azauracil resistant variants of cultured plant cells lack uracil phosphoribosyltransferase activity, *Plant Physiol.,* 75, 161, 1984.

128. **Jones, G. E. and Hann, J.,** *Haplopappus gracilis* cell strains resistant to pyrimidine analogues, *Theor. Appl. Genet.,* 54, 81, 1979.

129. **Kanabus, J., Pikaard, C. S., and Cherry, J. S.,** Heat shock proteins in tobacco cell suspension during growth cycle, *Plant Physiol.,* 75, 639, 1984.

130. **Kandra, G. and Maliga, P.,** Is bromodeoxyuridine resistance a consequence of cytokinin habituation in *Nicotiana tabacum, Planta,* 133, 131, 1977.

131. **Kao, F. T. and Puck, T. T.,** Genetics of somatic mammalian cells. VII. Induction and isolation of nutritional mutants in Chinese hamster cells, *Proc. Natl. Acad. Sci. U.S.A.,*. 60, 1275,1968.

132. **Kapista, O. S., Kulinich, A. V., and Vinetski, Y. P.,** Spontaneous lac $^+$ mutants of tobacco tissue culture cells and their use in transgenosis (translation), *Dokl. Akad. Nauk SSSR,* 235, 1426, 1977.

133. **Karp, A., Nelson, R. S., Thomas, E., and Bright, W. J.,** Chromosome variation in protoplast derived potato plants, *Theor. Appl. Genet.,* 63, 265, 1982.

134. **Kasperbauer, M. J., Sulton, T. J., Anderson, R. A., and Gilpton, C. L.,** Tissue culture of plants from a chimeral mutation of tobacco, *Crop Sci.,* 21, 588, 1981.

135. **Kato, H. and Yoshida, T. H.,** Isolation of aneusomic clones from Chinese hamster cell lines following induction of nondisjunction, *Cytogenetics,* 10, 392, 1971.

136. **Katz, A. and Tal, M.,** Salt tolerance in wild relatives of the cultivated tomato: proline accumulation in callus tissue of *Lycopersicon esculentum* and *L. peruvianum, Z. Pflanzenphysiol.,* 98, 429, 1980.

137. **Keil, R. L. and Chaleef, R. S.,** Genetic characterization of hydroxyurea-resistant mutants obtained from cell cultures of *Nicotiana tabacum, Mol. Gen. Genet.,* 192, 218, 1983.

138. **Kemble, R. J. and Shepard, J. F.,** Cytoplasmic DNA variation in a potato protoclonal population, *Theor. Appl. Genet.,* 69, 211, 1984.

139. **King, J.,** The response of *Datura innoxia* wild type and auxotrophic cells to several enrichment agents, *Plant Cell Rep.,* 4, 123, 1985.

140. **King, J., Horsch, R. B., and Savage, A. D.,** Partial characterization of two stable auxotrophic cell strains of *Datura innoxia, Planta,* 149, 480, 1980.

141. **King, J. and Khanna, V.,** A nitrate reductase-less variant isolated from suspension cultures of *Datura innoxia, Plant Physiol.,* 66, 632, 1980.

142. **King, J. and Maretzki, A.,** Isolation from sugarcane cultures of variants resistant to antimetabolites, *Physiol. Plant.,* 58, 457, 1983.

143. **Koblitz, H., Schumann, U., Bohm, H., and Franke, J.,** Tissue culture of alkaloid plants. IV. *Macleaya microcarpa* (Maxim), *Experientia,* 31, 768, 1975.

144. **Kochba, J., Ben-Hayyam, G., Spigel-Roy, P., Saad, S., and Neumann, H.,** Selection of stable salt tolerant callus cell lines and embryos in *Citrus sinensis* and *C. aurantium, Z. Pflanzenphysiol.,* 106, 111, 1982.

145. **Kramer, P. J.** Drought, stress and the origin of adaptations, in *Adaptation of Plants to Water and High Temperature Stress,* Turner, N. C. and Kramer, P. J., Eds., John Wiley & Sons, New York, 1980, 7.

146. **Krishnamurthy, M.,** *Sugarcane Breeders Newsletter,* 35, 24, 1974.

147. **Krumbiegel, G.,** Response of haploid and diploid protoplasts from *Datura innoxia* and *Petunia hybrida* to treatment with X-rays and a chemical mutagen, *Environ. Exp. Bot.,* 19, 99, 1979.

148. **Kueh, J. S. H. and Bright, S. W. J.,** Proline accumulation in a barley mutant resistant to trans-4-hydroxy-L-proline, *Planta,* 153, 166, 1981.

149. **Landsmann, J. and Uhrig, H.** Somaclonal variation in *Solanum tuberosum* detected at the molecular level, *Theor. Appl. Genet.,* 71, 500, 1985.

150. **Larkin, P. J. and Scowcroft, W. R.,** Somaclonal variation — a novel source of variability from cell cultures for plant improvement, *Theor. Appl. Genet.,* 60, 197, 1981.

151. **Larkin, P. J. and Scowcroft, W. R.,** Somaclonal variation and crop improvement, in *Genetic Engineering of Plants,* Kosuge, T., Meredith, C., and Hollaender, A., Eds., Plenum Press, New York, 1983, 257.

152. **LaRosa, P. C., Handa, A. K., Hasegawa, P. M., and Bressan, R. A.,** Abscisic acid accelerates adaptation of cultured tobacco cells to salt, *Plant Physiol.,* 79, 138, 1985.

153. **Lazar, G. B., Dudits, D., and Sung, Z. R.,** Expression of cycloheximide resistance in carrot somatic hybrids and their segregants, *Genetics,* 98, 347, 1981.

154. **Lebrun, L., Rajasekaran, K., and Mullins, M. G.,** Selection in vitro for NaCl tolerance in *Vitis rupestris*, *Ann. Bot.*, 56, 733, 1985.

155. **Lescure, A. M.** Mutagenese de cellules vegetales cultives in vitro: methods et resultats, *Soc. Bot. Fr. Mem.*, 353, 1970.

156. **Lescure, A. M.,** Selection of markers of resistance to base analogues in somatic cultures of *Nicotiana tabacum*, *Plant Sci. Lett.*, 1, 375, 1973.

157. **Lescure, A. M. and Peaud-Lenoel, C.,** Production par traitment mutagene de ligness cellulaires d' *Acer pseudoplatanus* anergiees a l'auxine, *C. R. Acad. Sci.*, 265, 1803, 1967.

158. **Levitt J.,** *Responses of Plants to Environmental Stresses*, Academic Press, New York, 1980.

159. **Lewis, W. H. and Wright, J. A.,** Altered ribonucleotide reductase activity in mammalian tissue culture cells resistant to hydroxyurea, *Biochem. Biophys. Res. Commun.*, 60, 926, 1974.

160. **Limberg, M., Cress, D., and Lark, K. G.,** Variant of soybean cells which grow in suspension with maltose as a carbon energy source, *Plant Physiol.*, 63, 718, 1979.

161. **Ling, D. H., Vidhyascharan, P., Borromeo, E. S., Zapata, F. J., and Mew, T. W.,** In vitro screening of rice germplasm for resistance to brown spot disease using phytotoxin, *Theor. Appl. Genet.*, 71, 131,1985.

162. **Little, M. C. and Preston, J. F.,** Sensitivity of carrot cell cultures and RNA polymerase II to amatoxins, *Plant Physiol.*, 77, 443, 1985.

163. **Liu, M. C. and Chen, W. H.,** Tissue and cell culture as aids to sugarcane breeding. I. Creation of genetic variation through callus culture, *Euphytica*, 25, 394, 1976.

164. **Liu, M. C.,** In vitro methods applied to sugarcane improvement in *Plant Tissue Culture*, Thorpe, T. A., Ed., Academic Press, New York, 1981, 299.

165. **LoSchiavo, F., Grovinazzo, G., and Terzi, M.,** 8-Azaguanine resistant carrot cell mutants and their use as universal hybrdizers, *Mol. Gen. Genet.*, 192, 326, 1983.

166. **LoSchiavo, F., Mela, L., Ronchi, N. V., and Terzi, M.,** Use of HAT system on mutants isolated from cell cultures of *Medicago sativa*, in *Plant Cell Cultures: Results and Perspectives*, Sala, F., Parisi, B., Cella, R., and Ciferri, O., Eds., Elsevier, Amsterdam, 1980, 127.

167. **Malemberg, R. L.,** Temperature sensitive variants of *Nicotiana tabacum* isolated from somatic cell culture, *Genetics*, 92, 215, 1979.

168. **Malemberg, R. L. and McIndoo, J.,** Abnormal floral development of a tobacco mutant with elevated polyamine level, *Nature (London)*, 305, 623, 1983.

169. **Malepszy, S., Grunewaldt, J., and Maluszynski, M.,** Uber die Skeletion von Mutanten in Zellkulturen aus Haploider *Nicotiana*, *Z. Pflanzenzuecht.*, 79, 160, 1977.

170. **Maliga, P.,** in *Cell Genetics in Higher Plants*, Dudits, D., Farkas, G. L., and Maliga, P., Eds., Akademiat Kiado, Budapest, 1976, 59.

171. **Maliga, P.,** Isolation, characterization and utiilzation of mutant cell lines in higher plants, in *Perspectives in Plant Cell and Tissue Culture*, Vasil, I. K., Ed., Academic Press, New York, 1980, 225.

172. **Maliga, P.,** Streptomycin resistance is inherited as a recessive mendelian trait in a *Nicotiana sylvestris* line, *Theor. Appl. Genet.*, 60, 1, 1981.

173. **Maliga, P., Breznovits, Sz., and Marton, L.,** Streptomycin resistant plants from callus cultures of haploid tobacco, *Nature, (London) New Biol.*, 244, 29, 1973.

174. **Maliga, P., Breznovits, A., Marton, L., and Joo, F.,** Nonmendelian streptomycin resistant tobacco mutant with altered chloroplasts and mitochondria, *Nature (London)*, 255, 401, 1975.

175. **Maliga, P., Kiss, Z. R., Dix, P. J., and Lazar, G.,** A streptomycin resistant line of *Nicotiana sylvestris* unable to flower, *Mol. Gen. Genet.*, 172, 13, 1979.

176. **Maliga, P., Lazar, G., Svab, Z., and Nagy, F.,** Transient cycloheximide resistance in a tobacco cell line, *Mol. Gen. Genet.*, 149, 267, 1976.

177. **Maliga, P., Marton, L., and Breznovits, Sz.,** 5-Bromodeoxyuridine-resistant cell lines from haploid tobacco, *Plant Sci. Lett.*, 1, 119, 1973.

178. **Maliga, P., Menczel, L., Sidorov, V., Marton, L., Cespelo, A., Medgyesy, P., ManhDung, T., Lazar, G., and Nagy, F.,** Cell culture mutants and their use, in *Plant Improvement and Somatic Cell Genetics*, Vasil, I., Scowcroft, W. R., and Frey, K. L., Eds., Academic Press, New York, 1982, 221.

179. **Maretzki, A. and Tom, M.,** Characteristics of a galactose adapted cell line grown in suspension culture, *Plant Physiol.*, 61, 544, 1978.

180. **Marton, L., ManhDung, T., Mendel, R. R., and Maliga, P.,** Nitrate reductase deficient cell lines from haploid protoplast culture of *Nicotiana plumbaginifolia*, *Mol. Gen. Genet.*, 186, 301, 1982.

181. **Marton, L., Nagy, F., Gupta, K. C., and Maliga, P.,** 5-Bromodeoxyuridine resistant tobacco cells incorporating the analogue, *Plant Sci. Lett.*, 12, 333, 1978.

182. **Mastrangello, I. A. and Smith, H. H.,** Selection and differentiation of aminopterin resistant cells of *Datura innoxia*, *Plant Sci. Lett.*, 10, 171, 1977.

183. **Mathur, A. K., Ganapathy, P. S., and Johri, B. M.,** Isolation of sodium chloride tolerant plantlets of *Kickxia ramosissima* under in vitro conditions, *Z. Pflanzenphysiol.*, 99, 287, 1980.

184. **McCoy, T. J., Phillips, R. L., and Rines, H. W.,** Cytogenetic analysis of plants regenerated from oat (*Avena sativa*) tissue cultures: high frequency of partial chromosome loss, *Can. J. Genet. Cytol.*, 24, 37, 1982.
185. **Meins, F.,** 5-Bromodeoxyuridine: a specific inhibitor of cytokinin-habituation in tobacco cell culture, *Planta*, 129, 239, 1976.
186. **Meins, F.,** Heritable variation in plant cell culture, *Annu. Rev. Plant Physiol.*, 34, 327, 1983.
187. **Meins, F. and Binns, A.,** Epigenetic variation of cultured somatic cells. Evidence for gradual changes in the requirements for factors promoting cell division, *Proc. Natl. Acad. Sci. U.S.A.*, 74, 2928, 1977.
188. **Meins, F. and Lutz, J.,** Tissue specific variation in the cytokinin habituation of cultured tobacco cells, *Differentiation*, 15, 1, 1979.
189. **Meins, F. and Lutz, J.,** The induction of cytokinin habituation in primary pith explants of tobacco, *Planta*, 149, 402, 1980.
190. **Menczel, L., Nagy, F., Kiss, Z. R., and Maliga, P.,** Streptomycin resistant and sensitive somatic hybrids of *Nicotiana tabacum* and *N. knightiana*. Correlation of resistance to *N. tabacum* plastids, *Theor. Appl. Genet.*, 59, 191, 1981.
191. **Mendel, R. R. and Muller, A. J.,** A common genetic determinant of xanthine dehydrogenase and nitrate reductase in *Nicotiana tabacum*, *Biochem. Physiol. Pflanz.*, 170, 538, 1976.
192. **Mendel, R. R. and Muller, A. J.,** Repair in vitro of nitrate reductase deficient tobacco mutants (cnx A) by molybdate and by molybdenum cofactor, *Planta*, 163, 370 1985.
193. **Meredith, C. P.,** Response of cultured tomato cells to aluminium, *Plant Sci. Lett.*, 12, 17, 1978.
194. **Merrick, M. M. A. and Collin, H. A.,** Asulam resistance in embryoids of celery tissue cultures, *New Phytol.*, 92, 435, 1982.
195. **Miller, O. K. and Hughes, K. W.,** Selection of paraquat-resistant variants of tobacco from cell culture, *In Vitro*, 16, 1085, 1980.
196. **Minton, K. W., Karmin, P., Hahn, G. M., and Minton, A. P.,** Nonspecific stabilization of stress-susceptible proteins by stress resistant proteins: a model for the biological role of heat shock proteins, *Proc. Natl. Acad. Sci. U.S.A.*, 79, 7107, 1982.
197. **Mok, M. C., Gableman, W. H., and Skoog, F.,** Carotenoid synthesis in carrot tissue cultures, *Plant Physiol.*, 56, (Suppl.), 29, 1976.
198. **Muller, A. J. and Grafe, R.,** Isolation and characterization of cell lines of *Nicotiana tabacum* lacking nitrate reductase, *Mol. Gen. Genet.*, 161, 67, 1978.
199. **Müller, A. J. and Grafe, R.,** Reconstitution of NADH-nitrate reductase in vitro from nitrate reductase deficient *Nicotiana tabacum* mutants, *Mol. Gen. Genet.*, 161, 77, 1978.
200. **Müller, A. J. and Grafe, R.,** Nitrate reductase deficient mutant cell lines of *Nicotiana tabacum*, *Mol. Gen. Genet.*, 177, 145, 1979.
201. **Müller, J. F., Goujaud, J., and Caboche, M.,** Isolation in vitro of naphthaleneacetic acid tolerant mutants of *Nicotiana tabacum* which are impaired in root morphogenesis, *Mol. Gen. Genet.*, 199, 194, 1985.
202. **Murphy, T. M. and Imbrie, C. W.,** Induction and characterization of chlorate resistant strains of *Rosa damascena* cultured cells, *Plant Physiol.*, 67, 910, 1981.
203. **Murphy, T. M. and Imbrie, C. W.,** Repair in vitro of nitrate reductase deficient tobacco mutants (cnx A) by molybdate and by molybdenum cofactor, *Planta*, 163, 370, 1985.
204. **Murphy, T. M., Wrona, A. F., and Wycoff, K.,** Chlorate-resistant rose cells: influx, efflux and reduction of (^{36}Cl)-chlorate, *Physiol. Plant.*, 64, 339, 1985.
205. **Nabros, M. W., Daniels, A., Nadolny, L., and Brown, C.,** Sodium chloride tolerant lines of tobacco cells, *Plant Sci. Lett.*, 4, 155, 1975.
206. **Nabors, M. W., Gibbs, S. E., Bernstein, C. S., and Meis, M. E.,** NaCl-tolerant tobacco plants from cultured cells, *Z. Pflanzenphysiol.*, 97, 13, 1980.
207. **Nabors, M. W., Kroskey, C., and McHugh, D. M.,** Green spots are predietors of high callus growth rates and shoot formation in normal and salt stressed tissue culture of oat, *Z. Pflanzenphysiol.*, 105, 341, 1982.
208. **Naef, J. and Turian, G.,** Sur les carotenoides du tissue cambial de racine de carotte cultive in vitro, *Phytochemistry*, 2, 173, 1963.
209. **Nafziger, E. D., Widholm, J. M., Steinrucken, H. C., and Killmer, J. L.,** Selection and characterization of a carrot cell lines tolerant to glyphosate, *Plant Physiol.*, 76, 571, 1984.
210. **Nakamura, C., Keller, W. A., and Fedak, G.,** In vitro propagation and chromosome doubling of a *Triticum crassum* × *Hordeum vulgare* intergeneric hybrid, *Theor. Appl. Genet.*, 60, 89, 1981.
211. **Nakamura, C., Mori, N., Nakata, M., and Ono, H.,** 2,4-D resistance in a tobacco cell culture variant. II. Effects of 2,4-D on nucleic acid and protein synthesis and cell respiration, *Plant Cell Physiol.*, 27, 243, 1986.
212. **Nakamura, C., Nakata, M., Shioji, M., and Ono, H.,** 2,4-D resistance in a tobacco cell culture variant: cross resistance to auxins uptake, efflux and metabolism of 2,4-D, *Plant Cell Physiol.*, 26, 271, 1985.
213. **Nanney, D. L.,** Epigenetic control systems, *Proc. Natl. Acad. Sci. U.S.A.*, 44, 712, 1958.

214. **Negrutiu, I., Cattoir-Reynearts, A., Verbruggen, I., and Jacobs, M.,** Lysine overproducer mutants with an altered dehydropiconilate synthase from protoplasts of *Nicotiana sylvestris, Theor. Appl. Genet.,* 68, 11, 1984.

215. **Negrutiu, I., DeBrouwer, D., Dirks, R., and Jacobs, M.,** Amino acid auxotrophs from protoplast cultures of *Nicotiana plumbaginifolia, Mol. Gen. Genet.,* 199, 330, 1985.

216. **Negrutiu, I., Dirks, R., and Jacobs, M.,** Regeneration of fully nitrate reductase deficient mutants from protoplast culture of *Nicotiana plumbaginifolia, Theor. Appl. Genet.,* 66, 341, 1983.

217. **Negrutiu, I., Jacobs, M., and Caboche, M.,** Advances in somatic cell genetics of higher plants — the protoplast approach in basic studies on mutagenesis and isolation of biochemical mutants, *Theor. Appl. Genet.,* 67, 289, 1984.

218. **Negrutiu, I., Jacobs, M., and Cattoir, A.,** *Arabidopsis thaliana* espece modele en genetique cellulaire, *Physiol. Veg.,* 16, 365, 1978.

219. **Negrutiu, I. and Muller, J. F.,** Culture conditions of protoplast-derived cells of *Nicotiana sylvestris* for mutuant selection, *Plant Cell Rep.,* 1, 14, 1981.

220. **Nickell, L. G. and Heinz, D. J.,** Potential of cell and tissue culture techniques as aids in economic plant improvement, in *Genes, Enzymes and Population,* Srb., A. M., Ed., Plenum Press, New York, 1973, 109.

221. **Nielsen, E., Selva, E., Sghirinzetti, C., and Devreux, M.,** The mutagenic effect of gamma rays on leaf protoplasts of haploid and dihaploid *Nicotiana plumbaginifolia* estimated by valine resistance mutation frequencies, *Theor. Appl. Genet.,* 70, 259, 1985.

222. **Nijkamp, H. J. J., Colijin, C. M., and Kool, A. J.,** in *Abstr. 4th Int. Congr. Plant Tissue and Cell Culture, IAPTC,* Calgary, Canada, 1978, 145.

223. **Novak, F. J.,** Phenotype and cytological status of plants regenerated from callus cultures of *Allium sativum, Z. Pflanzenzuecht.,* 84, 250, 1980.

224. **Nyman, L. P., Gonzales, C. J., and Arditti, J.,** In vitro selection for salt tolerance of taro (*Colocasia esculenta* var. *antiquorum*), *Ann. Bot.,* 229, 1983.

225. **Ogijhara, Y.,** Tissue culture in *Howarthia.* IV. Genetic characterization of plants regenerated from callus, *Theor. Appl. Genet.,* 60, 353, 1981.

226. **Ohyama, K.,** Properties of 5-bromodeoxyuridine-resistant lines of higher plant cells in liquid culture, *Exp. Cell. Res.,* 89, 31, 1974.

227. **Ohyama, K.,** A basis for bromodeoxyuridine resistance in plant cells, *Environ. Exp. Bot.,* 16, 209, 1976.

228. **Oinuma, T. and Yoshida, T.,** Genetic variation among doubled haploid lines of burley tobacco varieties, *Jpn. J. Breed.,* 24, 211, 1974.

229. **Ojima, K. and Ohira, K.,** Characterization of aluminium and manganese tolerant cell lines selected from carrot cell cultures, *Plant Cell Physiol.,* 24, 789, 1983.

230. **Ojima, K. and Ohira, K.,** Release of citric acid into the medium by aluminium tolerant carrot cells, *Plant Cell Physiol.,* 25, 855, 1984.

231. **Ojima, K. and Ohira, K.,** Reduction of aluminium toxicity by addition of a conditioned medium from aluminium tolerant cells of carrot, *Plant Cell Physiol.,* 26, 281, 1985.

232. **Ono, H.,** Genetical and physiological investigations of a 2,4-D resistant cell line isolated from tissue cultures of tobacco. I. Growth responses to 2,4-D and IAA, *Sci. Rep. Fac. Agric. Kobe Univ.,* 13, 272, 1979.

233. **Ono, H. and Nakano, M.,** The regulation of expression of cellular phenotypes in cultured tissues of tobacco, *Jpn. J. Genet.,* 53, 241, 1978.

234. **Oono, K.,** Test tube breeding of rice by tissue culture, *Trop. Agric. Res.,* 11, 109, 1978.

235. **Oono, K.,** In vitro methods applied to rice, in *Plant Tissue Culture,* Thorpe, T. A., Ed., Academic Press, New York, 1981, 273.

236. **Owens, L. D.,** Characterization of kanamycin resistant cell lines of *Nicotiana tabacum, Plant Physiol.,* 67, 1166, 1981.

237. **Palmer, J. E. and Widholm, J. M.,** Characterization of carrot and tobacco cell cultures resistance to p-fluorophenylalanine, *Plant Physiol.,* 56, 233, 1975.

238. **Pandey, R. and Ganapathy, P. S.,** Isolation of sodium chloride tolerant callus line of *Cicer arietinum, Plant Cell Rep.,* 3, 45, 1984.

239. **Pandey, R. and Ganapathy, P. S.,** Effects of sodium chloride stress on callus cultures of *Cicer arietinum.* Growth and ion accumulation, *J. Exp. Bot.,* 35, 1194, 1984.

240. **Pandey, R. and Ganapathy, P. S.,** The proline enigma: NaCl-tolerant and NaCl-sensitive callus lines of *Cicer arietinum, Plant Sci.,* 40, 13, 1985.

241. **Polacco, J.,** Arsenate as a potential negative selection agent for deficiency variants in cultured plant cells, *Planta,* 146, 155, 1979.

242. **Polacco, J. C. and Polacco, M. L.,** Inducing and selecting valuable mutants in plant cell culture: a tobacco mutant resistant to carboxin, *Ann. N.Y. Acad. Sci.,* 287, 385, 1978.

243. **Radin, D. N. and Carlson, P. S.,** Herbicide tolerant tobacco mutants selected in situ recovered via regeneration from cell culture, *Genet. Res.,* 32, 85, 1978.

244. **Rains, D. W., Stavarek, S. J., and Croughan, T. P.**, Selection of salt tolerant plants using tissue culture, in *Genetic Engineering of Osmoregulation Impact on Plant Productivity for Food, Chemicals and Energy,* Rains, D. W., Valentine, R. C., and Hollaender, A., Eds., Plenum Press, New York, 1980, 279.

245. **Ranch, J. P., Rick, S., Brotherton, J. E., and Widholm, J. M.**, Expression of 5-methyltryptophan resistance in plants regenerated from resistant cell lines of *Datura innoxia, Plant Physiol.,* 71, 136, 1983.

246. **Rangan, T. S. and Vasil, I. K.**, Sodium chloride tolerant embryogenic cell lines of *Pennisetum americanum, Ann. Bot.,* 52, 59, 1983.

247. **Reisch, B. and Bingham, E. T.**, Plants from ethionine resistant alfalfa tissue culture: variation in growth and morphological characteristics, *Crop. Sci.,* 21, 783, 1981.

248. **Riccardi, G., Cella, R., Camerino, G., and Ciferri, O.**, Resistance to azetidine-2-carboxylic acid and sodium chloride tolerance in carrot cell cultures and *Spirulina platensis, Plant Cell Physiol.,* 24, 1073, 1983.

249. **Rines, H. W. and Luke, H. H.**, Selection and regeneration of toxin-insensitive plants from tissue cultures of oats (*Avena sativa*) susceptible to *Helminthosporium victorae, Theor. Appl. Genet.,* 71, 16, 1985.

250. **Rosen, A. and Tal, M.**, Salt tolerance in the wild relatives of cultivated tomato: responses of naked protoplasts isolated from leaves of *Lycopersicon esculentum* and *L. peruvianum* to NaCl and proline, *Z. Pflanzenphysiol.,* 102, 91, 1981.

251. **Roth, E. J. and Lark, K. G.**, Isopropyl-N(3-chlorophenyl) carbamate (CIPC) induced chromosomal loss in soybean: a new tool for plant somatic cell genetics, *Theor. Appl. Genet.,* 68, 421, 1984.

252. **Roth, E. J., Weber, G., and Lark, K. G.**, Use of isopropyl-N(3-chlorophenyl) carbamate (CIPC) to produce partial haploid cells from suspension cultures of soybean (*Glycine max*), *Plant Cell Rep.,* 1, 205, 1982.

253. **Rubin, J. L., Gaines, C. G., and Jensen, R. A.**, Glyphosate inhibition of 5-enolpyruvylshikimate-3-phosate synthase from suspension cultures of *Nicotiana sylvestris, Plant Physiol.,* 75, 839, 1984.

254. **Rutger, J. N., Peterson, M. L., and Hu, C. H.**, Registration of calrose 76, *Crop Sci.,* 17, 978, 1977.

255. **Sacristan, M. D.**, Resistance responses to *Phoma lingam* of plants regenerated from selected cell and embryogenic cultures of haploid *Brassica napus, Theor. Appl. Genet.,* 61, 193, 1982.

256. **Sacristan, M. D. and Melchers, F.**, The karyological analysis of plants regenerated from tumorous and other callus cultures of tobacco, *Mol. Gen. Genet.,* 105, 317, 1969.

257. **Salgado-Garciglia, R., Lopez-Guthprez, F., and Ochoaalejo, N.**, NaCl-resistant variant cells isolated from sweet potato cell suspensions, *Plant Cell Tissue Organ Culture,* 3, 3, 1985.

258. **Savage, A. D., King, J., and Gamborg, O. L.**, Recovery of pantothenate auxotroph from a cell suspension culture of *Datura innoxia, Plant Sci. Lett.,* 16, 367, 1979.

259. **Schaeffer, G. W.**, Recovery of heritable variability in anther-derived double-haploid rice, *Crop Sci.,* 22, 1160, 1982.

260. **Schaeffer, G. W. and Sharpe, F. T.**, Lysine in seed protein from S-aminoethyl-L-cysteine resistant anther-derived tissue culture of rice, *In Vitro,* 17, 345, 1981.

261. **Schaeffer, G. W. and Sharpe, F. T.**, Mutations and cell selections: genetic variation for improved protein in rice, in *Genetic Engineering: Applications to Agriculture,* Owens, L. D., Ed., Roman and Allanheld, London, 1983, 237.

262. **Schieder, O.**, Isolation of mutants with altered pigmentation after irradiation of haploid protoplasts from *Datura innoxia* with X-rays, *Mol. Gen. Genet.,* 149, 251, 1976.

263. **Scott, A. I., Mizukami, H., and Lee, S. L.**, Characterization of a 5-methyltryptophan resistant strain of *Catharanthus roseus* cultured cells, *Phytochemistry,* 18, 795, 1979.

264. **Shepard, J. F.**, Protoplasts as sources of disease resistance in plants, *Annu. Rev. Phytopathol.,* 19, 145, 1981.

265. **Shepard, J. F., Bidney, D., and Shahin, E.**, Potato protoplasts in crop improvement, *Science,* 208, 17, 1980.

266. **Shimamoto, K. and King, P. J.**, Isolation of a histidine auxotroph of *Hyoscyamus muticus* during attempts to apply BUdr-enrichment, *Mol. Gen. Genet.,* 189, 69, 1983.

267. **Shimamoto, K. and Nelson, O. E.**, Isolation and characterization of aminopterin resistant cell lines in maize, *Planta,* 153, 436, 1981.

268. **Sidorov, V., Menczel, L., and Maliga, P.**, Isoleucine-requiring *Nicotiana* plant deficient in threonine deaminase, *Nature (London),* 294, 87, 1981.

269. **Sidorov, V., Menczel, M., Nagy, F., and Maliga, P.**, Chloroplast transfer in *Nicotiana* based on metabolic complementation between irradiated and iodoacetate treated protoplasts, *Planta,* 152, 341, 1981.

270. **Siminovitch, L.**, On the nature of heritable variation in cultured somatic cells, *Cell,* 7, 1, 1976.

271. **Singer, S. R. and McDaniel, C. N.**, Selection of amitrole tolerant tobacco calli and the expression of this tolerance in regenerated plants and progeny, *Theor. Appl. Genet.,* 67, 427, 1984.

272. **Singer, S. R. and McDaniel, C. N.**, Selection of glyphosate tolerant tobacco calli and the expression of this tolerance in regenerated plants, *Plant Physiol.,* 78, 411, 1985.

273. **Singh, N. K., Handa, A. K., Hasegawa, P. M., and Bressman, R. A.,** Proteins associated with adaptation of cultured tobacco cells to NaCl, *Plant Physiol.,* 79, 126, 1985.

274. **Skirvin, R. M.,** Natural and induced variation in tissue culture, *Euphytica,* 27, 241, 1978.

275. **Skirvin, R. M. and Janick, J.,** Tissue culture induced variation in scented *Pelargonium* spp., *J. Am. Soc. Hortic. Sci.,* 101, 281, 1976.

276. **Smith, R. H., Bhaskaran, S., and Schertz, K.,** Sorghum plant regeneration from aluminium selection media, *Plant Cell Rep.,* 2, 129, 1983.

277. **Smith, M. K. and McComb, J. A.,** Effect of NaCl on the growth of whole plants and their corresponding callus cultures, *Aust. J. Plant Physiol.,* 8, 267, 1981.

278. **Smith, M. K. and McComb, J. A.,** Selection for NaCl tolerance in cell cultures of *Medicago sativa* and recovery of plants from a NaCl tolerant cell line, *Plant Cell Rep.,* 2, 126, 1983.

279. **Sree Ramulu, K., Dijkhuis, P., and Roest, S.,** Phenotypic variation and ploidy level of plants regenerated from protoplasts of tetraploid potato *Solanum tuberosum, Theor. Appl. Genet.,* 65, 329, 1983.

280. **Sree Ramulu, K., Dijkhuis, P., Cate, H. T., and DeGroot, B.,** Patterns of DNA and chromosome variation during in vitro growth in various genotypes of potato, *Plant Sci.,* 41, 69, 1985.

281. **Sree Ramulu, K., Dijkhuis, P., Roest, S., Bokelmann, G. S., and DeGroot, B.,** Early occurrence of genetic instability in protoplast cultures of potato, *Plant Sci. Lett.,* 36, 79, 1984.

282. **Steffen, A. and Schieder, O.,** Biochemical and genetical characterization of nitrate reductase deficient mutants of *Petunia, Plant Cell Rep.,* 3, 134, 1984.

283. **Stewart, C. R.,** Effects of proline and carbohydrates on the metabolism of exogenous proline by excised bean leaves in the dark, *Plant Physiol.,* 50, 551, 1972.

284. **Steward, C. R. and Lee, J. A.,** The role of proline accumulation in Halophytes, *Planta,* 120, 279, 1974.

285. **Strauss, A., Bucher, F., and King, P. J.,** Isolation of biochemical mutants using haploid mesophyll protoplasts of *Hyoscyamus muticus, Planta,* 153, 75, 1981.

286. **Strauss, A., Frankhauser, H., and King, P. J.,** Isolation and cryopreservation of O-methylthreonine resistant *Rosa* cell lines altered in the feed back sensitivity of L-threonine deaminase, *Planta,* 163, 554, 1985.

287. **Strom, A. R., LeRudulier, D., Jakowec, M. W., Bunnell, R. C., and Valentine, R. C.,** Osmoregulatory genes and osmoprotective compounds, in *Genetic Engineering of Plants,* Kosuge, T., Meredith, C., and Holleander, A., Eds., Plenum Press, New York, 1983, 38.

288. **Sugano, N., Miya, S., and Nishi, A.,** Carotenoid synthesis in a suspension culture of carrot cells, *Plant Cell Physiol.,* 12, 525, 1971.

289. **Sung, Z. R.,** Mutagenesis of cultured plant cells, *Genetics,* 84, 51, 1976.

290. **Sung, Z. R.,** Relationship of indole-3-acetic acid and tryptophan concentrations in normal and 5-methyl-tryptophan resistant cell lines of wild carrots, *Planta,* 145, 339, 1979.

291. **Sung, Z. R. and Jacques, S.,** 5-Fluorouracil resistance in carrot culture. Its use in studying the interaction of pyrimidine and arginine pathways, *Planta,* 148, 389, 1980.

292. **Sung, Z. R., Liu, S. T., and Sell, S.,** in *Abstr. 4th Int. Congr. Plant Tissue and Cell Culture, IAPTC,* Calgary, Canada, 1978, 138.

293. **Swartzberg, D., Izhar, S., and Beckmann, J. S.,** Tobacco callus line tolerant to amitrole, selection, regeneration of plants and genetic analysis, *J. Plant Physiol.,* 121, 29, 1985.

294. **Tabata, M., Ogino, T., Yoshioka, K., Yoshikawa, N., and Hiroaka, N.,** Selection of cell lines with higher yields of secondary products, in *Frontiers of Plant Tissue Culture,* Thorpe, T. A., Ed., IAPTC, Calgary, Canada, 1978, 213.

295. **Thomas, B. R. and Pratt, D.,** Isolation of paraquat-tolerant mutants from tomato cell cultures, *Theor. Appl. Genet.,* 63, 169, 1982.

296. **Thomas, E., Bright, S. W. J., Franklin, J., Lancaster, V. A., Miflin, B. J., and Gibson, R.,** Variation amongst protoplast-derived potato plants (*Solanum tuberosum* cv. Maris Bard), *Theor. Appl. Genet.,* 62, 65, 1982.

297. **Tlaskal, J. G.,** The Cytology of Sugarcane, M.Sc. thesis, University of South Wales, 1978.

298. **Tyagi, A. K., Rashid, A., and Maheshwari, S. C.,** Sodium chloride resistant cell line from haploid *Datura innoxia.* A resistance trait carried from cell to plantlet and vice versa in vitro, *Protoplasma,* 165, 327, 1981.

299. **Tyagi, A. K., Rashid, A., and Maheshwari, S. C.,** Streptomycin resistance of a cell line from haploid *Datura innoxia* is transferred from cell to plantlet and back in vitro, *J. Exp. Bot.,* 35, 756, 1984.

300. **Umiel, N. and Goldner, R.,** Effects of streptomycin on diploid tobacco callus cultures and the isolation of resistant mutants, *Protoplasma,* 89, 83, 1976.

301. **Van Harten, A. M., Bouter, H., and Broertjes, C.,** In vitro adventitious bud techniques for vegetative propagation and mutation breeding of potato (*Solanum tuberosum*). II. Significance for mutation breeding, *Euphytica,* 30, 1, 1981.

302. **Wakasa, K.,** Variation in plants differentiated from tissue culture of pineapple, *Jpn. J. Breed.,* 29, 13, 1979.

303. **Wakasa, K., Kobayashi, M., and Kamada, H.,** Colony formation from protoplasts of nitrate reductase deficient rice cell lines, *J. Plant Physiol.*, 117, 223, 1984.

304. **Watad, A. A., Lerner, H. R., and Reinhold, L.,** Stability of salt-resistance character in *Nicotiana* cell lines adapted to grow in high NaCl concentration, *Physiol. Veg.*, 23, 887, 1985.

305. **Watad, A. A., Reinhold, L., and Lerner, H. R.,** Comparison between a stable NaCl-selected *Nicotiana* cell line and wild type, *Plant Physiol.*, 73, 624, 1983.

306. **Weber, G., DeGroot, E., and Schweiger, H. G.,** Methotrexate-resistant somatic cells of *Glycine max*: selection and characterization, *J. Plant Physiol.*, 117, 339, 1985.

307. **Weber, G. and Lark, K.,** An efficient plating system for rapid isolation of mutants from plant cell suspensions, *Theor. Appl. Genet.*, 55, 81, 1979.

308. **Weber, G. and Lark, K.,** Quantitative measurements of the ability of different mutagens to induce an inherited change in phenotype to allow maltose utilization in suspension cultures of sobyean *Glycine max*, *Genetics*, 96, 213, 1980.

309. **Wenzel, G., Schieder, O., Przewozny, T., Sopory, S. K., and Melchers, G.,** Comparison of single cell culture derived *Solanum tuberosum* plants and a model of their application in breeding programme, *Theor. Appl. Genet.*, 55, 49, 1979.

310. **Werry, P. and Stoffelson, K. M.,** The effect of ionizing radiation on the survival of plant cells cultivated in suspension cultures, *Int. J. Radiat. Biol.*, 35, 293, 1979.

311. **Werry, P. and Stoffelson, K. M.,** Theoretical and practical aspects of radiation-induced mutagenesis in plant cells, in *Plant Cell Cultures: Results and Perspectives*, Sala, F., Parisi, B., Cella, R., and Cifferi, O., Eds., Elsevier, Amsterdam, 1980, 157.

312. **Widholm, J. M.,** Cultured *Nicotiana tabacum* cells with an altered anthranilate synthetase which is less sensitive to feedback inhibition, *Biochim. Biophys. Acta*, 261, 52, 1972.

313. **Widholm, J. M.,** Anthranilate synthetase from 5-methyltryptophan-susceptible and -resistant cultured *Daucus carota* cells, *Biochim. Biophys. Acta*, 279, 48, 1972.

314. **Widholm, J. M.,** The use of fluorescein diacetate and phenosafranine for determination of viability of cultured plant cells, *Stain Technol.*, 47, 189, 1972.

315. **Widholm, J.M.,** Selection and characterization of cultured carrot and tobacco cells resistant to lysine, methionine and proline analogs, *Can. J. Bot.*, 54, 1523, 1976.

316. **Widholm, J. M.,** Selection and characterization of amino acid analog resistant plant cell cultures, *Crop Sci.*, 17, 597, 1977.

317. **Williams, K. L. and Barrand, P.,** Parasexual genetics in the cellular slime mold *Dictyostelium discoideum*: haploidization of diploid strains using benlate, *FEM Microbiol. Lett.*, 4, 155, 1978.

318. **Wong, J. R. and Sussex, I. M.,** Isolation of abscisic acid resistant variants from tobacco cell cultures. I. Physiological bases for selection, *Planta*, 148, 97, 1980.

319. **Wong, J. R. and Sussex, I. M.,** Isolation of abscisic acid resistant variants from tobacco cell cultures. II. Selection and characterization of variants, *Planta*, 148, 103, 1980.

320. **Wood, J. S.,** Mitotic chromosome loss induced by methylbenzidizole-2-yl-carbamate as a rapid mapping method in *Saccharomyces cerevisiae*, *Mol. Cell. Biol.*, 2, 1080, 1982.

321. **Wu, Min-Tze and Walner, S. J.,** Heat stress responses in cultured plant cells, *Plant Physiol.*, 75, 778, 1984.

322. **Yeo, A. R.,** Salinity resistance: physiologies and prices, *Physiol. Plant.*, 58, 214, 1983.

323. **Yoshida, S. and Miki, T.,** Cell membrane permeability and respiratory activity in chilling stressed callus, *Plant Cell Physiol.*, 20, 1237, 1979.

324. **Yurina, N. P., Odinstova, M. S., and Maliga, P.,** An altered chloroplast ribosomal protein in a streptomycin resistant tobacco mutant, *Theor. Appl. Genet.*, 52, 125, 1978.

325. **Zamski, E. and Umiel, N.,** Streptomycin resistance in tobacco. II. Effects of the drug on the ultrastructure of plastids and mitochondria in callus cultures, *Z. Pflanzenphysiol.*, 88, 317, 1978.

326. **Zamski, E. and Umiel, N.,** Streptomycin resistance in tobacco. IV. Effects of drug on the ultrastructure of plastids and mitochondria in cotyledons of germinating seeds, *Z. Pflanzenphysiol.*, 105, 143, 1982.

327. **Zelitich, I. and Berlyn, M. B.,** Altered glycine decarboxylation inhibition in isonicotinic acid hydrazide-resistant mutant callus lines and regenerated plants and seed progeny, *Plant Physiol.*, 69, 198, 1982.

328. **Zenk, M. H.,** Haploids in physiological and biochemical research, in *Haploids in Higher Plants: Advances and Potential*, Kasha, K.J., Ed., University of Guelph Press, Guelph, Canada, 1974, 339.

Chapter 3

CELL FUSION

I. INTRODUCTION

Cell fusion in animals is an important analytical method, but cell fusion in plants could not be considered until large-scale isolation[30] and regeneration[118] of plant protoplasts was possible. Due to the absence of a cell wall, two protoplasts can be made to fuse together and form a heterokaryon. While cell fusion in animals has remained an analytical tool, protoplast fusion in plants followed by recovery of plants from a fusion product, due to the inherent totipotent nature of plant cells, has opened new vistas in plant improvement.

At present, this promising technique of protoplast fusion has been applied to a small number of plants and its potential has been demonstrated. For routine application there is a need to learn more about the process of fusion, selection of heterokaryon, and regeneration of plants from fusion products. This chapter concerns the current state of knowledge about these aspects of protoplast technology. The achievements of somatic cell fusion are highlighted and information about genetic analysis of somatic cell hybrids is given in brief. The possibilities of new gene combinations through protoplast fusion are also discussed.

II. A NEW SYSTEM FOR GENETIC ENGINEERING OF HIGHER PLANTS

Although agriculture is the oldest industry, production has not kept pace with increasing population. The first green revolution, which was based on conventional agriculture, was able to save the human race from hunger. Future adequate food supply is expected through another green revolution which will not be with conventional agriculture because of its limitations and the constant reduction in land area available for cultivation. Self-sufficiency is possible only through novel plants which will be high yielding, disease resistant, and energy efficient.

In conventional agriculture a time-honored approach for crop improvement has been sexual hybridization. However, sexual hybridization is limited to closely related species, and crossability barriers reduce the utility of this technique in increasing the pool of genetic variability needed for crop improvement. As an alternative, protoplast fusion is considered to be a new approach to overcome these barriers and result in novel plants not previously possible by gametic fusion. In protoplast fusion, in addition to the possibility of nuclear mix from diverse resources, there is also a possibility of cytoplasmic mix, which is certainly novel. The cytoplasms of two cells differing in either or both mitochondria and plastids can be brought together in the plasmalemma of one cell, thus producing a heteroplasmodic cell. In a heteroplasmon or a cybrid one can obtain information not only about chloroplast and mitochondrial assortment and transmission, but also about the nuclear-cytoplasmic genomic relationship. Proper cellular manipulation can result in cybrid or hybrid (Figure 1) plants. The challenge is to exploit somatic cell fusion for crop improvement.

III. FUSION OF PROTOPLASTS

A. Spontaneous

Spontaneous fusion of two or more protoplasts has been recorded by many workers in the process of enzymatic isolation. This is interpreted[163] due to expansion and coalescence of plasmodesmatal connections between cells and can be prevented by preplasmolysis of tissue. Spontaneous fusion is more prevalent when protoplasts[113] are prepared from actively

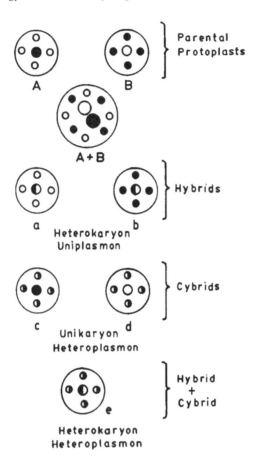

FIGURE 1. Diagrammatic representation of possibilities on protoplast fusion. Nuclear and organelle genomes are represented by large and small circles, respectively. An heterokaryon and uniplasmon results when there is nuclear fusion but unidirectional sorting of cytoplasmic genome takes place. Unikaryon and heteroplasmon results when either nuclear fusion fails or chromosomes of one of the parents are selectively eliminated but fusion of organelle genome takes place. Heterokaryon as well as heteroplasmon are possible when fusion occurs beween nuclear as well as cytoplasmic genomes.

dividing cells from a culture. Protoplasts prepared from microsporocytes[79] fused spontaneously at an efficiency of 50 to 70%. The preembryonic nature of these cells suggests that preembryonic membranes are predisposed to fusion or are fusogenic. This is consistent with the fact that fertilization is the best example of naturally occurring fusion in plants and animals. The possibilty of fusogenic plant membranes is indicated by recent investigation on protoplasts isolated from cell suspension of carrot which were embryogenic and had high potential for fusion.[20] The frequency of fusion in these protoplasts could be up to 60% when 10 mM calcium at pH 6 was present and this could be prevented by adding EGTA. However, induced fusion of protoplasts is more significant.

B. Induced

Induced fusion of isolated plant protoplasts for the production of somatic hybrids has attracted the attention of many plant scientists. Since the first report of induced fusion of protoplasts[136] a number of methods have been devised.

Plant protoplasts prepared from somatic tissues are negatively charged;[67] they repulse each other, thus preventing spontaneous fusion. Therefore, in any treatment which brings about protoplast fusion the first step is agglutination of protoplasts whereby protoplasts are brought in close contact. This is followed by membrane adhesion/fusion forming cytoplasmic bridges between two protoplasts. The expansion of cytoplasmic bridges results in a dumble-shaped structure. Finally, there is rounding off of two cytoplasms with two nuclei. Fusion of nuclei is an altogether different event and practically nothing is known about it. Therefore, for all practical purposes, protoplast fusion refers to fusion of two cytoplasms. The large number of agglutinating agents employed so far are the following:

1. Agglutinating Agents

1.   Sodium nitrate[136]
2.   Mechanical adhesion[140]
3.   Artificial seawater[39]
4.   Lysozyme[131]
5.   Virus[162]
6.   Gelatin[83]
7.   High calcium, high pH[91]
8.   Antibodies[75]
9.   Salt mixture[18]
10. Polyethylene glycol[89,157]
11. Polyvinyl alcohol[116]
12. Electrical stimulus[147,169]
13. Dextran[84]
14. Artificial lipid vesicles[155]

Of these various agents, fusion of protoplasts and recovery of somatic hybrid cells/plants has been accomplished with $NaNO_3$, high calcium, polyethylene glycol, and electrical stimulus. Each of these four agents is discussed individually.

a. Sodium Nitrate

Using sodium nitrate (0.25 M), fusion was demonstrated between protoplasts from oat root.[136] Later this fusogen was used to recover a somatic cell hybrid of *Nicotiana glauca* + *N. langsdorffii*.[24] However, this method is marked by a low frequency of fusion and is toxic to vacuolate protoplasts from mesophyll cells.[24,136] The fusion was ascribed to changes in electrical properties of membranes by sodium ions.

b. High Calcium

Mesophyll protoplasts of tobacco readily fused in the presence of high calcium (0.05 M $CaCl_2$) and high pH (10.5) on incubation at 37°C. The fused protoplasts formed colonies. Using this method, intraspecific and interspecific hybrids were possible in tobacco.[105,106] However, the frequency of fusion was not high. High pH also affects the survival of protoplasts. This method is also marked by production of large clumps comprising many protoplasts.[23]

How calcium brings about fusion remains unknown. Divalent cations (especially Ca^{++}) modify the electrophoretic mobility of protoplasts.[128] $CaCl_2$ at 100 mM also decreases the membrane potential. It is about 0 mV whereas it is -25 to -35 mV in normal medium.[117] An insight into the role of calcium[20] in fusogenesis has been possible using highly fusogenic protoplasts obtained from cell suspension of carrot which were capable of forming embryos. These protoplasts fused at a frequency of over 50% with only a mild calcium stimulus. Only calcium was required; not other cations stimulated fusion, and EGTA (a calcium chelator) strongly inhibited protoplast fusion. Furthermore, lowering of intracellular calcium lowered fusion potential while raising intracellular stores of calcium-enhanced fusion potential.[66] Irrespective of the amount of calcium sequestered in a store, mobilization of calcium with A23187 increased fusion frequency within 10 min, and antagonists of calmodulin were potent inhibitors of protoplast fusion.

The involvement of calcium in fusion of the bilayer membrane is seen in plant as well as animal systems. Both extracellular and intracellular calcium have been implied. Calcium is described as altering the physical structure of the bilayer and this alteration enhances fusion. Intracellular calcium has been implicated in Sendai virus-induced red blood cell fusion and myoblast fusion, but the biochemical mechanism remains unclear. Particularly in myoblasts, prior to fusion there is a transient increase in intracellular calcium which when blocked results in inhibition of fusion.

c. Polyethylene Glycol (PEG)

High-frequency fusion of plant protoplasts was possible when PEG was discovered as a new fusogen by two groups of workers.[33,89,157] This compound not only supports high-frequency fusion, but it also favors the formation of binucleate fusion products and is unspecific. It is effective for fusion of animal cells[2,130] and plant cells.[3]

PEG is a high-molecular-weight polymer (1500 to 6000) and 25 to 50% of its solution causes an instantaneous shock to protoplasts; they immediately shrink on exposure to PEG. This is followed by agglutination of protoplasts (Figure 2); the plasmalemma of two or more protoplasts is tightly appressed over a significant portion. After keeping the protoplast in PEG solution it is gradually removed. Numerous factors affecting the frequency of agglutination and subsequent fusion are:

1. Molecular weight and concentration of PEG solution (Figure 3).
2. Aggregation is more rapid when freshly isolated protoplasts are employed.
3. Protoplasts from young leaves and actively growing tissue are more responsive.
4. The enzymes employed for isolation of protoplasts also affect the response; protoplasts formed by dresilase were more responsive to PEG, but dresilase itself may be harmful.
5. Protoplast aggregation is dependent on density.
6. Protoplast aggregates more preferentially when in culture media than in enzyme solution (Figure 4).
7. Aggregation is promoted in the presence of divalent cations Ca^{++} and Mg^{++}; monovalent cations Na^+ and K^+ are ineffective; divalent cation-mediated aggregation can be promoted by an increase in pH.
8. Gradual dilution of PEG is helpful; elution of PEG with high calcium and high pH solution is especially helpful.
9. High temperature, up to 37°C, favors increased frequency of aggregation, but in itself may be harmful; therefore, treatment at 25°C is recommended.
10. Prolonged incubation in PEG should be avoided.
11. Fusion should be performed on a flat surface in droplets of protoplast solution.

This method has turned out to be very effective for fusion of protoplasts and somatic hybrids have been recovered in many systems. However, the method has some drawbacks:

1. There is deformation of mitochondria following PEG treatment.[16]
2. Fusion products remain bound to the fusion vessel
3. Very few of the fused protoplasts divide on culture.

Therefore, as an improvement[112] fusion could be induced in freely suspended protoplasts

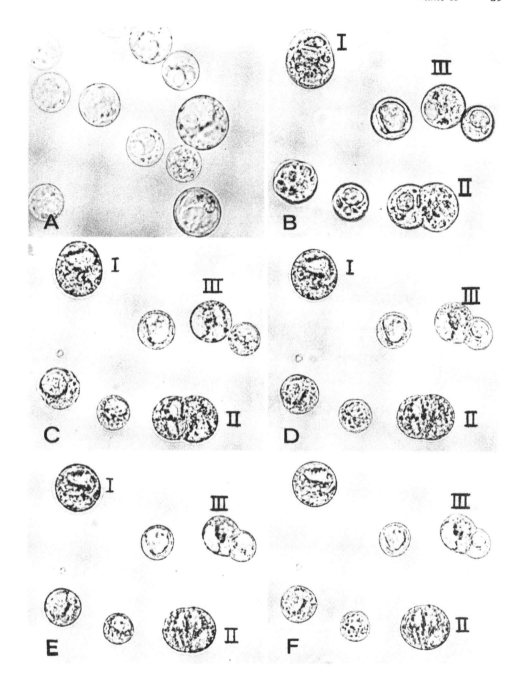

FIGURE 2. PEG-induced aggregation and fusion of carrot protoplasts; 1 mℓ of protoplast suspension in nutrient medium was mixed with 1 mℓ of 56% PEG w/w. (A) Protoplasts at the time of isolation. (B) After a minute of PEG treatment, the protoplasts are shocked, appear to shrink, and form aggregates; fusion in aggregates I and II is in progress. (C) After 5 min of treatment, fusion in aggregate I is completed and fusion in aggregate II is at an advanced stage. (D) After 7 min treatment, fusion has begun in aggregate III. (E) After 10 min, fusion in aggregate III is at advanced stage. (F) After 40 min, fusion in aggregate III is over. (From Wallin, A., Glimelius, K., and Eriksson, T., *Z. Pflanphysiol.*, 74, 64, 1974. With permission.)

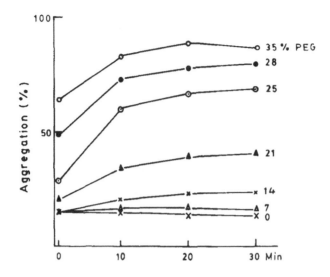

FIGURE 3.   Effect of different concentrations of PEG on aggregation frequency of protoplasts of *Daucus carota*. (From Wallin, A., Glimelius, K., and Eriksson, T., Z. *Pflanzenphysiol.*, 74, 64, 1974. With permission.)

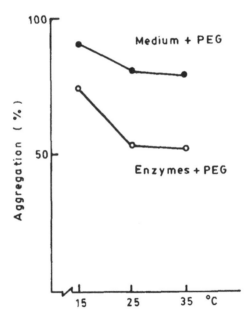

FIGURE 4.   Effect of temperature and enzyme on PEG-induced aggregation of protoplasts of *Daucus carota*. (From Wallin, A., Glimelius, K., and Eriksson, T., Z. *Pflanzenphysiol.*, 74, 64, 1974. With permission.)

using a mixture of 10% PEG and 10% DMSO at high pH (0.1 *M* glycine NaOH buffer). In this process, since the fusion products are not bound to surface, they can be easily picked up. Also, more than 50% of the fusion products divided after 2 days, indicating that the fusion process did not affect their viability.

The mechanism by which PEG brings about protoplast fusion is unknown. However, some explantations advanced are the following:

1. PEG appears to act as a molecular bridge between the surfaces of adjacent protoplasts either directly or indirectly through Ca^{++}. Fusion presumably results from disturbance and redistribution of electric charge[88] when PEG molecules are washed away.
2. Hydrogen bonding affinity of PEG for water and osmotic properties of PEG solution may be related to its effectiveness in membrane fusion.[157] Osmotic properties include the establishment of a two-phase system involving the polymer and protoplasts or direct effects of PEG on the membrane resulting from localized dehydration at the membrane interface. Also, reduction of water concentration at the protoplast membrane would change membrane protein concentration or configuration and thereby change membrane catalytic activities. Alternatively, strong membrane dehydration may shift the fluid state of lipids to a more fluid state. The lipids in natural membranes need to be in a fluid condition to undergo fusion.
3. Similar to PEG, polyvinylalcohol (PVA)[116] and dextran[84] have been shown to have fusogenic properties. The weak nonionic surfactant properties of these chemicals are possibly involved in inducing membrane fusion.
4. Membrane fluidity[146] is important in regulating fusion because fusion of plant protoplasts is greatly enhanced in the presence of PEG or PVA.

Despite these explanations, much remains to be learned about the mode of PEG-induced fusion. In this context an important finding is the loss[78] of agglutination on purification of commercial grade PEG. When an antioxidant, such as α-tocopherol, was added to commercial grade PEG it promoted fusion in the presence of PEG.

d. Electrical Stimulation

A new note in plant protoplast fusion was introduced when protoplasts of *Rauwolfia serpentina*, seen at the stage of point-adherance, were fused by the application of electric current.[147] When two protoplasts at the stage of point-adherance were given an electrical impulse of 5 to 12 μamp lasting 1 to 5 msec, the protoplasts at first appeared shocked and then fused instantaneously. On application of electric current the protoplasts shrank slightly, indicating a transient change in their membrane state.

Prior to this, in an animal system, it was shown that fusion occurs when a cell suspension in an electrolytic discharge chamber is subjected to an external electrical field pulse of sufficient strength to induce breakdown of the cell membrane.[168] The low frequency of fusion was explained in terms of electrical repulsion among the cells caused by negative surface charge. Also, from an animal system it was known that cell repulsion could be avoided in a nonuniform and alternating electrical field. Because of the alternating electrical field, the net surface charge is masked and the nonuniform field induces a dipole in the cell. This divergent nature of field causes one end of the dipole to be of higher field strength than the other and consequently the dipole or a cell moves in the direction of increasing field. The movement of the cell within the field is a function of field strength, cell size, and the mobility of charge within the cell. This effect is called dielectrophoresis. It is possible only in a nonuniform field and has been used for collecting live cells by means of an electric field.[129] Cells undergoing dielectrophoresis are often mutually attracted when the polarizability of the cells is higher than the surrounding medium. Consequently, the cells are often clumped together or bind together and form pearl chains when moving in the direction of higher field intensity. This effect is called mutual dielectrophoresis. On removal of the electrical field the pearl chain disintegrates and, due to Brownian movement, the cells separate from each other.

High-frequency fusion[169] of plant protoplasts was demonstrated on application of electric field. It was possible on inducing the protoplasts to (1) come together and establish close membrane contact, forming pearl chains, in the process of dielectrophoresis and (2) fuse together by reversible electrical breakdown of cell membranes.

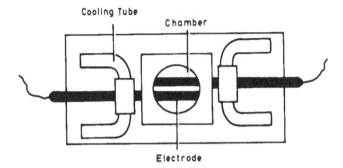

FIGURE 5.   Diagrammatic representation of dielectrophoresis chamber for fusion of plant protoplasts. The perspex chamber (capacity 1 mℓ) with built-in electrodes is mounted on a slide and covered with a cover slip during the experiment. The minimum distance between the electrodes is 200 μm. (From Zimmermann, U. and Scheurich, P., *Planta*, 151, 26, 1981. With permission.)

Dielectrophoresis is achieved in a highly inhomogenous alternating electric field (sine wave 5 to 10 V, peak-to-peak value 500 kHz, electrode distance 200 μm). Under these conditions protoplasts form bridges of four to six protoplasts or pearl chains between electrodes. This pearl chain arrangement is stable for the duration of the applied field only. For dielectrophoretically induced adhesion of protoplasts the field strength chosen should be such that breakdown of membrane is avoided. It is possible to control dielectrophoretic adhesion of protoplasts in such a way as to produce aggregates of only two protoplasts. The size of aggregates is dependent on the density of protoplast suspension and on the amplitude and frequency of alternating voltage.

Fusion is achieved within the aggregate or bridges by the application of a high single-field pulse (square wave 15 V, 50 μsec) which results in breakdown of the membrane. The fusion product forms a sphere within a few minutes. The field intensity must be sufficiently high to induce reversible breakdown in the area of close membrane contact.[156] This electrically stimulated fusion of protoplasts is possible at room temperature and without the use of chemical agents.

The nature of electrically induced structural changes in the membrane at the point of contact is unknown. The fluidity of membrane may be changed when material is removed from it by electromechanical compression and subsequent pore formation (electroporation).[167] The technique has also been used for cell transformation.

Electrofusion of plant protoplasts[169] as originally described is essentially a microtechnique (Figures 5 and 6). In a small perspex chamber of 1-mℓ capacity, two cylindrical brass rods coated with gold serve as electrodes. This chamber is mounted on a glass slide and covered with a cover glass. A function generator serves as a source of alternating fields between the two electrodes and a pulse generator is employed for injecting square pulses into the suspension to induce breakdown of cell membranes. The function generator is connected to electrodes by means of a resistor while the pulse generator is linked directly. The frequency of voltage required for dielectrophoresis is monitored on an oscilloscope along with the amplitude and duration of pulse required for fusion. The temperature in the perspex chamber is maintained at 25°C by a thermistor located close to the electrodes. A modification of this method has been proposed,[159] in which the principal drawback is removed so that a relatively large volume of the culture, up to 2 mℓ, could be uniformly used for electrofusion with no tendency for the protoplasts to attach to electrodes and fusion products could be recovered aseptically at high frequency.

Electrofusion of protoplasts is considered to be of great importance,[120] since it avoids the

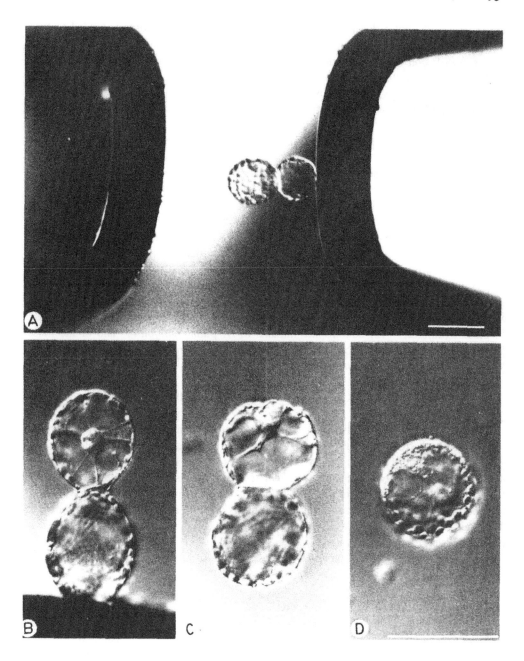

FIGURE 6. Electrofusion of plant protoplasts of *Nicotiana tabacum*. (A) Pair of protoplasts in the electrode gap; bar represents 50μm. (B, C, D) Details of fusion. (From Koop, H-U. and Schweiger, H-G., *Eur. J. Cell Biol.*, 39, 46, 1985. With permission.)

use of potentially toxic chemicals. Study of factors affecting fusion and innovation of the technique have turned out to be rewarding. Protoplasts from mesophyll cells of *N. tabacum* could be fused with protoplasts from suspension cells of *N. plumbaginifolia*, in a simple-to-use large-scale fusion chamber which could accommodate 0.5 mℓ of suspension at a density of 10^6 units/mℓ, at efficiencies approaching 40%.[11,93] Optimum fusion[11] was possible with an alternating current field of 150 V/cm and direct current square wave pulses of 1000 V/cm. Half of the fusion products were heterokaryons, indicating that fusion is random; 60% were bi- or trinucleate. The frequency of fusion is dependent on protoplast origin, size and chain length, pulse duration, and voltage. Leaf mesophyll protoplasts tended to fuse more readily than suspension culture protoplasts.[152] For both types there was a correlation of size with fusion frequency. Large protoplasts tended to fuse more readily than smaller protoplasts. In short chains (five protoplasts) fusion frequency was lower, but the proportion of one to one fusion was higher than in long chain (ten protoplasts). In fusion of mesophyll and suspension cell protoplasts the fusion frequency response curves reflected homofusion of mesophyll protoplasts rather than suspension cell protoplasts. Based on these results it was possible to obtain high-frequency fusion of one to one protoplast (mesophyll + suspension) by having a low density of mesophyll protoplasts in a high density of suspension cell protoplasts and by using a short fusion pulse.

When using electrofusion of protoplasts one must consider their future survival. During this process it is very likely that along with cell membrane there is breakdown of tonoplast membranes, resulting in leakage of toxic substances from vacuoles which could affect protoplast viability. Therefore, a method has been devised to fractionate evacuolate protoplasts.[71] These protoplasts also readily fuse by the same method.

Electrofusion has been successfully used to obtain intraspecific (*Brassica napus* and *B. napus*), interspecific (*B. napus* and *B. campestris*), and intergeneric (*B. napus* and *Primula acaulis*) cell hybrids.[165] It has also been used to recover somatic cell hybrids which are complementation hybrids of (1) nitrate reductase (NR)-deficient strains of *N. tabacum*[92] and *N. plumbaginifolia*,[137] (2) auxotrophic mutants of *N. plumbaginifolia*[120] and of the moss *Physcomitrella*,[158] and (3) hybrids between *N. plumbaginifolia* and *N. tabacum*[12] and *Solanum brevidens* + *N. sylvestris* or *Datura innoxia*.[152]

IV. SELECTION OF A HYBRID (HETEROKARYON)

From the above account it becomes clear that more than one method is available which results in a high frequency of fusion of protoplasts. However, the difficulty involved in successful isolation of somatic cell hybrids is the means to identify the hybrid and separate it from the parental protoplasts. An efficient recovery process requires selective cultural conditions which will favor growth of hybrid cells only and inhibit or preclude growth of parental types. This is important because protoplasts are cultured at high density to enable them to divide and form a callus, and unless the hybrid cell is at an advantage, the parental types will overgrow and mask it.

Recovery of hybrid cells so far has been possible due to complementation of characters, either genetic or physiological or even visual.

A. Complementation of Visual Characters

The first somatic cell hybrid between two *Nicotiana*[24] spp., *N. glauca* + *N. langsdorffii*, was based on auxin autotrophy of hybrid cells and auxin heterotrophy of parental cells. On a medium lacking auxin, only the hybrid cells proliferated and parental cells were precluded. However the design of the selection system was greatly facilitated due to availability of (1) sexual hybrid between the two species and (2) growth characteristics of hybrid and parental cells. Also, the authors had a prior knowledge that on a particular NT medium[118] the protoplasts of the parental types did not divide. On subjecting the protoplast mixture from two parents to a fusogen (NaNO₃) the mixed protoplasts were cultured on NT medium. Of the 33 calli recovered, all grew vigorosuly on medium lacking hormone. Plants regenerated from three isolates were of a hybrid nature, the leaves were like that of the sexual hybrid and were different from either parent. The density of trichomes on hybrid plants was intermediate between *N. langsdorffii* and *N. glauca*. The somatic hybrid had the capacity of spontaneous tumor formation, similar to that of the sexual hybrid. Also, the chromosome number (42) was the sum of diploid *N. glauca* (24) and *N. langsdorffii* (18). The leaf peroxidase banding pattern represented the combined isozyme bands of the two parents. However, the somatic hybrid differed from the sexual hybrid in being self-fertile.

Additional evidence for the hybrid nature of these plants was provided by the polypeptide composition of large and small subunits of the enzyme ribulose bisphosphate (RUBP)-carboxylase. On electrofocusing in 8 *M* urea, the nucleus-encoded small subunit and the chloroplast-encoded large subunit of this enzyme can be resolved into their respective polypeptides. The patterns obtained are characteristic of each species and can serve as markers of chloroplast and nuclear genomes. The somatic hybrid contained small subunit polypeptides of both parents (*N. glauca* as well as *N. langsdorffii*), but only the large subunit polypeptide of *N. glauca*.[96] This indicates that chloroplasts of *N. langsdorffii* either were not present or were not expressed in the hybrid plant.

Confirmation of these results, in principle, was possible when PEG was used for protoplast fusion.[29,150] The hybrid cells were auxin autotrophic, but plants recovered from them had higher chromosome numbers, ranging from 56 to 64. Further, the hybrid plants had small units of RUBP-carboxylase from both parents (*N. glauca* and *N. langsdorffii*) and only one plant had large subunits of this enzyme from both parents; the rest had a large subunit either of *N. glauca* or *N. langsdorffii*. Progeny from one plant with large and small subunits of both parents showed the large subunit of *N. glauca* only, and regenerates from the callus culture produced from leaves of this plant showed a large subunit of *N. langsdorffii*.[27] This indicated that the nuclear genome remained static and cytoplasmic components segregated during sexual and vegetative propagation.

An increase in frequency of fusion[22] in this sytem (*N. glauca* + *N. langsdorffii*) was possible when protoplasts were treated with PEG and Ca^{++}. About 5% of the calli were found to be hybrid.

B. Genetic Complementation

Genetic complementation[106] was used to recover somatic hybrids when green callus formation was a marker for fusion of two homozygous recessive albino mutants of *N. tabacum*. Two recessive chlorophyll-deficient mutants of tobacco sublethal (S) and virescent (V) were used. These mutants were also light sensitive; they grew slowly under direct illumination, but were normal in weak diffuse light. Protoplasts isolated from haploid plants were fused and green calli were recovered on exposure of plates to light. The plants regenerated from these green calli were hybrids — morphologically similar to plants produced in sexual cross.[107]

An advantage of using mutants in the selection process is that the hybrid can be confirmed and analyzed genetically. Recovery of both classes among the selfed progeny of the hybrid can be taken as unequivocal proof of their hybrid nature. Indeed, on self-fertilization of one hybrid plant, V and S phenotypes were identified.[106] Later this method was used to recover the interspecific hybrid between *N. tabacum* + *N. sylvestris*[105] using a S mutant of the former and a chlorophyll-deficient mutant of the latter.

Chlorophyll-deficient albino mutants of *Datura innoxia*[141] AI/5a and A7/LS, obtained by X-ray treatment of protoplasts, were fused and the green callus was recovered as a putative hybrid (Figure 7). Callus had various ploidies, but tetraploid callus readily regenerated to form shoots. Similarly, interspecific hybrids.[34] were raised following fusion of chlorotic protoplasts of *N. rustica* with albino protoplasts of *N. tabacum*.

Since chlorophyll synthesis is a function of nuclear as well as chloroplast genomes, mutation in either genome can cause chlorophyll deficiency. A complementation of two genomes was attempted using a chloroplast mutant and a semidominant nuclear mutant — sulfur (Su) — of tobacco.[52] The Su mutation is lethal in the homozygous state (Su/Su); the plants are yellow and unable to photosynthesize, but the seedlings can be grown on nutrient medium supplied with sucrose. By contrast, heterozygous plants (Su/+) are yellow-green

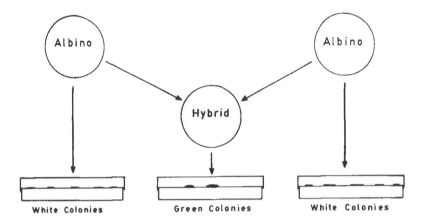

FIGURE 7.   Utility of albino mutants in selection of cell hybrid on protoplast fusion.

and photosynthetic. On fusing protoplasts of an albino Su mutant having normal chloroplasts with protoplasts from albino sectors of variegating green plant (chloroplast mutant), yellow-green and green plants were recovered. These results can be interpreted on the basis of formation of heterokaryons, their fusion, and subsequent segregation into separate cells. The cytoplasmic components could also segregate independent of nuclear events.

For the selection of a somatic hybrid, a single recessive character such as albinism can also be combined with either a morphological character or growth response. A combination of albinism and morphological character[35] was possible when albino protoplasts from *Daucus carota* were fused with normal protoplasts of *D. capillifolius*. Since regeneration of plants from protoplasts of *D. capillifolius* is at a very low frequency, green calli recovered were putative hybrids later confirmed by morphological characters and chromosome number of plant produced. Also, fusion of albino protoplasts of *D. carota* was done with green protoplasts of *Aegopodium podagraria*.[36] Since protoplasts of *Aegopodium* cannot divide, green callus was the putative hybrid. Plants regenerated from this callus had normal carrot chromosomes and some from *Aegopodium*.

Using protoplasts of a semidominant albino mutant several interspecific hybrids have been recovered in *Nicotiana*. The shoots regenerated can be identified easily. The albino protoplasts produce albino shoots,[49] whereas protoplasts from normal *Nicotiana* produce green plants.[40] Light green shoots were verified to be somatic hybrids.[43] Specifically, when protoplasts of a Su mutant of tobacco were fused with green protoplasts of *N. glauca*, the inability of the protoplasts of the latter to divide helped in the identification of the somatic hybrids. The regenerated plants were light green, which was expected of a heterozygote for Su allele, and produced anthocyanin at the base of the leaf petiole, which is a feature of *N. glauca*.

A combination of albinism and growth characteristic was used to construct an interspecific hybrid of *Petunia*.[31,134] Green protoplasts from *P. parodii* were fused with albino protoplasts from other species (*P. hybrida, P. inflata,* and *P. parviflora*). After fusion, when the protoplast mixture was plated on MS medium, callus formation from the protoplasts of *P. parodii* did not proceed beyond a small colony stage. Albino calli were produced from other protoplasts. Green and growing callus was considered to be the product of hybridization.

When albino protoplasts from chlorophyll-deficient *Datura innoxia* were fused with green protoplasts of *Atropa belladonna*,[94] pubescence of *Datura* callus served as marker. Green pubescent callus was considered to be due to somatic cell fusion. The plants regenerated from this callus had chromosomes from *Atropa* as well as *Datura*.

A better combination[63] of albinism and growth characteristic was possible when protoplasts of an albino mutant of *N. tabacum,* characterized by a deficient chloroplast genome, were

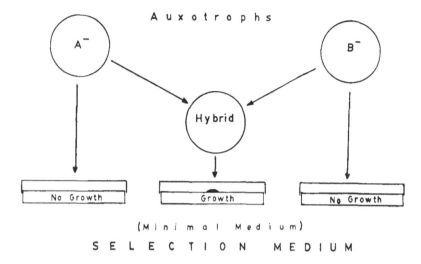

FIGURE 8.   Utility of auxotrophic mutants in selection of cell hybrid on protoplast fusion.

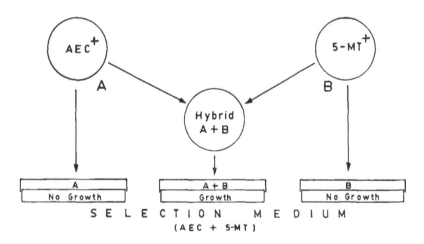

FIGURE 9.   Utility of resistant mutants in selection of cell hybrid on protoplast fusion.

fused with the NR-deficient mutant of tobacco. In a somatic hybrid the NR deficiency was complemented by albino mutant and deficiency of albino mutant was complemented by chloroplasts of NR mutant.

C. Complementation of Metabolic Characters

Selection schemes based on complementing albino mutants are based on the green color of the hybrid cell, but its ability to synthesize chlorophyll does not confer any growth advantage over the parental cell lines. Rather, a hybrid between distantly related species may be less vigorous and is likely to be overgrown by parental cell lines. Therefore, selection based on metabolic mutants is desirable because it permits the growth of only hybrid cells.

Two types of metabolic markers used for selection are resistant and auxotrophic mutants. The auxotrophic mutants on fusion complement are recessive so that the hybrid cells alone are autotrophic and capable of growth on unsupplemented medium that does not support the growth of auxotrophs (Figure 8). The resistant markers should be dominant. In the presence of both markers, hybrids alone will be able to grow, due to complementation (Figure 9), and parental cell types resistant to only one metabolite will be inhibited by the other.

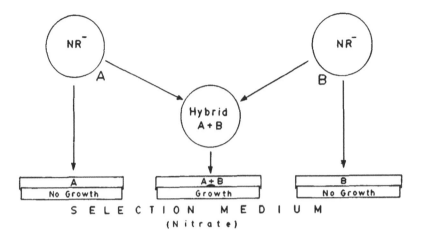

FIGURE 10.   Metabolic complementation, employing NR-deficient mutants, in selection of cell hybrid on protoplast fusion.

The metabolic mutant,[100] without any complementation, was used to recover hybrid cells when chlorophyll-deficient, kanamycin-resistant protoplasts of *N. sylvestris* were fused with green mesophyll protoplasts of *N. knightiana*. Colonies were allowed to develop on a nonselective medium and then were transferred to another medium which did not support the growth of *N. knightiana* protoplasts, and protoplasts of *N. sylvestris* formed colorless colonies. Green colonies on this medium were putative hybrids. Two such colonies on transfer to kanamycin medium were found to be resistant and hybrids. Similarly, a streptomycin-resistant mutant of *N. tabacum*[164] was used to recover an intraspecific hybrid and interspecific hybrids *N. sylvestris*[104] and *N. knightiana*.[111] The green colonies which were resistant to streptomycin were taken as putative hybrids. Falling in the same category are protoplasts from the fusion of cycloheximide (CH)-resistant cell lines of *Daucus carota* and albino protoplasts of *D. carota*. The somatic hybrids could be identified as green and CH resistant.

Also, protoplasts from a *D. carota*[85] cell line showing double resistance to 5-methyltryptophan (5-MT) and azetidine-2-carboxylate (A2CA) were used to recover an interspecific hybrid with *D. capillifolius*. Selection of hybrid cells was based on resistance to 5-MT.

Complementation of metabolic markers[65] in the selection of hybrid cells was used when protoplasts from two NR-deficient mutants of *N. tabacum* were fused and grown on nitrate medium. The hybrids could complement and grow on nitrate medium (Figure 10) whereas auxotrophs could not grow. Complementation and dominant expression of amino acid analogue resistant markers was possible when protoplasts from two cell lines of carrot resistant to the amino acid analog AEC (*S*-2-aminoethyl-L-cysteine) or 5-MT were fused. On a selective medium containing 5-MT and AEC, only hybrids survived[74] (Figure 9). No colonies appeared in controls including homologous fusion and co-culture. Auxotrophic complementation[132] was used to recover intergeneric cell hybrids when protoplasts of nicotinamide-deficient *Hyoscyamus muticus* nic⁻ were fused with protoplasts of *N. tabacum* which were nitrate reductase deficient (NR⁻). Recently, extensive fusion experiments have been performed with various auxotrophic mutants of *N. plumbaginifolia*.[120]

A unique means of metabolic complementation is possible using the irreversible biochemical inhibitors iodoacetate or diethylpyrocarbonate. Iodoacetate treatment[104] was used to recover hybrids between *N. sylvestris* and *N. tabacum*, *N. tabacum* and *N. plumbaginifolia*,[148] and *Pennisetum americanum* + *P. maximum*.[122a.] The parental protoplasts treated with iodoacetate were unable to divide while hybrid protoplasts divided. Streptomycin resistance in *N. tabacum* was transferred to *N. sylvestris* by fusing iodoacetate-treated non-

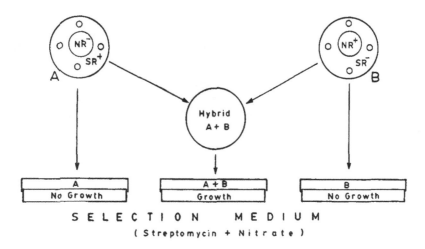

FIGURE 11. Utility of a double mutant, which could serve as universal hybridizer, in selection of cell hybrid on protoplast fusion.

dividing protoplasts of *N. tabacum* with streptomycin-sensitive protoplasts of *N. sylvestris*. Similarly, somatic hybrids of carrot[25] were recovered following fusion of iodoacetate-inactivated A2CA-resistant and -sensitive protoplasts.

Of great importance in selection schemes are double mutants (Figure 11). For example, streptomycin resistant (SR) and NR deficient. Such a double mutant in tobacco could be obtained[125] by sexually crossing SR *N. tabacum* (♀) with NR-deficient *N. tabacum* nia 130 (♂). In a double mutant, the location of streptomycin resistance is in the chloroplast, while the location of NR deficiency is in the nucleus. Instead of resorting to sexual fusion, combination of NR-deficient nuclear genome and streptomycin-resistant chloroplast genome is also possible by protoplast fusion.[70] The utility of this type of double mutant for somatic hybridization experiment with species lacking any marker is shown in experiments on protoplast fusion of *N. tabacum* with *N. rustica*.[126] Somatic hybrids were recovered on medium containing nitrate as the sole nitrogen source and also containing streptomycin. Similarly, leaf mesophyll protoplasts of *N. tabacum* deficient in NR and having streptomycin resistance were fused with protoplasts from a cell suspension of wild-type *Petunia hybrida*. Somatic hybrid cell colonies were recovered for streptomycin resistance and NR proficiency.[127]

In a double selection scheme[1] a metabolic character and visual character can be employed. In this way, interspecific somatic hybrids between *Lycopersicon* spp. were recovered. Regeneration ability of *L. peruvianum* and resistance to antibiotic G 418 (deoxystreptamine) of *L. pennelii* cell line were the characters. First, the putative hybrids were grown on an antibiotic medium and then transferred to a regeneration medium containing antibiotic. Similarly, protoplasts of a 6-azauracil (AU)-resistant cell line of *Solanum melongena* were fused with protoplasts of *S. sisymbrifolium* to create a somatic hybrid between these sexually incompatible species.[62] Following fusion, colonies were selected which were capable of growing on 1 *m* AU. These colonies were transferred to medium containing zeatin which had been shown to stimulate anthocyanin production during shoot formation in *S. sisymbrifolium*, but not *S. melongena*. From anthocyanin-producing colonies, plants regenerated were intermediate in morphological characters (flower color, leaf shape, and trichome density) to those of the parents.

D. Visual Characters — Sole Basis of Selection

For want of a proper selection scheme, due to a paucity of mutants, selections based on visual characters have also been used.

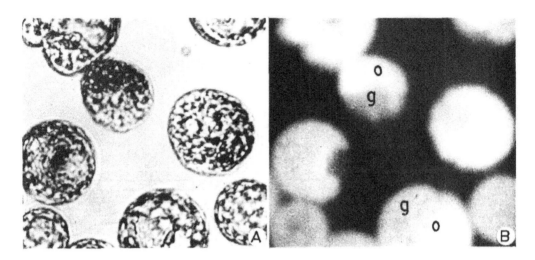

FIGURE 12. Fluorescence-activated sorting of hybrid protoplasts. Heterokaryons of *Nicotiana tabacum* and *N. sylvestris* isolated by fluorescence-activated sorting (A) as seen in light microscopy and (B) as seen with fluorescence microscopy (here, shown in black and white). The majority of cells exhibit both fluorescein fluorescence (g) and chlorophyll autofluorescence (o), indicating that they are hybrid protoplasts. (Photographs courtesy of Professor D. W. Galbraith.)

Hybrid vigor is such a system. Interspecific hybrids of *Datura*[142,143] formed a more vigorous callus than the parental species. This was also true of intraspecific hybrids of *Solanum tuberosum*.[5]

When green vacuolated mesophyll protoplasts are fused with colorless cytoplasmic protoplasts containing starch granules originating from cells in culture, it is possible to identify the fusion product for some time in culture.[87] Soon after fusion the hybrid protoplasts contain chloroplasts in one half and starch granules in the other half. During the first cell division the chloroplast can be seen clustering around the nucleus. Using this visual marker intergeneric hybrids originating from mesophyll protoplasts of *Brassica campestris* and protoplasts from cells in culture of *Arabidopsis thaliana* were picked up mechanically using a micropipette and cultured in small volumes of a cuprak dish.[54,55] Individual heterokaryons of *Nicotiana* after microisolation were also cultured with the nurse technique.[110] Single heterokaryons were placed on a droplet of albino protoplasts capable of fast growth. The heterokaryon was identified as the green colony among colorless ones from albino protoplasts.

Heterokaryons have been isolated by means of a micromanipulator[124] using a capillary pipette coupled to a specially constructed syringe.

With the success in electrofusion of protoplasts, visual characters (green mesophyll protoplasts and colorless suspension cell protoplasts) or fluorescence followed by microisolation are increasingly employed for selection of heterokaryons. When protoplasts from cell suspension and mesophyll cell origin are stained with fluorescein isothiocyanate and fused, the heterokaryon shows apple green cytoplasmic fluorescence of cell suspension and red fluorescence from mesophyll chloroplast.

Microisolation of heterokaroyons is a tedious and slow process. Therefore, a new development[46] in this field is fluorescence-activated cell sorting (FACS). In this process two parent protoplasts are separately labeled with different vital stains: fluorescein isothiocyanate (FITC, showing green fluorescence) and rhodamine isothiocyanate (RITC, showing red fluorescence). The hybrids are identified on the basis of showing two types of fluorescences (Figure 12) and can be selected using a cell sorter. Although it is a sophisticated technique, it is employed successfully to characterize and select plant heterokaryons.[4] The fluorescent compounds described above are not soluble in water. Therefore, the nontoxic, water-soluble, highly fluorescent compounds[86] scopoletin and carobxyfluorescein have been used to label both tissue culture and leaf mesophyll protoplasts. These compounds localize within the vacuoles of a cell in about 15 hr. They remain in the vacuole during cell wall digestion; when morphologically indistinguishable protoplasts were labeled and treated with PEG, multicolored fluorescent fusion products were observed. This provides a novel method for monitoring protoplast fusion.

V. CONFIRMATION OF A HYBRID

The confirmation of a hybrid is possible when it is intermediate between the two parents in terms of morphological characters, vegetative or floral. Among the vegetative characters, leaf shape of somatic hybrid was intermediate between *N. rustica* and *N. tabacum* when these were employed as parents.[114] Somatic hybrids between *D. carota* and *D. capillifolius* were intermediate[35] in leaf area. An intermediate trichome density and intermediate trichome length were taken to be criteria to distinguish somatic hybrids of *N. glauca* and *N. langsdorffii*.[24] To confirm the somatic hybrids, floral characters used were corolla length[150] (*N. glauca* and *N. langsdorffii*) and corolla morphology[142] (*D. innoxia* and *D. discolour*, and *D. innoxia* and *D. stramonium*). Occasionally, a character behaving as single gene dominant trait is present in only one parent, but is also expressed in a somatic hybrid. An example is stem anthocyanin of *N. glauca* appearing on a somatic hybrid of *N. tabacum* and *N. glauca*.[43] However, reliance on morphological characters alone is possible when sexual hybrids are available and a comparison can be made. Whenever possible, additional evidence should be provided to support the hybrid nature.

The chromosome number of the hybrid should conform to the sum of the two parents. However, how misleading this can be is seen from three independent experiments on the same system of *N. glauca* and *N. langsdorffii*. A small population of hybrid plants in the first experiment[24] was, as expected, amphiploid. By contrast, in the second experiment[150] all hybrid plants were aneuploid. In a third experiment[29] both aneuploid and amphidiploid plants were obtained. The variation in chromosome number can be explained in terms of (1) fusion of more than two protoplasts, (2) cytological variation arising out of culture and its duration and, (3) incompatibility of cytoplasmic and nuclear genes brought together.

Isoenzyme analysis is another source to verify the hybrid nature of plants or tissues. This can be applied at the tissue level when plants are not formed. The enzymes that have been shown to be of unique banding pattern for somatic hybrid vs. parental species are isoperoxidase[24] (*N. glauca* and *N. langsdorffii*), amylase[97] (interspecific hybrids of *Datura*), aspartate aminotransferase[43] (interspecific hybrids of *Nicotiana*, *N. tabacum*, and *N. glauca*), and alcohol dehydrogenase[122a] (*Pennisetum americanum* + *Panicum maximum*). However, it is important that the zymogram be prepared and interpreted cautiously, particularly when tissue[160] alone is employed for analysis. When hybrid plants are to be screened, the same tissue should be taken from each plant and at a similar stage of development.

Isoelectrofocusing of small and large subunits of RUBPCAse, which can be resolved into individual polypeptides, has also been used for confirmation of the hybrid nature in interspecific somatic fusion of *N. glauca* and *N. langsdorffii*.[96]

Species-specific repetitive DNA[138a] can also be used for identification of somatic hybrids. Plant DNA clones containing repetitive DNA sequence isolated from *Hyoscyamus muticus* and *N. tabacum* were hybridized with DNA isolated from a presumably somatic hybrid. This provided unequivocal proof of their hybrid nature.

In brief, it can be said that it is safe to combine a number of characters to confirm a hybrid.

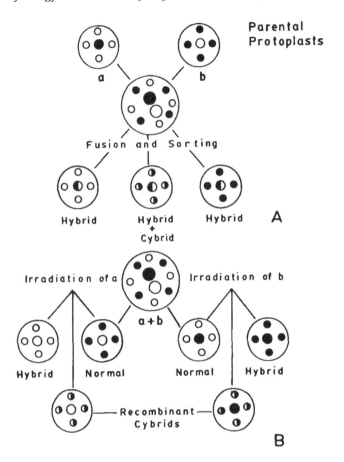

FIGURE 13.   Diagrammatic representation of protoplast fusion and fate of nuclear and cytoplasmic genomes. (A) Nuclear fusion and fusion or sorting of organelle genome. (B) Selective transfer of cytoplasmic character, on elimination of nuclear fusion.

VI. SELECTION OF A CYBRID

In protoplast fusion it is possible to have cytoplasmic hybrids or cybrids with chloroplasts and/or mitochondria from the "donor" and a nucleus from the "recipient" (Figure 13). Obtaining a cybrid, elimination of the "donor" nucleus can be facilitated by irradiation of donor protoplasts with X- or γ-rays before fusion. When protoplasts of *N. sylvestris* were fused with X-irradiated protoplasts of *N. tabacum*,[166] only cell cybrids could grow on a particular medium because protoplasts of *N. sylvestris* do not grow on this medium and protoplasts of *N. tabacum* were inactivated due to irradiation. An alternate way of preventing hybrid formation is to take enucleated protoplasts or cytoplasts.[101]

Direct selection of cybrids[21] was possible on simultaneous selection for "donor" chloroplasts and recipient "nuclei" when mesophyll protoplasts of two tobacco mutants — SRI (streptomycin resistant) and val-2 (valine resistant) — were fused. SR resistance is a maternally inherited chloroplast trait while valine resistance is a Mendelian character. The colonies formed on fusion were selected for resistance to valine and streptomycin.

VII. ACHIEVEMENTS OF CELL FUSION

Cell fusion is in its infancy. Large-scale isolation of plant protoplasts[30] was possible in the early 1960s and the first heterokaryon was demonstrated a decade later;[136] the first somatic hybrid[24] plant was possible 2 years later. Since then, a spate of papers have appeared confirming the feasibility of this technique and its potential contribution in crop improvement and in basic research on development and differentiation.

A. Intraspecific Hybrids

Although intraspecific hybrids have been possible in the genera *Nicotiana, Datura, Petunia,* and *Daucus,* most have been in *Nicotiana.* Basically these are attempts to test the feasibility of the technique and test the selection system. Recently, intraspecific fusion has also been shown in *Solanum tuberosum,*[5-7] indicating that somatic fusion may be useful for the transfer of traits within the group.

B. Interspecific Hybrids

Interspecific hybrids have been possible in five genera: *Nicotiana, Datura, Petunia, Solanum,* and *Daucus.* These include incompatible crosses between different species of *Nicotiana, Petunia,* and *Solanum.* For example, as a result of fusion of sexually incompatible mesophyll protoplasts of *P. parodii* with albino protoplasts from a cell suspension of *P. parviflora,*[134] it was possible to obtain an amphidiploid somatic hybrid plant with a new character — "hanging basket-type petunia". This indicates the importance of somatic cell fusion for improvement of horticultural plants. In *Datura,* it is not clear whether interspecific fusion is between incompatible species. With the exception of *Daucus,* all interspecific fusions are limited to the family of Solanaceae. In the genus *Solanum,*[10] somatic hybrids were obtained on fusing protoplasts of an albino *S. tuberosum* with protoplasts of *S. brevidens,* a nontuber-bearing species which is sexually incompatible with the tuber-bearing potato. This offers the possibility of introgressing valuable germplasm. Using the same system, disease resistance[6] (to potato leaf roll virus) could be transferred from *S. brevidens* to *S. tuberosum.*

C. Intergeneric Hybrids

Of great importance are intergeneric hybrids. Plants have been possible on intergeneric[108] fusions of *Lycopersicon esculentum* + *S. tuberosum,*[108] *Arabidopsis thialiana* + *Brassica campestris,*[54-56] *B. napus* + *Raphanus sativus,* and *B. campestris* + *R. sativus,*[28] *Citrus sinensis* + *Poncirus trifoliate,*[122] *N. tabacum* + *Petunia hybrida,*[127] *Lycopersicon esculentum* + *S. lycopersicoides,*[72] and *L. esculentum* + *S. rickii.*[121]

Intergeneric cell hybrids where plants have not been possible are between *Parthenocissus* + *Petunia,*[133] *Glycine max* + *N. glauca,*[87] *Daucus carota* + *Aegopodium,*[35] and *Petunia* + *Vicia.*[1e] Of these, only in cell hybrids between *Petunia* and *Vicia* is there chromosome stability; otherwise, in the rest the chromosomes of one of the parents are progressively lost. Recently, it has been possible to have intergeneric cell fusion in the Graminae *Pennisetum americanum* + *Panicum maximum.*[122a]

D. Transfer of Male Sterility

Using protoplast fusion it has been possible to transfer cytoplasmic male sterility,[14] thus initiating an altogether new line of investigation. When protoplasts of male sterile *N. techne* (TS, a hybrid with the nucleus of *N. tabacum* in the cytoplasm of *N. dabneyi*) were fused with protoplasts of male fertile *N. tabacum* cv. *xanthi,* 936 plants were recovered. These plants were classified on the basis of leaf and floral morphology and sterility/fertility. Of these, 654 proved to be similar to the parental types and were suspected to be derived from unfused protoplasts, since no selection pressure for hybrids was applied; 57 were taken to be hybrids because of intermediate leaf and floral morphology. The remaining 225 plants had leaves identical to one of the parents, but intermediate in floral characters and were taken to be cybrids. Incidentally, most were fertile *techne* and sterile *xanthi*, suggesting their cybrid nature. Further,[15] it was resolved that these cybrids in their mitochondrial genome, according to mt-restriction patterns, were not identical to either of the parental mt-DNA, confirming their cybrid nature.

A clear demonstration of transfer of cytoplasmic male sterility,[9,166] using selection pressure, was possible when mesophyll protoplasts of *N. sylvestris* were fused with protoplasts of male sterile *N. tabacum* (having nuclear genome of *N. tabacum* and cytoplasm from an unknown parent). The protoplasts of *N. sylvestris* did not divide on mannitol medium and protoplasts from male sterile *N. tabacum* were given a heavy dose of X-rays to prevent their division; only the fusion products could grow on this medium. Seven calli were possible and the plants regenerated were cybrids because they depicted the morphological features of either *N. sylvestris* or both parents, but were male sterile — a characteristic conditioned by the cytoplasm of the male sterile line. A diagrammatic representation of protoplast fusion and fates of nuclear and cytoplasmic genomes is provided in Figure 13.

Similarly, transfer of male sterility by somatic cell hybridization was demonstrated in *Petunia* spp.[80,81] (cms *P. hybrida* and fertile *P. axillaris*). In this system transfer of cms is also possible through sexual reproduction.

The above results indicate that transfer of cytoplasmic male sterility from one species to another is possible in one cycle of hybridization in comparison to lengthy recurrent backcrossing needed for transfer of this trait through conventional breeding.

Following cell fusion there is reorganization[119] of mitochondrial DNA, which has been shown in another system (*N. tabacum* and *N. knightiana*,) confirming the observation of earlier workers.[159]

E. Transfer of Disease Resistance

A semidominant albino mutation (Su) has been successfully employed to select somatic hybrids in the genus *Nicotiana*. The light green, when selfed, produces dark green, light green, and albino seedlings in a ratio of 1:2:1. A cell suspension from albino serves as a source of protoplasts to be fused with protoplasts of wild tobacco in crosses which do not occur in nature and are not possible by conventional means. Disease resistance has been incorporated into *N. tabacum* by fusing its protoplast with that of wild *N. nesophila*,[41] demonstrating the utility of this method in crop improvement.

Also, in a fusion of protoplasts of *N. tabacum* + *N. glutinosa*, leaves of hybrid plants which were intermediate in morphology between the parents showed resistance towards tobacco mosaic virus.[153]

It has also been possible to transfer resistance to potato leaf roll virus (PLRV) from *S. brevidens* + *S. tuberosum* by somatic fusion.[6] *S. brevidens* is resistant to PLRV and frost, but is difficult to cross sexually with cultivated potato.

F. Restoration of Regeneration Potential

A new aspect that has come to light through cell fusion is the induction or restoration of regeneration potential.[100] Kanamycin-resistant cell lines of *N. tabacum* lost the potential for plant regeneration, as did protoplasts of *N. knightiana*. However, when the protoplasts of the two were fused, the somatic hybrid regenerated into plants. Thus, lack of morphogenic potential could be complemented by genetic complementation.

Similarly, two NR-deficient cell lines could not be induced to undergo organogenesis, but somatic hybrids[65] from these lines complemented not only for NR, but also for organogenesis. It remains to be seen whether NR is critical for organogenesis, because these lines also failed to undergo organogenesis on amino acid medium, indicating that it is probably the endogenous balance of reduced and oxidized nitrogen which controls organogenesis. Also, in an analysis of F_1 progeny of somatic hybrids obtained from two lines of *N.*

plumbaginifolia (green NR deficient and regeneration defective + NR⁺ and pigment deficient), a cosegregation of NR⁻ and an absence of regeneration capacity indicate that nitrate assimilation is a determinate of shoot regeneration ability.[102]

VIII. GENETIC ANALYSIS OF CELL HYBRIDS

For effective use of cell fusion the first and foremost requirement is that hybrid cells be able to regenerate into plants and in regenerated plants the transferred genes should be stably inherited. Information that is helpful in genetic analysis of somatic hybrid is the frequency of (1) cell and nuclear fusion, (2) nuclear segregation and recombination, and (3) organelle segregation and recombination. The phenotype of somatic hybrids may further be affected by cultural conditions which bring about chromosome instability, gene mutation, and epigenetic changes. Therefore, there is a tremendous potential for variability in somatic hybrid plants. In fact, populations of plants regenerated from cell fusion are more variable than comparable populations raised from sexual fusion. The variations recorded in phenotypic characters are plant height[114,115] (21 to 113 cm), leaf shape, leaf size and petiole length,[42,156,166] flower length,[150] flower color,[99] and pollen viability.[114]

In a recent hybridization between *S. tuberosum* and *S. brevidens*, a wide range of variation in morphological characters was observed in plants of somatic hybrid origin.[7]

A. Nuclear Fusion in Cell Hybrids

In most experiments on cell fusion, diploid protoplasts are fused, resulting in an amphidiploid condition, indicating that nuclear fusion has occurred. Aneuploidy is also not rare and is due to chromosome elimination after nuclear fusion. Multiple cell fusion also cannot be ruled out. In protoplast fusion of *N. glauca* (2n = 24) and *N. langsdorffii* (2n = 18) none of the plants produced had 42 chromosomes;[150] instead the number ranged from 56 to 63. It is proposed that hybrid plants were derived from triple fusion having either two glauca (GG + GG + LL = 66) or two langsdorffii (LL + LL + GG = 60) genomes.

Multiple cell fusion is also a factor in somatic cell genetics. In intraspecific fusion of *Nicotiana*, fusion products isolated mechanically yielded both hybrid and parental type cells. This could be due to multiple cell fusion in which two nuclei fused and a third unfused nucleus segregated during mitotic division.[50,51] Similar results were also obtained in *N. tabacum* and *N. plumbaginifolia*.[57,148] Thus, nonfusion and segregation of nuclei is possible in fusion products.

After cell fusion, nuclear fusion need not follow; this is evidenced by the high frequency of nuclear segregants.[50] In contrast, there is high rate of nuclear fusion,[148] as high as 100%. This frequency could be reduced to 60% when protoplasts of one of the parents was preirradiated with ⁶⁰Co rays (60 J/kg). The frequency[109] further declined to 22% by increasing the radiation dose to 210 J/kg. It is worth recalling that preirradiation of protoplasts[166] was used to transfer male sterility from *N. tabacum* and *N. sylvestris*.

Genetic analysis of interspecific hybrids completed in *Datura* and *Nicotiana* indicates that there is segregation of albinism, the nuclear genetic marker. In $R_1R_2R_3$ progeny of *D. innoxia* + *D. discolour* and *D. innoxia* + *D. stramonium* the number of chlorophyll-deficient seedlings increased from 0 to 10%.[143a] This is expected because of limited chromosome pairing during meiosis in interspecific hybrids. The same was true of interspecific hybrids of *Nicotiana*, *N. otophora* + *N. tabacum*.[41] On anther culture of somatic hybrids it was possible to recover plants with the same chromosome number as sexual hybrid; 45 wild type and 17 albino plants were recovered following anther culture of interspecific *Datura* somatic hybrids.[143a] In contrast, in interspecific hybrid *N. otophora* + *N. tabacum*, all plants with 2N = 72 were light green Su/Su/+/+. The former, Su/Su, represented by tobacco genome and homologous allelic pair + + of otophora. During microsporogenesis the pollen produced is of Su/+ and hence all pollen plants produced were light green.

B. Cytoplasmic Fusion in Cell Hybrids

The fate of mixed cytoplasm has been followed in many somatic hybrids. The presence of mixed cytoplasm was inferred from variegated plants obtained following fusion of two *N. tabacum*[60] lines; cytoplasmic heterogeneity was confirmed by transmission of the variegated trait to sexual progeny. Protoplasts of a *N. tabacum* plastome mutant were also fused with protoplasts from a cms line (which contained *N. tabacum* genome in *N. debneyi* cytoplasm) and *N. debneyi*.[60] In both cases variegated plants were recovered. Plants were cytoplasmic heterogeneic because large subunit of RUBP-carboxylase comprised polypeptides of both *N. tabacum* and *N. debneyi*. Similarly, heterogeneity of plastome genes causing chlorophyll deficiency is seen in cell hybridization of tobacco.[63,148]

Segregation of cytoplasmic characters can be random or nonrandom. In cell hybrids of *N. glauca* + *N. langsdorffii* segregation of chloroplasts occurred prior to meiosis and was random to one or the other type of cytoplasm. This was inferred from a follow up of large subunit of fraction I protein.[27] The nonrandom segregation reported[44] may be due to nucleocytoplasmic incompatibility or due to the cell system used for fusion. Chloroplast segregation occurs shortly after cell fusion in intraspecific somatic hybrids[145] of *N. debneyi*.

Recombination of chloroplast DNA was not observed in the cybrids *N. tabacum* + *N. debneyi*,[13] *N. tabacum* + *N. knightiana*,[111] *N. tabacum* + *N. suaveolens*,[64] and *N. tabacum* + *N debneyi*.[145] However, there is an indication of chloroplast DNA recombination in *N. glauca* + *N. langsdorffii* somatic hybrids.[32] Also, mitochondrial recombination is seen in several cytoplasmic hybrids: *N. tabacum* + *N. debneyi*,[15] *N. tabacum* + *N. sylvestris*,[48] *N. tabacum* + *N. knightiana*,[119] *Brassica napus* + *Raphanus sativus*, and *B. campestris* + *R. sativus*.[128]

C. Intergeneric Cell Hybrids

In the first intergeneric[90] cell hybrids of *N. glauca* + *Glycine max*, which were isolated mechanically and cultured individually, chromosome analysis of first cell division after fusion indicated that chromosomes of *N. glauca* had a tendency to condense and fuse together, superlong chromosomes (megachromosomes); ring chromosomes and multiconstrictioned chromosomes were also observed. These features persisted for up to 1 to 2 months of culture. During subsequent months glauca chromosomes were gradually reduced. The isoenzyme study correlated with the chromosome study. The isoenzymes of both parents were seen in hybrids up to 2 to 4 months and by the 8th month isoenzymes of glauca were reduced. Reintroduction of glauca information was possible by back-fusion of *N. glauca* protoplasts to hybrid cells[161] after 27 months of initial fusion. This information was retained for up to 2 years after back-fusion, indicating that a recombination between glauca and soybean chromosome occurs, resulting in select information. Chromosome elimination also occurs in *Vicia faba* + *Petunia hybrida* cell hybrids.[19] In some hybrid cell lines there is loss of *Petunia* and in others there is loss of *Vicia* chromosomes.

In somatic hybrids of *N. tabacum* + *Glycine max*,[26] some lines lost almost all tobacco chromosomes while in others about half of the chromosomes were retained. Similarly, in cell hybrids of *Pisum sativum* and *N. chinensis*,[58] chromosomes and species-specific multiple molecular forms of the enzymes of both parents were retained. However, the chromosomes of the two parent groups separated in a common metaphase, indicative of spatial separation of the two genomes. These examples provide evidence that chromosomal stability could be maintained in distant hybridization. However, plant regeneration is not possible.

In a somatic hybrid of *Arabidopsis thaliana* + *Brassica campestris*,[54-56,77] cytological analysis revealed reconstructed chromosomes and mitotic abnormalities (anaphase bridges, multichoromosome chains, etc.). The chromosomes and multiple molecular forms of enzymes, esterases, lactase, alcohol dehydrogenase, and peroxidase of both parents were retained for up to 7 months of fusion. The plants regenerating from chromosomal stable cell lines were highly homogenous and morphologically intermediate between *Arabidopsis* and *Brassica*, but contained *Brassica* plastids. These regenerates however did not flower and had abnormal roots. The cell line which has lost most of the turnip chromosomes produced anomalous as well perfect flowering plants[77] similar to *Arabidopsis*. However, these plants had a small amount of turnip genetic material and turnip characters such as yellow petals, simple nonbranching form of trichomes, and plastids derived from turnip. These plants were sterile.

Somatic hybrid of *Daucus carota* (2n = 18) + *Aegopodium podagraria* (2n = 42) produced plants[36] which were similar to carrot and had a chromosome number of 18. In a similar experiment on a somatic hybrid of *D. carota* + *Peteroselinium hortense* (2n = 22) the regenerated plants had 19 chromosomes, indicating an incomplete elimination of the parsley genome.[37] However, the presence of parsley isoenzymes and correction of albinism of carrot indicate transfer of the parsley genome to carrot.

Anomalous shoots have been possible from cell hybrids *N. tabacum* + *Solanum tuberosum*, *N. tabacum* + *S. sucrense*,[149] *Atropa belladonna* + *N. chinensis*.[59] and *S. tuberosum* + *L. exculentum*.[108] From these reports the following conclusions are derived:[53,60]

1. The hybrid state is relatively stable.
2. Genetic rearrangements (reconstruction and partial elimination of parental chromosomes) is extensive.
3. Diverse genetic lines are recovered from fusion products.
4. Intertribal hybrids are capable of regeneration.

For the success of experiments on cell fusion, information is needed about the processes that affect the fate of parental genetic material in fused cells.

IX. GENE TRANSFER THROUGH CELL FUSION

Cell fusion can be utilized in two major ways to effect transfer of genes in plants for crop improvement. One is the merging of two or more species into a synthetic species and the other approach involves transfer of a few characters from one species to another (controlled interogression). These aspects are outlined below.

A. Gene Transfer in Whole-Cell Fusion

The recent success of protoplast fusion in bringing together the genomes of two incompatible species has greatly stimulated plant breeders and biotechnologists to use this technique for distant hybridization and raising hybrid seeds. One of the objectives of fusion of protoplasts of widely divergent species is to incorporate into recipient species some limited attributes of a donor species. In mammalian cells this was possible by X-ray irradiation of parental cells and other chromosome-destabilizing processes to introduce directional chromosome elimination. Also, in plants such a situation can be seen in pollination of *Nicotiana* with irradiated pollen which results in transfer of some genes without fertilization.[82,123] In protoplasts fusion directional chromosome elimination can be seen when protoplasts of carrot (2n = 18) were fused with irradiated protoplasts of parsley (2n = 22). Among the selected regenerated plants, there were some with 2n = 19 chromosomes and with some genetic marker from the irradiated donor.[38] Transfer of nuclear genes through irradiated protoplasts can also be seen in the fusion of an X-irradiated (100 krad) protoplast of *N. tabacum* cv. *xanthi* with protoplasts of *N. tabacum* lacking NR and incapable of utilizing nitrate as the sole source of nitrogen. Colonies recovered showed NR proficiency, suggesting that gene transfer has occurred (Cooper-Bland, S., University of Nottingham, U.K., unpublished). More direct evidence to this effect was provided when γ-irradiated barley protoplasts were fused with NR-deficient protoplasts of *N. tabacum*. The former were incapable of division, but later became NR proficient as a result of fusion.[150a]

Also, direct evidence of limited gene transfer[68] is seen in fusion of protoplasts of *Physalis minima* and *Datura innoxia*. In these species no hybrids are possible either by sexual means or protoplast fusion. However, when X-irradiated protoplasts of *P. minima* were fused with albino protoplasts of *D. innoxia*, 1.15% of the heterokaryon showed chlorophyll synthesis. The introduction and expression of the *P. minima* genome was demonstrated by the presence of additional nuclear DNA and the expression of various isoenzymes.

B. Gene Transfer in Cytoplasmic Fusion

In cell fusion, cytoplasmic mix is certainly novel and provides a means for heterogeneity of extrachromosomal genes.

In the chloroplast genome, the two types of chloroplasts in a somatic hybrid eventually sort out into one or the other or, in some cases, only one of the parental types. This was seen in the very first somatic hybrid, *N. glauca + N. langsdorffii*.[27,96] In somatic hybrids of *Petunia (P. parodii + P. hybrida, P. inflata, P. parviflora)* there was unidirectional sorting out in favor of *P. parodii*.[95] Also, in somatic hybrids of *Lycopersicon esculentum + L. pennelli* the callus clones carried predominantly either one parental plastid DNA or mixtures of both types. However, in intergeneric cell fusion between *L. esculentum + Solanum rickii* somatic hybrids plants invariably had chloroplast DNA from *S. rickii*.[121]

Chloroplast sorting is common, but not exclusive.[47] This was inferred from experiments on fusion of *N. sylvestris* protoplasts (recipient) with X-irradiated *N. tabacum* streptomycin-resistant protoplasts. Of the 38 cybrids (all with *N. sylvestris* nuclei) regenerated, 8 were homoplastomic and streptomycin resistant and their progeny did not segregate, 16 plants were homoplastomic and streptomycin sensitive, and 14 plants were heteroplastomic and their progeny segregated into resistant and sensitive plants. Also, in other fusion experiments delayed sorting out of chloroplasts was seen; the result of the delay was persistent variegation.

Cell fusion is a way to investigate interactions between organelle and nuclear genome. It is possible to expose either or both mitochondria and chloroplasts of a given species in an alien nuclear environment.[44,103] When protoplasts of a mutant line of *N. tabacum* having a maternally transmitted chlorophyll deficiency were fused with protoplasts of two alloplasmic male sterile *Nicotiana* lines by the "donor-recipient" technique, in both fusion experiments variegated plants were regenerated which contained cytoplasm of mixed chloroplast nature. Thus, it is possible to have mixed cytoplasm of genetically different chloroplasts.[45] In another experiment,[8] cybrid plants having the nuclear genome of one species and either or both plastome and chondriome of another species were obtained by fusing protoplasts of *N. sylvestris* as recipient with irradiated protoplasts of *N. rustica* as donor of chloroplasts and mitochondria. As expected, all 49 plants analyzed had *N. sylvestris* morphology. The chloroplast restriction pattern indicated that 8 had *N. rustica* and 41 had *N. sylvestris* plastomes. Chondriome analysis revealed that all were similar to *N. sylvestris*. Further, it appears that the nuclear genome does not influence the expression of cytoplasmic male sterility which is chondriome encoded. This was seen when progeny of a somatic hybrid between male sterile *N. tabacum (N. debneyi* cytoplasm) and *N. glutinosa* were back-crossed with pollen of either *N. tabacum* or *N. glutinosa*. In these crosses male sterility was consistently maintained.[154]

Opportunities for increasing cytoplasmic variability by protoplast fusion are greater with mitochondria. Mitochondrial DNA recombination[15,119] occurs in cell fusion. A correlation between mitochondrial genome and cytoplasmic male sterility is seen[15,47] and can be illustrated by interspecific protoplast fusion between a cms line and a normal line. The donor X-irradiated protoplasts were from a cms *N. tabacum* line (L 92) and recipient protoplasts were normal *N. sylvestris* protoplasts. The protoplasts of *N. tabacum* were incapacitated in division due to X-irradiation, and protoplasts of *N. sylvestris* were incapable of division in the standard mannitol medium. Only the hybrids could grow and the plants regenerated from these hybrids were of *N. sylvestris* morphology, but were male sterile. Chloroplasts of these plants in three tests (tentoxin sensitivity, electrofocusing of large subunits of RUBPcase polypeptides, and chloroplast restriction DNA patterns) resembled chloroplasts of L 92 plants but were male sterile. Therefore, it can be inferred that these plants acquired the chloroplast characters and male sterility characters from L 92 plants.

Mitochondrial DNA polymorphism is seen in crucifer plants regenerated from somatic cell fusion of *Brassica napus* + *Raphanus sativa* and *B. campestris* + *R. sativus*,[28] *Daucus* sp.,[93a,102a] and *Petunia* sp.[138] Restriction fragment analysis of mitochondrial DNA indicates that these genomes differ from each other and from both parents.

C. Gene Transfer in Microplast Fusion

Isolation of microplasts or subprotoplasts[17,98] has opened a new possibility for fusion of protoplasts and microplasts. This is likely to be similar to microcell-mediated transfer of chromosomes in mammalian cells, a technique which has helped resolve the chromosome assignment of genes. A beginning in this direction is the demonstration of the transfer of isolated chromosomes into protoplasts of wheat, maize, and parsley.[151]

D. Gene Transfer in Plant and Microbial Cell Fusion

A new development is the fusion of plant protoplasts and bacterial spheroplasts. Using this approach it was possible to transform *Vinca rosea* cells into auxin autonomy by fusing their protoplasts with the spheroplasts of *Agrobacterium tumefaciens*.[76] These results were confirmed on fusion of *Agrobacterium* and *Escherichia coli* spheroplasts with protoplasts of tobacco.[69] Residual contaminating spheroplasts and bacterial cells are eliminated by lysozyme and antibiotic treatment. The frequency of transformation was higher compared to transformation by transfer of plasmid DNA. This is consistent with the increased frequency of transformation of monkey cells[139] on fusing with spheroplasts of *E. coli* harboring a recombinant plasmid of SV 40.

The association between bacterial spheroplasts and plant protoplasts has been followed employing electron microscopy.[73] This study demonstrates that fusion between spheroplasts and protoplasts does not occur. The spheroplasts attach to protoplasts via a plasma membrane protrusion after high pH high Ca^{++} treatment. By contrast, PEG, high pH, and high Ca^{++} promoted endocytosis of spheroplasts into a plasma membrane-bound vesicle. Following fusogenic treatment it is possible to detect bacterial antigens associated with protoplasts using immunofluorescence. The antigens are dispersed within the peripheral cytoplasm.

X. LIMITATIONS OF CELL FUSION

Although protoplasts from a number of species have been fused and plants regenerated from heterokaryons, there are a number of limitations for widespread application of this technique for crop improvement.

Unfortunately, it has not been possible to regenerate plants from protoplasts of our important crop plants — cereals and seed legumes. This limitation precludes the potential of this technique for improvement of staple diet crops. At present, the technique is limited to the family Solanaceae and other model plants. Even among Solanaceae, chromosome instability is a marked feature of somatic hybrids.[107] The intergeneric somatic hybrid *Arabidopsis* + *Brassica* is aneuploid; it is neither sterile nor capable of flowering.[56] Species limitations for successful somatic hybridization are yet to be outlined, based on extensive experimentation. For example, somatic hybrids between the reproductively isolated tomato species *L. peruvianum* + *L. pennelli* were sterile and subvital. This questions the value of somatic hybridization as a useful breeding approach in *Lycopersicon*.[1] Also, hybrid plants of somatic cell origin between *Solanum melongena* + *S. sisymbrifolium*, the two sexually incompatible species, were sterile.[62]

Nucleocytoplasmic genomic incompatibility can also be a limitation in obtaining functional somatic cell hybrids. In interspecific somatic hybrids plants of *Petunia*,[144] *P. parodii* + *P. inflata*, even cytologically stable somatic hybrids displayed aberrant reproductive and floral morphologies, whereas cytologically unstable somatic hybrids showed various degrees of aneuploidy coupled with corolla splitting and irregularities in reproductive organs such as double stigmas and styles and reduced pollen viability. Explanations for this aberrant behavior include nuclear-cytoplasmic genomic incompatibility, chromosomal loss, or a phenomenon similar to hybrid dysgenesis occurring as a result of somatic fusion. Genomic incompatibility was also observed in intergeneric cell hybrids *N. tabacum* + *P. hybrida*,[127] resulting in loss of tobacco genome.

In brief, the inability to regenerate plants from protoplasts, species barriers to cell fusion, chromosome instability of somatic hybrids and aneuploidy, and nucleocytoplasmic genomic incompatibility are the principal limitations of cell fusion. These limitations call for intensive and extensive experimentation concerning various aspects of protoplast technology and cell fusion.

XI. PROSPECTS

Using protoplast fusion it is possible to obtain plants having (1) a nucleus from one parent and cytoplasm from another parent and (2) chloroplast genes of one parent and mitochondria from another parent. Certain of these nuclear cytoplasmic combinations are difficult or impossible in sexual hybridization. Also of frequent occurrence in cell fusion is chromosome elimination, and hence it is possible to have nuclear recombination involving portions of a nucleus. In addition, mitotic recombination is possible in organelle DNA. Since chloroplasts and mitochondria are both concerned with energy processes, it is likely that additional agriculturally useful traits encoded by these organelles will be found. Traits concerning cytoplasmic male sterility, herbicide resistance, disease resistance, response to toxins, and antibiotic resistance are found in these organelles. As our knowledge of organelle genetics increases, it is likely that organelles may be used as vehicles for transfer of genes.

One of the important aspects of cell fusion is distant hybridization, not possible by conventional means. Some amphiploid hybrids have been possible in some species, which are not possible by conventional breeding, but hybrids of more distant genera are aneuploid and sterile. However, limited coordination of the two genomes in distant hybridization in terms of transfer of a small amount of genetic material has offered a new approach to crop improvement.

REFERENCES

1. **Adams, T. L. and Quiros, C. F.**, Somatic hybridization between *Lycopersicon peruvianum* and *L. pennellii*. Regenerating ability and antibiotic resistance as selection systems, *Plant Sci.*, 40, 209, 1985.
2. **Ahkong, Q. F., Fischer, D., Tempion, W., and Lucy, J. A.**, Mechanism of cell fusion, *Nature (London)*, 253, 194, 1975.
3. **Ahkong, Q. F., Howell, J. I., Lucy, J. A., Safwat, F., Davey, M. R., and Cocking, E. C.**, Fusion of hen erythrocytes with yeast protoplasts induced by polyethylene glycol, *Nature (London)*, 255, 66, 1975.
4. **Alexander, R. G., Cocking, E. C., Jackson, P. J., and Jett, J. H.**, The characterization and isolation of plant heterokaryons by flow cytometry, *Protoplasma*, 128, 52, 1985.

5. **Austin, S., Baer, M., Ehlenfeldt, M., Kazamierezak, P. J., and Helgeson, J. P.**, Intraspecific fusions in *Solanum tuberosum, Theor. Appl. Genet.*, 71, 172, 1985.
6. **Austin, S., Baer, M., and Helgeson, J. P.**, Transfer of resistance of potato leaf roll virus from *Solanum brevidens* into *S. tuberosum* by somatic fusion, *Plant Sci.*, 39, 75, 1985.
7. **Austin, S., Ehlenfeldt, M. K., Baer, M. A., and Helgeson, J. P.**, Somatic hybrids produced by protoplast fusion between *S. tuberosum* and *S. brevidens:* phenotypic variation under field conditions, *Theor. Appl. Genet.*, 71, 682, 1986.
8. **Aviv, D., Bleichman, S., Arzee-Gonen, P., and Galun, E.**, Intersectional cytoplasmic hybrids in *Nicotiana, Theor. Appl. Genet.*, 67, 499, 1984.
9. **Aviv, D., Fluhr, B., Edelman, M., and Galun, E.**, Progeny analysis of the interspecific somatic hybrids: *Nicotiana tabacum* (cms) + *N. sylvestris* with respect to nuclear and chloroplast markers, *Theor. Appl. Genet.*, 56, 145, 1980.
10. **Barsby, T. L., Shepard, J. F., Kemble, R. J., and Wong, R.**, Somatic hybridization in the genus *Solanum: S. tuberosum* and *S. brevidens, Plant Cell Rep.*, 3, 165, 1984.
11. **Bates, G. W.**, Electrical fusion for optimal formation of protoplast heterokaryons in *Nicotiana, Planta,* 165, 217, 1985.
12. **Bates, G. W. and HasenKampf, C. A.**, Culture of plant somatic hybrids following electrical fusion, *Theor. Appl. Genet.*, 70, 227, 1985.
13. **Belliard, G. and Pelletier, G.**, Morphological characteristics and chloroplast DNA distribution in different cytoplasmic parasexual hybrids of *Nicotiana tabacum, Mol. Gen. Genet.*, 165, 231, 1978.
14. **Belliard, G., Pelletier, G., and Ferault, M.**, Fusion de protoplastes de *Nicotiana tabacum* a cytoplasmes differents: etude des hybrides cytoplasmiques neoformes, *C. R. Acad. Sci. Ser. D*, 284, 749, 1977.
15. **Belliard, G., Vedel, F., and Pelletier, G.**, Mitochondrial recombination in cytoplasmic hybrids of *Nicotiana tabacum* by protoplast fusion, *Nature (London)*, 281, 401, 1979.
16. **Benbadis, A. and de Virville, J. D.**, Effects of polyethylene glycol treatment used for protoplast fusion and organelle transplantation on the functional and structural integrity of mitochondria isolated from spinach leaves, *Plant Sci. Lett.*, 26, 257, 1982.
17. **Bilkey, P. C., Davey, M. R., and Cocking, E. C.**, Isolation, origin and properties of enucleate plant microplasts, *Protoplasma*, 110, 147, 1982.
18. **Binding, H.**, Fusionsversuche mit isolierten Protoplasten von, *Petunia hybrida, Z. Pflanzenphysiol.*, 72, 421, 1974.
19. **Binding, H. and Nehls, R.**, Somatic cell hybridization of *Vicia faba* and *Petunia hybrida, Mol. Gen. Genet.*, 164, 137, 1978.
20. **Boss, W. F., Grimes, H. D., and Brightman, A.**, Calcium-induced fusion of fusogenic wild carrot protoplasts, *Protoplasma*, 120, 207, 1984.
21. **Bourgin, J. P., Missonier, C., and Goujaud, J.**, Direct selection of cybrids by streptomycin and valine resistance in tobacco, *Theor. Appl. Genet.*, 72, 11, 1986.
22. **Brar, D. S., Ono, M., Kobayashi, S., Uchimija, H., and Harada, H.**, Analysis of chromosomes and ribosomal RNA genes in parasexual hybrids of *Nicotiana, Protoplasma*, 121, 228, 1984.
23. **Burgess J. and Fleming, E. N.**, Ultrastructural studies of the aggregation and fusion of protoplasts, *Planta,* 118, 183, 1974.
24. **Carlson, P. S., Smith, H. H., and Dearing, R. D.**, Parasexual interspecific plant hybridization, *Proc. Natl. Acad. Sci. U.S.A.*, 69, 2292, 1972.
25. **Cella, R., Carbonera, D., and Iadarola, P.**, Characterization of intraspecific somatic hybrids of carrot obtained by fusion of iodoacetate-inactivated A2CA-resistant and -sensitive protoplasts, *Z. Pflanzenphysiol.*, 112, 449, 1983.
26. **Chien, Y. C., Kao, K. N., and Wetter, L. R.**, Chromosomal and isozyme studies of *Nicotiana tabacum-Glycine max* hybrid cell lines, in *Proc. 5th Int. Congr. Plant Tissue and Cell Culture*, Fujiwara, A., Ed., Maruzan, Tokyo, 1982, 633.
27. **Chen, K., Wildmann, S. G., and Smith, H. H.**, Chloroplast DNA distribution in parasexual hybrids as shown by polypeptide composition of fraction I protein, *Proc. Natl. Acad. Sci. U.S.A.*, 74, 5109, 1977.
28. **Chetrit, P., Mathieu, C., Vedel, F., Pelletier, G., and Primard, C.**, Mitochondrial DNA polymorphism induced by protoplast fusion in Cruciferae, *Theor. Appl. Genet.*, 69, 361, 1985.
29. **Chupeau, Y. C., Missonier, C., Hommel, M. C., and Goujaud, L.**, Somatic hybrids of plant by fusion of protoplasts, *Mol. Gen. Genet.*, 165, 239, 1978.
30. **Cocking, E. C.**, A method for the isolation of plant protoplasts and vacuoles, *Nature (London)*, 187, 962, 1960.
31. **Cocking, E. C., George, D., Price-Jones, M. J., and Power, J. B.**, Selection procedures for the production of interspecies somatic hybrids of *Petunia hybrida* and *P. parodii.* II. Albino complementation selection, *Plant Sci. Lett.*, 10, 7, 1977.
32. **Conde, M. R.**, Chloroplast DNA recombination in *Nicotiana* somatic parasexual hybrids, *Genetics*, 97, 26, 1981.

33. **Constabel, F. and Kao, K. N.**, Agglutination and fusion of plant protoplasts by polyethylene glycol, *Can. J. Bot.*, 52, 1603, 1974.

34. **Douglas, G. C., Wetter, L. R., Keller, W. A., and Setterfield, G.**, Somatic hybridization between *Nicotiana rustica* and *N. tabacum*. IV. Analysis of nuclear and chloroplast genome expression in somatic hybrids, *Can. J. Bot.*, 59, 1509, 1981.

35. **Dudits, D., Haldlaczky, G. Y., Levi, E., Fejer, O., Haydu, Z., and Lazar, G.**, Somatic hybridization of *Daucus carota* and *D. capillifolius* by protoplast fusion, *Theor. Appl. Genet.*, 51, 127, 1977.

36. **Dudits, D., Hadlaczky, G. Y., Bajazar, G. Y., Konencz, C. S., and Lazar, G.**, Plant regeneration from intergeneric cell hybrids, *Plant Sci. Lett.*, 15, 101, 1979.

37. **Dudits, D., Hadlaczky, G. Y., Lazar, G., and Haydu, Z.**, Increase in genetic variability through somatic cell hybridization of distantly related plant species, in *Plant Cell Cultures: Results and Perspectives*, Sala, F., Parisi, B., Cella, R., and Ciferi, O., Eds., Elsevier/North-Holland Biomedical Press, Amsterdam, 1980, 207.

38. **Dudits, D., Fejer, O., Hadlaczky, G. Y., Konencz, C., Lazar, G. B., and Horvath, G.**, Intergeneric gene transfer mediated by protoplast fusion, *Mol. Gen. Genet.*, 179, 283, 1980.

39. **Eriksson, T.**, Isolation and fusion of plant protoplasts, in *Les Cultures de Tissus de Plantes*, No. 193, Colloq. Int. C.N.R.S. Paris, 1970, 297.

40. **Evans, D. A.**, Chromosome stability of plants regenerated from mesophyll protoplasts of *Nicotiana* species, *Z. Pflanzenphysiol.*, 95, 459, 1979.

41. **Evans, D. A., Flick, C. E., and Jensen, R. A.**, Disease resistance incorporation into sexually incompatible somatic hybrids of genus *Nicotiana*, *Science*, 213, 907, 1981.

42. **Evans, D. A., Flick, C. E., Kut, S. A., and Reed, S. M.**, Comparison of *Nicotiana tabacum* and *N. nesophila* hybrids produced by ovule culture and protoplast fusion, *Theor. Appl. Genet.*, 62, 193, 1982.

43. **Evans, D. A., Wetter, L. R., and Gamborg, O. L.**, Somatic hybrid plants of *Nicotiana glauca* and *N. tabacum* obtained by protoplast fusion, *Physiol. Plant.*, 48, 225, 1980.

44. **Flick, C. E. and Evans, D. A.**, Evaluation of cytoplasmic segregation in somatic hybrids in the genus *Nicotiana*: tentoxin sensitivity, *J. Hered.*, 73, 264, 1982.

45. **Fluhr, R., Aviv, D., Galun, E., and Edelman, M.**, Generation of heteroplastidic *Nicotiana* cybrids by protoplast fusion: analysis for plastid recombinant types, *Theor. Appl. Genet.*, 67, 491, 1984.

46. **Galbraith, D. W. and Mauch, T. J.**, Identification of fusion of plant protoplasts. II. Conditions for the reproducible fluorescence labelling of protoplasts derived from mesophyll tissue, *Z. Pflanzenphysiol.*, 98, 129, 1980.

47. **Galun, E.**, Somatic cell fusion for inducing cytoplasmic exchange: a new biological system for cytoplasmic genetics in higher plants, in *Plant Improvement and Somatic Cell Genetics*, Vasil, I. K., Scowcroft, W. B., and Frey, K. L., Eds., Academic Press, New York, 1982, 205.

48. **Galun, E., Arzee-Gonen, P., Fluhr, R., Edelman, M., and Aviv, D.**, Cytoplasmic hybridization in *Nicotiana*: mitochondrial DNA analysis in progenies resulting from fusion between protoplasts having different organelle constitutions, *Mol. Gen. Genet.*, 186, 50, 1981.

49. **Gamborg, O. L., Shyluk, J. P., Fowke, L. C., Wetter, L. R., and Evans, D. A.**, Plant regeneration from protoplasts and cell cultures of *Nicotiana tabacum* sulfer mutant (Su/Su), *Z. Pflanzenphysiol.*, 95, 255, 1979.

50. **Gleba, Yu. Yu.**, Transmission genetics of *Nicotiana* hybrids produced by protoplast fusion. I. Genetic constitution of the intergeneric *N. tabacum* progeny obtained by cloning individual fusion products, *J. Hered.*, in press.

51. **Gleba, Yu. Yu. and Berlin, J.**, Somatic hybridization by protoplast fusion in *Nicotiana*. Fate of nuclear genetic determinants, in *Int. Protoplast Symp.*, Szeged, Hungary, 1979, 73.

52. **Gleba, Yu. Yu., Butenko, R. G., and Sytnik, K. M.**, Fusion of protoplasts and parasexual hybridization in *Nicotiana tabacum*, *Dokl. Akad. Nauk. USSR*, 221, 1196, 1975.

53. **Gleba, Yu. Yu. and Evans, D. A.**, Genetic analysis of somatic hybrid plants, in *Handbook of Plant Cell Culture*, Vol. 1, Evans, D. A., Sharp, W. R., Ammirato, P. V., and Yamada, Y., Eds., Macmillan, New York, 1983, 322.

54. **Gleba, Yu. Yu. and Hoffmann, E.**, Hybrid cell lines *Arabidopsis thaliana* and *Brassica campestris*. No evidence for specific chromosome elimination, *Mol. Gen. Genet.*, 165, 257, 1978.

55. **Gleba, Yu. Yu. and Hoffmann, F.**, "*Arabido brassica*" plant-genome engineering by protoplast fusion, *Naturwissenschaften*, 66, 547, 1979.

56. **Gleba, Yu. Yu., and Hoffmann, F.**, "*Arabido brassica*" a novel plant obtained by protoplast fusion, *Planta*, 149, 112, 1980.

57. **Gleba, Yu. Yu., Kanevsky, L., and Cherep, N. N.**, Transmission genetics of *Nicotiana* hybrids produced by protoplast fusion, *Plant Cell Tissue Organ Culture*, 4, 19, 1985.

58. **Gleba, Yu. Yu., Momot, V. P., Okolot, A. N., and Cherep, N. N.**, Isolation of somatic cell hybrid *Pisum sativum* and *Nicotiana chinensis*, *Z. Pflanzenphysiol.*, in press.

59. **Gleba, Yu. Yu., Momot, V. P., Okolot, A. N., Cherep, N. N., and Skarzhynskaya, M. V.,** Interstitial hybrid cell lines of *Atropa* and *Nicotiana*. I. Genetic constitution of cells of different clonal origin grown in vitro, *Plant Cell Tissue Organ Culture,* in press.

60. **Gleba, Yu. Yu., Piven, N. M., Komarnitsky, I. K., and Sytnik, K. M.,** Cytoplasmic hybrids (cybrids) *Nicotiana tabacum* and *N. debneyi* obtained by protoplast fusion, *Dokl. Akad. Nauk USSR,* 240, 225, 1978.

61. **Gleba, Yu. Yu. and Sytnik, K. M.,** *Protoplast Fusion and Genetic Engineering of Higher Plants,* Naukova Dumka Publishers, Kiev, 1982.

62. **Gleddie, S., Keller, W. A., and Setterfield, G.,** Production and characterization of somatic hybrids between *Solanum melongena* and *S. sisymbirifolium, Theor. Appl. Genet.,* 71, 613, 1986.

63. **Glimelius, K. and Bonnett, H. T.,** Somatic hybridization in *Nicotiana:* restoration of photoautotrophy to an albino mutant with defective plastids, *Planta,* 153, 497, 1981.

64. **Glimelius, K., Chen, K., and Bonnett, H. T.,** Somatic hybridization in *Nicotiana.* Segregation of organellar traits among hybrid and cybrid plants, *Planta,* 153, 504, 1981.

65. **Glimelius, K., Eriksson, T., Grafe, R., and Muller, A.,** Somatic hybridization of nitrate reductase deficient mutants of *Nicotiana tabacum* by protoplast fusion, *Physiol. Plant.,* 44, 273, 1978.

66. **Grimes, D. H. and Boss, W. F.,** Intracellular calcium and calmodulin involvement in protoplast fusion, *Plant Physiol.,* 79, 253, 1985.

67. **Grout, B. W. W., Willison, J. H. M., and Cocking, E. C.,** Interaction at the surface of plant cell protoplasts: an electrophoretic and freeze-etch study, *J. Bioenerg.,* 4, 311, 1972.

68. **Gupta, P. P., Schieder, O., and Gupta, M.,** Intergeneric nuclear gene transfer between somatically and sexually incompatible plants through asymmetric protoplast fusion, *Mol. Gen. Genet.,* 197, 30, 1984.

69. **Hain, R., Steinbiss, H. H., and Schell, J.,** Fusion of *Agrobacterium* and *E. coli* spheroplasts with *Nicotiana tabacum* protoplasts. Direct gene transfer from microorganisms to higher plant, *Plant Cell Rep.,* 3, 60, 1984.

70. **Hamill, J. D., Pental, D., and Cocking, E. C.,** The combination of a nitrate reductase deficient nuclear genome with a streptomycin resistant chloroplast genome in *Nicotiana tabacum* by protoplast fusion, *J. Plant Physiol.,* 115, 253, 1984.

71. **Hampp, R., Steingraber, M., Mehrle, W., and Zimmermann, U.,** Electric-field-induced fusion of evacuolated mesophyll protoplasts of oat, *Naturwissenchaften,* 72, 91, 1985.

72. **Handley, L. W., Nickels, R. L., Cameron, M. W., Moore, P. P., and Sink, K. C.,** Somatic hybrid plants between *Lycopersicon esculentum* and *Solanum lycopersicoides, Theor. Appl. Genet.,* 71, 691, 1986.

73. **Harding, K. and Cocking, E. C.,** The interaction between *E. coli* spheroplasts and plant protoplasts: a proposed procedure to deliver foreign genes into plant cells, *Protoplasma,* 130, 153, 1986.

74. **Harms, C. T., Potrykus, I., and Widholm, J. M.,** Complementation and dominant expression of amino acid analogue resistance markers in somatic hybrid clones from *Daucus carota* after protoplast fusion, *Z. Pflanzenphysiol* 101, 377, 1981.

75. **Hartman, J. X., Kao, K. N., Gamborg, O. L., and Miller, R. A.,** Immunological methods for the agglutination of protoplasts from cell suspension culture of different genera, *Planta,* 132, 45, 1973.

76. **Hasezawa, S., Nagata, T., and Syono, K.,** Transformation of Vinca protoplasts mediated by *Agrobacterium* spheroplasts, *Mol. Gen. Genet.,* 182, 206, 1981.

77. **Hoffmann, F. and Adachi, T.,** "*Arabidobrassica*" chromosome recombination and morphogenesis in asymmetric intergeneric hybrid cells, *Planta,* 153, 586, 1981.

78. **Honda, K., Maeda, Y., Sasakawa, S., Ohno, H., and Tsuchida, E.,** The components contained in polyethylene glycol of commercial grade PEG 6000 as cell fusogen, *Biochem. Biophys. Res. Commun.,* 101, 165, 1981.

79. **Ito, M.,** Studies on the behaviour of meiotic protoplasts. II. Induction of a high fusion frequency in protoplasts from *Liliaceous* plants, *Plant Cell Physiol.,* 14, 865, 1973.

80. **Izhar, S. and Power, J. B.,** Somatic hybridization in *Petunia.* A male sterile cytoplasmic hybrid, *Plant Sci. Lett.,* 14, 49, 1979.

81. **Izhar, S. and Tabib, Y.,** Somatic hybridization in *Petunia.* II. Heteroplasmic state in somatic hybrids followed by cytoplasmic segregation into male sterile and male fertile lines, *Theor. Appl. Genet.,* 57, 241, 1980.

82. **Jinks, J. L., Caligari, P. D. S., and Ingram, N. R.,** Gene transfer in *Nicotiana rustica* using irradiated pollen, *Nature (London),* 291, 586, 1981.

83. **Kameya, T.,** The effect of gelatin on aggregation of protoplasts from higher plants, *Planta,* 115, 77, 1973.

84. **Kameya, T.,** Studies on plant cell fusion. Effect of dextran and pronase E on fusion, *Cytologia,* 44, 449, 1979.

85. **Kameya, T., Horn, M. E., and Widholm, J. M.,** Hybrid shoot formation from fused *Daucus carota* and *D. capillifolius* protoplasts, *Z. Pflanzenphysiol* 104, 459, 1981.

86. **Kanchanapoom, K., Brightman, A. O., Grimes, H. D., and Boss, W. F.,** A novel method for monitoring protoplast fusion, *Protoplasma,* 124, 65, 1985.

87. **Kao, K. N.**, Chromosomal behaviour in somatic hybrids of soybean and *N. glauca*, *Mol. Gen. Genet.*, 150, 225, 1977.

88. **Kao, K. N., Constabel, F., Michayluk, M. R., and Gamborg, O. L.**, Plant protoplast fusion and growth of intergeneric hybrid cells, *Planta*, 120, 215, 1974.

89. **Kao, K. N. and Michayluk, M. R.**, A method for high frequency intergeneric fusion of plant protoplasts, *Planta*, 115, 355, 1974.

90. **Kao, K. N. and Wetter, L. R.**, Advances in techniques of plant protoplast fusion and culture of heterokaryocytes, in *International Cell Biology*, Brinkley, B. R. and Porter, K. R., Eds., Academic Press, New York, 1976, 216.

91. **Keller, W. A. and Melchers, G.**, The effect of high pH and calcium on tobacco leaf protoplast fusion, *Z. Naturforsch.*, 28, 737, 1973.

92. **Kohn, H., Schieder, R., and Schieder, O.**, Somatic hybrids in tobacco mediated by electrofusion, *Plant Sci.*, 38, 121, 1985.

93. **Koop, H-U. and Schweiger, H-G.**, Regeneration of plants after electrofusion of selected pairs of protoplasts, *Eur. J. Cell Biol.*, 39, 46, 1985.

93a. **Kothari, S. L., Monte, D. C., and Widholm, J. M.**, Selection of *Daucus carota* somatic hybrids using drug resistance markers and characterization of their mitochondrial genomes, *Theor. Appl. Genet.*, 72, 494, 1986.

94. **Krumbiegel, G. and Schieder, O.**, Selection of somatic hybrids after fusion of protoplasts from *Datura innoxia* and *Atropa belladonna*, *Planta*, 145, 371, 1979.

95. **Kumar, A., Cocking, E. C., Bovenberg, W. A., and Kool, A. J.**, Restriction endonuclease analysis of chloroplast DNA in interspecies somatic hybrids of *Petunia*, *Theor. Appl. Genet.*, 62, 377, 1982.

96. **Kung, S. P., Gray, J. C., Wildman, S. G., and Carlson, P. S.**, Polypeptide composition of fraction I protein from parasexual hybrid plants in the genus *Nicotiana*, *Science*, 187, 353, 1975.

97. **Lonnendonker, N. and Schieder, O.**, Amylase isoenzymes of the genus *Datura* as a simple method for an early identification of somatic hybrids, *Plant Sci. Lett.*, 17, 135, 1980.

98. **Lörz, H., Paszhowski, J., Dierks-Ventling, C., and Potrykus, I.**, Isolation and characterization of cytoplasts and miniprotoplasts derived from protoplasts of cultured cells, *Physiol. Plant.*, 53, 385, 1981.

99. **Maliga, P., Kiss, Z. R., Nagy, A. H., and Lazar, G.**, Genetic instability in somatic hybrids of *Nicotiana tabacum* and *N. knightiana*, *Mol. Gen. Genet.*, 163, 145, 1978.

100. **Maliga, P., Lazar, G., Joo, F., Nagy, A. H., and Menczel, L.**, Restoration of morphogenic potential in *Nicotiana* by somatic hybridization, *Mol. Gen. Genet.*, 157, 291, 1977.

101. **Maliga, P., Lörz, H., Lazar, G., and Nagy, F.**, Cytoplast-protoplast fusion for interspecific chloroplast transfer in *Nicotiana*, *Mol. Gen. Genet.*, 185, 211, 1982.

102. **Marton, L., Biasini, G., and Maliga, P.**, Co-segregation of nitrate reductase activity and normal regeneration ability in selfed sibs of *Nicotiana plumbaginifolia* somatic hybrids, heterozygotes for nitrate reductase deficiency, *Theor. Appl. Genet.*, 70, 340, 1985.

102a. **Matthews, B. F., Widholm, J. M.**, Organelle DNA compositions and isozyme expression in an interspecific somatic hybrid of *Daucus*, *Mol. Gen. Genet.*, 198, 371, 1985.

103. **Medgyesy, P., Golling, R., and Nagy, F.**, A light sensitive recipient for the effective transfer of chloroplast and mitochondrial traits by protoplast fusion in *Nicotiana*, *Theor. Appl. Genet.*, 70, 590, 1985.

104. **Medgyesy, P., Menczel, L., and Maliga, P.**, The use of cytoplasmic streptomycin resistance chloroplast transfer from *Nicotiana tabacum* into *Nicotiana sylvestris* and isolation of their somatic hybrids, *Mol. Gen. Genet.*, 179, 693, 1980.

105. **Melchers, G.**, Kombination Somatischen und Konventionellergenetik fur die Pflanzenzuchtung, *Naturwissenschaften*, 64, 184, 1977.

106. **Melchers, G. and Labib, G.**, Somatic hybridization of plants by fusion of protoplasts. I. Selection of light resistant hybrids of haploid light sensitive varieties of tobacco, *Mol. Gen. Genet.*, 135, 277, 1974.

107. **Melchers, G. and Sacristan, M. D.**, Somatic hybridization of plants by fusion of protoplasts. II. The chromosome numbers of somatic hybrid plants of four different fusion experiments, in *La Culture des Tissus et des Cellules des Vegetaux*, Gautheret, G., Ed., Masson, Paris, 1977, 169.

108. **Melchers, G., Sacristan, M. D., and Holder, A. A.**, Somatic hybrid plants of potato and tomato regenerated from fused protoplasts, *Carlsberg Res. Commun.*, 43, 203, 1978.

109. **Menczel, L., Galiba, G., Nagy, F., and Maliga, P.**, Effect of radiation dose on efficiency of chloroplast transfer by protoplasts fusion in *Nicotiana*, *Genetics*, 100, 487, 1982.

110. **Menczel, L., Lazar, G., and Maliga, P.**, Isolation of somatic hybrids by cloning *Nicotiana* heterokaryons in nurse culture, *Planta*, 143, 29, 1978.

111. **Menczel, L., Nagy, F., Kiss, Z. R., and Maliga, P.**, Streptomycin resistant and sensitive somatic hybrids of *Nicotiana tabacum* and *N. knightiana*: correlation of resistance of *N. tabacum* plastids, *Theor. Appl. Genet.*, 59, 191, 1981.

112. **Menczel, L. and Wolf, K.**, High frequency of fusion induced in freely suspended protoplast mixtures by polyethylene glycol and dimethylsulfoxide at high pH, *Plant Cell Rep.*, 3, 196, 1984.

113. **Motoyoshi, F.**, Protoplasts isolated from callus cells of maize endosperm, *Exp. Cell Res.*, 68, 452, 1971.
114. **Nagao, T.**, Somatic hybridization by fusion of protoplasts. I. The combination of *N. tabacum* and *N. rustica, Jpn. J. Crop Sci.*, 47, 491, 1978.
115. **Nagao, T.**, Somatic hybridization by fusion of protoplasts. II. The combinations of *Nicotiana tabacum* and *N. glutinosa* and of *N. tabacum* and *N. alata, Jpn. J. Crop Sci.*, 48, 385, 1979.
116. **Nagata, T.**, A novel cell fusion method of protoplasts by polyvinyl alcohol, *Naturwissenschaften*, 65, 263,1978.
117. **Nagata, T. and Melchers, G.**, Surface charge of protoplasts and their significance in cell cell interactions, *Planta*, 142, 235, 1978.
118. **Nagata, T. and Takebe, I.**, Plating of isolated tobacco mesophyll protoplasts on agar medium, *Planta*, 99, 12, 1971.
119. **Nagy, F., Torok, I., and Maliga, P.**, Extensive rearrangements in the mitochondrial DNA in somatic hybrids of *Nicotiana tabacum* and *N. knightiana, Mol. Gen. Genet.*, 183, 437, 1981.
120. **Negrutiu, I., DeBrouwer, D., Watts, J. W., Sidorov, V. I., Dirks, R., and Jacobs, M.**, Fusion of plant protoplasts: a study using auxotrophic mutants of *Nicotiana plumbaginifolia, Theor. Appl. Genet.*, 72, 279, 1986.
121. **O'Connell, M. A. and Hanson, M. R.**, Regeneration of somatic hybrid plants formed between *Lycopersicon esculentum* and *Solanum rickii, Theor. Appl. Genet.*, 72, 59, 1986.
122. **Ohgawara, T., Kobayashi, S., Ohgawara, E., Ichimiya, H., and Ishii, S.**, Somatic hybrid plants obtained by protoplast fusion between *Citrus sinensis* and *Poncirus trifoliata, Theor. Appl. Genet.*, 71, 1, 1985.
122a. **Ozias-Akins, P., Ferl, R. J., and Vasil, I. K.**, Somatic hybridization in the Graminae: *Pennisetum americanum* (pearl millet) and *Panicum maximum* (Guinea grass), *Mol. Gen. Genet.*, 203, 365, 1986.
123. **Pandey, K. K.**, Sexual transfer of specific genes without gametic fusion, *Nature (London)*, 256, 310, 1975.
124. **Patnaik, G., Cocking, E. C., Hamill, J., and Pental, D.**, A simple procedure for the manual isolation and identification of plant heterokaryons, *Plant Sci. Lett.*, 24, 105, 1982.
125. **Pental, D., Cooper-Bland, S., Harding, K., Cocking, E. C., and Muller, A. J.**, Cultural studies on nitrate reductase deficient *Nicotiana tabacum* mutant protoplasts, *Z. Pflanzenphysiol* 105, 219, 1982.
126. **Pental, D., Hamill, J. D., and Cocking, E. C.**, Somatic hybridization using a double mutant of *Nicotiana tabacum, Heredity*, 53, 79, 1984.
127. **Pental, D., Hamill, J. D., Pirrie, A., and Cocking, E. C.**, Somatic hybridization of *Nicotiana tabacum* and *Petunia hybrida, Mol. Gen. Genet.*, 202, 342, 1986.
128. **Pilet, P. E. and Stenn, A.**, Effect du Ca^{++} et du K$^+$ sur la mobilite electrophoretique des protoplastes, *C.R. Acad. Sci.*, 278, 269, 1974.
129. **Phol, H. A.**, *Dielectrophoresis*, Cambridge University Press, Cambridge, 1978.
130. **Pontecorvo, G.**, "Alternatives to sex": genetics by means of somatic cells, in *Modification of the Information Content of Plant Cells*, North-Holland/American Elsevier, Amsterdam, 1975, 1.
131. **Potrykus, I.**, Isolation, fusion and culture of protoplasts of *Petunia*, in *Yeast, Mould and Plant Protoplasts*, Villanveva, J. R., Garcia-Acha, I., Gascon, S., and Urubuni, F., Eds., Academic Press, London, 1973, 319.
132. **Potrykus, I., Jia, J., Lazar, G. B. and Saul, M.**, *Hyoscyamus muticus* and *Nicotiana tabacum* fusion hybrids selected via auxotroph complementation, *Plant Cell Rep.*, 3, 68, 1984.
133. **Power, J. B., Hayward, C. and Cocking, E. C.**, Some consequences of the fusion and selective culture of *Petunia* and *Parthenocissus* protoplasts, *Plant Sci. Lett.*, 5, 97, 1975.
134. **Power, J. B., Berry, S. F., Chapman, J. V., Cocking, E. C., and Sink, K. C.**, Somatic hybridization of sexually incompatible petunias: *Petunia parodii* and *Petunia parviflora, Theor. Appl. Genet.*, 57, 1, 1980.
135. **Power, J. B. and Cocking, E. C.**, Fusion of plant protoplasts, *Sci. Prog.*, 59, 181, 1971.
136. **Power, J. B., Cummins, B. E., and Cocking, E. C.**, Fusion of isolated plant protoplasts, *Nature (London)*, 225, 1016, 1970.
137. **Puite, K. J., Van Wikselaar, P., and Verhoeven, H.**, Electrofusion, a simple and reproducible technique in somatic hybridization of *Nicotiana plumbaginifolia* mutants, *Plant Cell Rep.*, 4, 274, 1985.
138. **Rothenberg, M., Boeshore, M. L., Hanson, M. R., and Izhar, S.**, Intergenomic recombinations of mitochondrial genomes in a somatic hybrid plant, *Curr. Genet.*, 9, 615, 1985.
138a. **Saul, M. W. and Potrykus, I.**, Species-specific repetitive DNA used to identify interspecific somatic hybrids, *Plant Cell Rep.*, 3, 65, 1984.
139. **Schaffner, W.**, Direct transfer of cloned genes from bacteria to mammlian cells, *Proc. Natl. Acad. Sci. U.S.A.*, 77, 2163, 1980.
140. **Schenk, R. U. and Hildebrandt, A. C.**, Production, manipulation and fusion of plant cell protoplasts as steps towards somatic hybridization, in *Les Cultures de Tissus de Palantes (Colloq. Int. C.N.R.S.)*, No. 193, 1970, 319.

141. **Schieder, O.**, Hybridization experiments with protoplasts from chlorophyll-deficient mutants of some solanceous species, *Planta*, 137, 253, 1977.

142. **Schieder, O.**, Somatic hybrids of *Datura innoxia* and *Datura discolour* or of *D. innoxia* and *D. stramonium*. I. Selection and characterization, *Mol. Gen. Genet.*, 162, 113, 1978.

143. **Schieder, O.**, Somatic hybrids between a herbaceous and two tree *Datura* species, *Z. Pflanzenphysiol.*, 98, 119, 1980.

143a. **Schieder, O.**, Somatic hybrids of *Datura innoxia* + *D. discolor* and *D. innoxia* + *D. stramonium*. II. Analysis of progenies of three sexual generation, *Mol. Gen. Genet.*, 139, 1, 1980.

144. **Schnabelrauch, L. S., Kloc-Bauchan, F., and Sink, K. C.**, Expression of nuclear-cytoplasmic genomic incompatibility in interspecific *Petunia* somatic hybrid plants, *Theor. Appl. Genet.*, 70, 57, 1985.

145. **Scowcroft, W. R. and Larkin, P. J.**, Chloroplast-DNA assorts randomly in intraspecific somatic hybrids of *Nicotiana debneyi*, *Theor. Appl. Genet.*, 60, 179, 1981.

146. **Senda, M., Morikawa, H., Katagi, H., Takada, T., and Yamada, Y.**, Effect of temperature on membrane fluidity and protoplast fusion, *Theor. Appl. Genet.*, 57, 33, 1980.

147. **Senda, M., Takeda, J., Abe, S., and Nakamura, T.**, Induction of cell fusion of plant protoplasts by electrical stimulation, *Plant Cell Physiol.*, 20, 1441, 1979.

148. **Sidorov, V. A., Menczel, L., Nagy, F., and Maliga, P.**, Chloroplast transfer in *Nicotiana* based on metabolic complementation between irradiated and iodoacetate-treated protoplasts, *Planta*, 152, 341, 1981.

149. **Skarzhynskaya, M. V., Cherep, N. N., and Gleba, Yu. Yu.**, Potato and tobacco hybrid cell lines and plants obtained by cloning individual protoplast fusion products, *Sov. Cytol. Genet.*, 6, 42, 1982.

150. **Smith, H. M., Kao, K. N., and Combatti, N. C.**, Interspecific hybridization by protoplast fusion in *Nicotiana*, *J. Hered.*, 67, 123 1976.

150a. **Somers, D. A., Narayanan, K. R., Kleinhofs, A., Cooper-Bland, S., and Cocking, E. C.**, Immuno-logical evidence for transfer of the barley nitrate reductase structural gene to *Nicotiana tabacum* by protoplast fusion, *Mol. Gen. Genet.*, 204, 296, 1986.

151. **Szabados, L., Hadlaczky, G. Y., and Dudits, D.**, Uptake of isolated plant chromosomes by plant protoplasts, *Planta*, 151, 141, 1981.

152. **Tempelaar, M. J. and Jones, M. G. K.**, Fusion characteristics of plant protoplasts in electric fields, *Planta*, 165, 205, 1986.

153. **Uchimiya, H.**, Somatic hybridization between male sterile *Nicotiana tabacum* and *N. glutinosa* through protoplast fusion, *Theor. Appl. Genet.*, 61, 69, 1982.

154. **Uchimiya, H., Kobayashi, S., Ono, M., Brar, D. S., and Harada, H.**, Characterization of nuclear and cytoplasmic information in the progeny of a somatic hybrid between male sterile *Nicotiana tabacum* and *N. glutinosa*, *Theor. Appl. Genet.*, 68, 95, 1984.

155. **Uchimiya, H., Kudo, N., Ohgawara, T., and Harada, H.**, Aggregation of plant protoplasts by artificial lipid vesicles, *Plant Physiol.*, 69, 1278, 1982.

156. **Vienken, J., Ganser, R., Hampp, R., and Zimmermann, U.**, Electric field-induced fusion of isolated vacuoles and protoplasts of different developmental and metabolic provenience, *Physiol. Plant.*, 53, 64, 1981.

157. **Wallin, A., Glimelius, K., and Eriksson, T.**, The induction of aggregation and fusion of *Daucus carota* protoplasts by polyethylene glycol, *Z. Pflanzenphysiol* 74, 64, 1974.

158. **Watts, J. W., Doonan, J. H., Cove, D. J., and King, J. M.**, Production of somatic hybrids of moss by electrofusion, *Mol. Gen. Genet.*, 199, 349, 1985.

159. **Watts, J. W. and King, J. M.**, A simple method for large scale electrofusion and culture of plant protoplasts, *BioSci. Rep.*, 4, 335, 1984.

160. **Wetter, L. R. and Kao, K. N.**, The use of isoenzymes in distinguishing the sexual and somatic hybrids in callus culture derived from *Nicotiana*, *Z. Pflanzenphysiol.*, 80, 455, 1976.

161. **Wetter, L. R. and Kao, K. N.**, Chromosome and isoenzyme studies on cells derived from protoplast fusion of *Nicotiana glauca* with *Glycine max-Nicotiana glauca* cell hybrids, *Theor. Appl. Genet.*, 57, 273, 1980.

162. **Withers, L.**, Plant protoplast fusion: methods and mechanism in *Les Cultures de Tissus de Plantes (Colloq. Int. C.N.R.S.)*, No. 212, 1973, 212.

163. **Withers, L. and Cocking, E. C.**, Fine structural studies on spontaneous and induced fusion of higher plant protoplasts, *J. Cell Sci.*, 11, 59, 1972.

164. **Wullems, G. J., Molendijk, L., and Schilperoort, R. A.**, The expression of tumour markers in intraspecific somatic hybrids of normal and crown gall cells from *Nicotiana tabacum*, *Theor. Appl. Genet.*, 56, 203, 1980.

165. **Zachrisson, A. and Bornman, C. H.**, Application of electric field fusion in plant tissue culture, *Physiol. Plant.*, 61, 314, 1984.

166. **Zelcher, A., Aviv, D., and Galun, E.**, Interspecific transfer of cytoplasmic male sterility by fusion between protoplasts of normal *Nicotiana sylvestris* and X-ray irradiated protoplasts of male sterile *N. tabacum*, *Z. Pflanzenphysiol.*, 90, 397, 1978.

167. **Zimmerman, U. and Benz, R.,** Dependence of the electrical breakdown voltage on the charging time in *Valonia utricularis, J. Membr. Biol.,* 53, 33, 1980.
168. **Zimmermann, U. and Pilwat, G.,** *6th Int. Biophys. Congr.,* Abstr. IV-19-H, Kyoto, Japan,1978, 140.
169. **Zimmermann, U. and Scheurich,P.,** High frequency fusion of plant protoplasts by electric fusion, *Theor. Appl. Genet.,* 151, 26, 1981.

Chapter 4

CELL TRANSFORMATION

I. INTRODUCTION

At the end of this century about 8 billion people are expected to inhabit this planet. In order to meet their dietary requirements, world food production should double, if not triple, during the next decades. World food production has in fact, nearly doubled during the last decade. This was possible due to conventional agriculture that fueled the green revolution and produced high yielding crop varieties. Despite this progress, during the coming decades it will be more and more difficult to achieve increased yield. A possible recourse is non-conventional means. In this context, molecular biology complements conventional agriculture.

Molecular biology can be briefly described as the identification and isolation of useful genes, cloning of these genes, synthesis of new genes, and transfer of these new gene constructs to cells. With these advances, the desired genetic modifications are fast becoming a reality. Much enthusiasm has been generated by the fact that various eukaryotic genes can be expressed as biologically functional products in bacteria and various foreign prokaryotic and eukaryotic genes can also be transcribed and translated into the environment of animal cells. Also, plant cells, analogous to microbes, are seen as a system to translate the conceptual advances of molecular biology to plant biology and, ultimately, to human welfare. The stage is set for genetic engineering of plant cells. This has been possible due to (1) technological advances in molecular biology, (2) advances in the technology of cloning and multiplication of plant cells, and (3) differentiation of plants from cells. Once detected, a single transformed cell can be multiplied and many transformed plants are possible on regeneration of these cells derived from a single cell.

Cell transformation includes any change brought about by the introduction of foreign DNA into the cell. Cell transformation is commonly described as "genetic engineering". Genetic engineering of cells is possible by molecular means which include methods involving the use of DNA without any change, purified or protected, or recombinant DNA. Cellular means include methods that deal with transfer of nuclei or individual chromosomes, organelles, or even cell fusion. This chapter is primarily concerned with the molecular aspects of cell transformation. A brief account of cell transformation by isolated nuclei is also included.

Transformation of a cell leading to transformed plants is a complex process. For a systematic evaluation,[97] cell transformation has been divided into five biochemical steps: (1) uptake, (2) integration, (3) replication, (4) expression, and (5) transmission. Of these, all except integration are essential for obtaining a transformed cell and, in turn, a plant. Integration of the introduced genetic information is desirable (it is expected to stabilize and facilitate its replication) but not essential because plant cells possess functional but nonintegrated DNA in the form of chloroplast and mitochondrial DNA.

II. HISTORY

The fascinating field of genetic engineering of plant cells had an unusual beginning. The initial claims of genetic transformation of plants or cells are not accepted by the scientific community and are only of historical importance. However, a brief summary of the history is in order because it is instructive. For the sake of convenience, the reported transformations are reviewed on the basis of systems employed, i.e., attempted transformation of seeds, seedlings, pollen, and cells.

A. Uptake of DNA by Seeds and Seedlings

Experiments on transformation of entire plants were attempted by workers led by Ledoux in Belgium and Hess in Germany. The Belgian group claimed the uptake, replication, and integration of bacterial DNA emanating from *Micrococcus lysodeikticus* into the nuclei of *Hordeum* plants.[107] Disinfected barley seeds were germinated in the presence of labeled bacterial DNA. DNA was isolated from seedlings and subjected to cesium chloride density gradient centrifugation. The occurrence of a radioactive DNA band at the density of donor DNA was taken as evidence of uptake of DNA. The appearance of a radioactive DNA band, approximately intermediate in density[107] between the recipient and donor DNA, was taken as evidence of the integration of radioactive donor DNA with the nonradioactive host DNA. The intermediate density DNA band was further analyzed by sonication and denaturation. Shearing by sonication resulted in separation of two peaks corresponding to donor and recipient (host) DNA. Shearing shifts the radioactivity of the intermediate density band back to donor DNA density, whereas denaturation shifts the intermediate density DNA band to higher density, as would be expected for the shift from double-stranded to single-stranded DNA. However, there was no change in density on dentauration of intermediate density DNA. This indicated that donor and host DNA could not be released by denaturation and were not held as complementary partners of double-stranded DNA.

Similar experiments were used to indicate the integration of *Escherichia coli* and *Streptomyces coelicolor* DNA in *Arabidopsis thaliana*,[108] phage T4 DNA in *Matthiola incana*.[60,149] In *Arabidopsis* a claim was made to correct nutritional deficiencies. By the application of *Bacillus subtilis* DNA, thi⁻ and tryp⁻ mutants of *Arabidopsis*[109] were reported to change to tryp⁺ and thi⁺. The change was described to be permanent and heritable. These results were sensational enough to prompt others to initiate similar work.

Attempts by other workers to reproduce these results were unsuccessful. In experiments on *Hordeum*[79] as well as *Arabidopsis*[116] an intermediate density DNA band was not observed. Further, attempts to reproduce these results in Ledoux's laboratory,[96] using the same strain of barley seeds, were unsuccessful. An intermediate density DNA band was never observed, except when the seeds were contaminated. Also, in genetic analysis of corrected mutants of *Arabidopsis* obtained from Ledoux's laboratory, no evidence was found for DNA-mediated correction.[150] In an independent study of integration of *Pseudomonas* DNA by pea seedlings it was concluded that partially degraded bacterial DNA can be taken up by a plant, but it is not found in the cell nuclei.[6]

A report was also made about the transfer of the gene for anthocyanin by the supply of exogenous DNA from red-flowered to white-flowered mutant plants of *Petunia*.[71] The seedlings of white-flowered mutants were treated for 48 hr with DNA from red-flowered plants. A very high frequency (27%) of transformation was reported. However, when white-flowered petunia seedlings were treated with their own DNA 5 to 9% showed slight anthocyanin. According to the author, this was due to environmental factors, but it still cast doubt about the transformation. These results are discounted[8] mainly because color change is also possible due to nutritional conditions and viral infection.

In brief, it can be said that it is possible to demonstrate the uptake of exogenous DNA, but its integration and replication remain to be proved. The fate of exogenous DNA incubated with plant cells seems to be degradation rather than integration and expression. It is doubtful whether there is significant uptake of exogenous DNA. Meiotic cells of lily did not incorporate appreciable amounts of exogenous DNA, either homologous or heterologous.[79] Also, when protoplasts were isolated from DNA-treated cells of tobacco a considerable amount of DNA was found adsorbed to the cell wall and was eliminated during isolation of protoplasts.[74]

Despite these contradictions, sporadic reports of transformation continue to appear in the literature. For example, a recent study reports the formation of transformed plants[166] on injection of barley donor DNA into grains of barley recipient plants.

B. Uptake of DNA by Pollen

It has also been reported that DNA uptake is possible by pollen, and pollination by this pollen results in transformation. Initially, uptake of bacterial DNA from *Rhizobium leguminosorum* was demonstrated by autoradiography and cesium chloride density gradient centrifugation.[72] Also, it was reported that DNA uptake by pollen can result in transformation. When pollen of petunia treated with phage DNA ($\lambda \rho gal$ or $\lambda \rho lac$) was used for pollination, the seeds thus produced formed seedlings which could grow on galactose or lactose medium.[73]

In yet another experiment, transformation of pollen[73] was reported. Plants of *Nicotiana glauca* and *N. langsdorffii* do not form tumors, but hybrids of these plants do. When pollen of *N. glauca* was fed with DNA from *N. langsdorffii* and used for pollination of *N. glauca* plants, the resulting seedlings formed tumors. However, in view of earlier reports, this claim needs to be confirmed. Instead, in a recent study on pollen-mediated plant transformation using genomic donor DNA it was not possible to show evidence of transformation of plants in *Zea mays* and *Lycopersicon esculentum*.[156] The authors used well-defined multiple markers to screen for transformants.

C. Uptake of DNA by Cells

Instead of using whole bacterial DNA, an attempt was made to transform plant cells with phage DNA. Use of bacteriophages is distinctly advantageous because (1) it is possible to have increased concentration of a specific gene, (2) DNA is protected by the protein coat, and (3) plant cells are susceptible to viral infection. When haploid cells of *L. esculentum* and *A. thaliana* were exposed to phages carrying gal$^+$ and lac$^+$ genes, cells inoculated with $\lambda \rho$ gal$^+$ or $\phi 80 \rho$ lac$^+$ were able to grow on galactose or lactose medium, respectively. The transformation of gene Z coding for β-galactosidase was inferred from the observation that the activity of the enzyme β-galactosidase was 43 times higher than the untreated control cells.[38] However, the ability to grow on lactose or galactose medium was lost progressively on subculture.

Further, it was reported that transfer of mutant suppressor gene supF$^+$ is possible from *Escherichia coli* to plant cells using $\phi 80$ as a vehicle. When introduced in defective bacteria, this gene restores protein synthesis and the reverse happens when it is introduced in normal cells, resulting in cessation of growth. When cells were inoculated with phage DNA carrying $\phi 80$supF$^+$ they stopped growing and died, indicating that expression of bacterial genes is possible in plant cells.

To account for the these phenotypic changes of plant cells brought about by transducing phages, a new term, "transgenosis", was proposed.[37] It was also clarified that transgenosis should help define transfer, expression, and inheritance of genetic information between donor and recipient, widely separated in evolution.

The above experiments on transgenosis were repeated with suspension culture of *Acer pseudoplatanus*.[88] The cells fed with DNA from $\lambda \rho$ lac$^+$ were described as developing the ability to grow on lactose medium. However, later it was pointed out that the observed growth on lactose medium was due mainly to an increase in cell volume.[61,165] This was accompanied by an increase in β-galactosidase activity. However, on subculture to lactose medium these cells failed to divide and there was a decline in β-galactosidase activity.

These experiments on transgenosis aroused considerable excitement in scientific circles. Using haploid cells of *Datura innoxia* and phage λ as carrier of the gene lac$^+$, it was found that approximately 60% of the cultures inoculated with $\lambda \rho$ *lac* alone or in combination with $\lambda \rho$ gal$^+$ showed an average increase in level of β-galactosidase, but the cells failed to grow on lactose medium.[120] These results (an increased level of β-galactosidase, but inability of cells to grow on lactose medium) were incompatible. The mystery was resolved when β-galactosidase was not found to be *E. coli* specific since it failed to hydrolyze lactose in vitro. These results therefore raised doubts about the transfer of genes from bacteria to plant cells via transducing phages.

III. TRANSFORMATION BY FREE DNA

The unsuccessful attempts to transform plant cells led to the use of protoplasts as an alternate system for transformation. Due to the absence of a cell wall, a protoplast is considered an ideal system for cell transformation. Also, protoplasts, being free cells, can be easily cloned without contamination. Protoplasts have been shown to take up viral RNA from tobacco mosaic virus (TMV) and cowpea chlorotic mottle virus (CCMV). Multiplication of these viruses in protoplasts has been demonstrated.[4,128]

A. Uptake of DNA by Protoplasts

Protoplasts from a number of plants have been shown to take up free DNA, homologous or heterologous, by different groups of workers.[135] About 0.06 to 2.8% of the exogenous *E. coli* DNA was shown to be taken up by protoplasts of *Ammi visnaga*, *Glycine max*, and *Daucus carota*.[136,137] This was after DNAse treatment to remove unbound DNA. Of this, about 20% appeared to be acid soluble. Factors affecting uptake are time of incubation, DNA concentration, temperature of incubation, and protoplast density. However, more significant are the polycations DAE dextran, poly-L-lysine, and poly-L-ornithine, which markedly enhanced DNA uptake at 5 μg/mℓ. Uptake of single-stranded DNA from bacteriophage *fd* by tobacco protoplasts was greatly enhanced by Zn ions.[170] Also, in tobacco protoplasts, uptake of native DNA was facilitated by Cu^+ ions.[176] However, it is not clear how DNA is taken up. It is postulated to occur through pinocytosis.[136,176]

Uptake of exogenous DNA is also possible by isolated nuclei. Nuclei from protoplasts of *G. max* showed binding with homologous exogenous DNA within 20 min of incubation.[138] However, compared to double-stranded DNA from *Salmonella*, soybean nuclei had greater affinity for single-stranded DNA from phage *fd*.

The fate of foreign DNA is degradation by protoplasts. A reextraction of *E. coli* DNA from plant protoplasts and analysis by CsCl density gradient centrifugation resulted in a broadening of band, indicating its degradation. Also, degradation was recorded in respect to single-stranded *fd* DNA. However, plasmid DNA from *E. coli* is taken up by *Vigna sinensis* protoplasts, and 3% of it is closely and irreversibly associated with protoplast nuclei and it loses its circular configuration.[117]

B. Transformation of Protoplasts

Starting with the negative reports of no transformation, reported transformations of protoplasts by free DNA have appeared in the literature. Tobacco protoplasts from a cultivar susceptible to TMV could not be transformed into a resistant type on feeding DNA from a resistant cultivar.[176] Although donor DNA was taken up, plants regenerated from protoplasts were not resistant to TMV. Also, binding of *E. coli* plasmid DNA, carrying resistance to kanamycin, was demonstrated for tobacco protoplasts,[142] but the tissue regenerated did not show resistance to kanamycin.

However, contrary to these reports, transformation of petunia protoplasts has been possible with tumor-inducing (Ti) plasmid DNA from *Agrobacterium tumefaciens*, the causal organism of crown gall (tumor). Transformation[25] was inferred from growth of protoplasts on a hormone-free medium, synthesis of octopine, and the presence of lysopine dehydrogenase activity. Preliminary DNA hybridization data indicated the presence of tumor-DNA (T-DNA) in cells that were octopine synthesizing. This report was soon followed by transformation of tobacco[103] protoplasts by DNA from octopine-type plasmid of *A. tumefaciens*. The characteristics of transformants were hormone autonomy and opine synthesis.

Unequivocal evidence for the transformation[39] of petunia protoplasts was provided by the presence of DNA sequences homologous to Ti-plasmid in phytohormone-independent callus colonies resulting from transformed protoplasts and maintained on a hormone-free medium for 9 months. These results, however, indicated that only a small part of the plasmid, approximately 2 Mdaltons, was maintained in the callus. In contrast, analysis of calli resulting from *N. tabacum* leaf protoplasts transformed[103] with DNA from octopine-type plasmid had a larger part of the plasmid integrated into the host genome than on infection with whole bacterium. Further work is required to clarify this anomaly.

The frequency of transformation[185] is very low, 0.01% of the original number of protoplasts. One of the reasons for this could be the inefficient method of introducing DNA which is subjected to degradation by nucleases. High pH, $ZnSO_4$, and kinetin are described to protect the exogenously supplied DNA, but these treatments decrease wall formation by protoplasts.[43,83] Polycations, particularly poly-L-ornithine, appear to be of value in protecting DNA from DNAse and it is not harmful to protoplasts. However, much of the added DNA is degraded.[83]

Recently, it has been shown that as compared to unsynchronized protoplasts, synchronized protoplasts showed a clear increase in transformation[126] provided that the transformation is performed at S or M phase. After completion of M phase, transformation efficiencies dropped to the level of unsynchronized cells.

IV. TRANSFORMATION BY LIPOSOME-COATED DNA

As an alternative to naked DNA, liposomes (phospholipid vesicles) were used to introduce foreign DNA into plant cells. The possible advantages of liposome-mediated delivery of DNA are low toxicity and protection of phospholipid-coated DNA from degradation by nucleases present in the culture medium and cell. Therefore, it is considered to be more efficient than direct introduction of DNA.

Liposomes have been successfully used to transfer genes from human to mouse cells. Metaphase X-chromosomes from hypoxanthine guanine phosphoribosyl transferase-positive cells were encapsulated forming "lipochromosomes" and were fused with HGPRTase-negative cells.[129] Transformed cells were selected on survival on hypoxanthine aminopterin thymidine (HAT) medium. Frequency of transformation was tenfold higher than uncapsulated chromosomes. The efficiency of liposome-mediated delivery of DNA can also be seen in the transfer of *E. coli* plasmid pBR322 to competent *E. coli* cells even in the presence of high amounts of DNAse.[48] Transformation occurred only when liposome-coated plasmids were used and could be seen by formation of colonies on tetracycline medium.

A. Preparation of Liposomes
Liposomes prepared by several different techniques and with different composition of lipids have been introduced into plant protoplasts. The contents of liposomes are delivered to protoplasts by fusion and endocytosis. Liposomes fuse with the plasma membrane of a protoplast[119] and deliver their contents to the cytoplasm and nucleus. Endocytosis also plays an important role. In the presence of cytochalasin B,[123] an inhibitor of endocytosis in mammalian cells, the delivery of liposome contents to carrot protoplasts was reduced by 35 to 60%. Delivery of liposome contents can be monitored by fluorescent dyes such as fluorescein diacetate[16,175] ethidium bromide and 4,6-diamidino-2-phylindole (DAPI)[115] and acridine orange DNA complex.[118]

For the preparation of liposomes, lipids used are phosphatidylcholine, phosphatidylserine, stearylamine, β-sitosterol, and cholesterol.

Entrapment of DNA into liposomes is dependent upon the encapsulation technique. Of the various methods used for the formation of liposomes which differ in ease of formation, efficiency of encapsulation, size, volume, homogeneity, and phospholipid composition, the net result is the formation of either multilamellar vesicles (MLV) or unilamellar vesicles (ULV). MLVs are possible by rotary evaporation of lipids in an organic phase until the round bottom flask is evenly coated with a thin film of lipid with no trace of organic solvent. The material to be entrapped is then added in aqueous solution and the flask is shaken by hand or sonicated. The suspension is allowed to incubate at room temperature for 2 hr, resulting in MLVs. The heterogeneity of liposome size, multilamellar structure, and low encapsulation efficiency limit the utility of this method. However, an advantage of this method is that it avoids the contact of material with an organic phase and is suitable for encapsulation of large particulate matter such as chloroplasts, mitochondria, and chromosomes.

ULVs or oligolamellar vesicles are possible by a reverse phase evaporation (REV) technique[171] in which the aqueous material to be encapsulated is added to the organic phase containing phospholipids. The mixture is sonicated or shaken by hand to form an emulsion which is then rotary evaporated under nitrogen until liposomes are formed with no trace of organic solvent. This method is suitable for encapsulation of molecular matter, DNA, and chromosomes only and destroys organelles.[122]

Liposomes of unencapsulated DNA and RNA can be purified[48] by treatment with nucleases or, alternatively, they can be purified by high-speed centrifugation which pellets the liposomes and the rest remains in the supernatant.

B. Interaction of Liposomes and Protoplasts

Liposome-protoplast interaction is determined by (1) charge of liposomes, (2) condition of encapsulation, and (3) condition of liposome protoplast incubation. The overall charge of the liposome membrane can be made positive, negative, or neutral by changing the phospholoipid composition. It is generally believed that liposomes containing cholesterol are more stable and retain the encapsulated material longer. Stearylamine is used with neutral lipids, cholesterol and phosphatidylcholine to form positively charged vesicles, whereas dicetyl phosphate or phosphatidylserine helps form negatively charged vesicles.

Conditions of liposome-protoplast incubation affecting the delivery of liposomes into the protoplast include (1) chemical composition and pH of the incubation medium, (2) proportion of liposome to protoplasts, (3) temperature of the incubation medium, and (4) fusing agents. Choosing TMV RNA for encapsulation in liposomes, it was found that when petunia protoplasts and liposome-coated preparations were incubated in buffer for 30 min, there was no detection of virus production after 48 hr.[49] However, the polymers polyethylene glycol (PEG) and polyvinyl alcohol (PVA) stimulated delivery. Also, high levels of $CaCl_2$ stimulated delivery and viral production. This was accompanied by increased liposome binding to protoplasts. Maximal interaction occurred at neutral pH. The optimal incubation periods were 5 min for preincubation of protoplasts and liposomes and 20 to 30 min for exposure of protoplasts and liposomes to PEG solution.

C. Biological Activity of Liposome-Coated DNA/RNA

It was possible to infect 50% of the tobacco protoplasts by encapsulated TMV RNA in large unilamellar phosphatidylserine vesicles.[55] Production of TMV was monitored using fluorescent antibody against the virus. Using the same method of virus detection, but encapsulating the virus by REV, infection of more than 80% of the protoplasts was achieved.[47] Also, CCMV RNA and tobacco vein mottling (TVMV) RNA, encapsulated by REV, could infect, respectively, cowpea[22] and tobacco[197] mesophyll protoplasts.

Similar results were obtained on infection of tobacco protoplasts with TMV RNA encapsulated in negatively charged liposomes.[153] Other substances found to affect liposome-protoplast interaction are lectins. Targeting of large liposomes with lectin increases their binding to protoplasts.[162] A differential effect on positively and negatively charged liposome binding to protoplasts is recorded. Positively charged vesicles bound best at pH 5.6, whereas negative ones bound best at pH 9. High-salt medium impaired the binding of positively charged liposomes, whereas it did not affect binding by the other.[162] A quantitation of liposomal contents into *Catharanthus roseus*[24] protoplasts indicated that highest uptake (1.09% of vesicle contents per 10^7 protoplasts) was possible from positively charged vesicles in the presence of 10% PEG 4000, whereas low uptake (0.1% of vesicle contents per 10^7 protoplasts) was possible with negatively charged vesicles in the presence of 25% PEG 1540. No uptake of neutral liposomes was detected. A more recent success[35a] is the liposome-mediated transfer of the neomycin-phosphotransferase II gene from bacteria to tobacco cells, conferring kanamycin resistance; the inheritance of this gene was shown to be Mendelian.

The results described above are quite encouraging and indicate the utility of liposomes in the transformation of plant cells. However, much remains to be learned about preparation of liposomes and their delivery into protoplasts because infection of tobacco protoplasts has been accomplished by liposome-mediated delivery of DNA from octopine-type Ti-plasmid, but at a very low frequency (10^{-6}).[31a,43] Hormone autonomous calli produced synthesized octopine and had T-DNA sequences.

In a recent study[53] a comparison of different methods for delivery of plasmid DNA to plant protoplasts indicated that treatments which increased association of plasmid with protoplasts decreased protoplast viability. Optimum association of plasmid with protoplasts, in the context of acceptable loss of viability, was achieved when protoplasts interacted with either naked plasmid or liposome-coated DNA in the presence of 15% w/v PEG 6000. Divalent cations did not stimulate significant plasmid delivery without an unacceptable loss of protoplast viability.

V. TRANSFORMATION BY BACTERIAL COCULTIVATION

The reasons for a low frequency of transformation of protoplasts by T-DNA from *A. tumefaciens*, either free or liposome encapsulated, could be many. The two important ones are (1) inefficient delivery and failure of T-DNA to integrate and (2) unsuitable conditions for identification of rare hormone-independent transformants among the majority of untransformed cells.

However, high-frequency transformation of protoplasts was possible on coculture[121] of tobacco protoplasts and intact cells of *A. tumefaciens*. Similarly, protoplasts of *Hyoscyamus*[67] and *Petunia*[87] could also be transformed by *A. tumefaciens*. The frequency of transformation could be dramatically increased[49,50] incorporating information on efficient culture of protoplasts (rapid decrease of osmoticum) and bacterial cocultivation. It was possible to have transformation frequency of 10^{-1} routinely and at times as high as 80%. For this, the protoplasts of *Petunia* were plated at a density of 10^5 intact protoplasts per milliliter. On the second day *A. tumefaciens* cells were added to a titer of 10^7 to 10^8/mℓ. The cultures were incubated for 24 to 28 hr and washed by centrifugation to remove free bacteria. The cells were plated on a medium with 500 μg/mℓ of carbenicillin and then replated on feeder plates with reduced hormones to begin selection for hormone-autonomous cells. After 10 days, the microcolonies on filter paper were transferred to hormone-free medium.

Still higher levels of transformants[145] were achieved when the concentration of carbenicillin was reduced; small microscopic protoplast-derived colonies were more effectively transformed (up to 70%) by *Agrobacterium*.

The protoplasts of many plants are relatively refractory to regeneration and protoplasts are killed by *Agrobacterium*. Therefore, an alternative is cocultivation of cells[3] (Figure 1) with *Agrobacterium*. Tobacco cells could be transformed up to a frequency of 50% by cocultivation with *A. tumefaciens* having a kanamycin-resistant marker. Frequency of transformation was dependent on physiology of cells; exponentially growing cells gave maximum frequency of transformation, suggesting that active growth is an essential aspect of cell transformation.

The mechanism by which *Agrobacterium* integrates its DNA into plant cells is not clear. However, the process of infection can be subdivided into different steps:

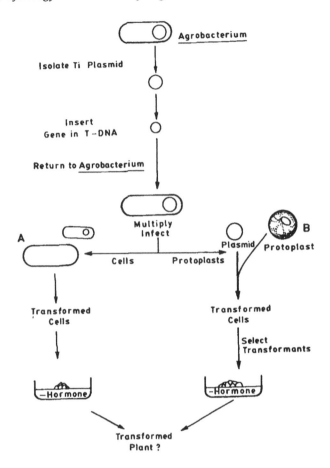

FIGURE 1.   Diagrammatic representation of strategies for genetic engineering of higher plant cells using *A. tumefaciens* as a vector. (A) Cell transformation on cocultivation of bacteria and cells. (B) Transformation of protoplast by plasmid DNA of *A. tumefaciens*.

1.   The attachment of bacterium to plant cell
2.   Induction of virulent (vir) region, which encodes functions required for the transfer of T-DNA to plant cell
3.   Transfer of T-DNA out of bacterium
4.   Transfer of T-DNA to plant nucleus
5.   Integration of T-DNA to plant genome
6.   Expression of T-DNA genes

The attachment of bacteria to plant cells is mediated by an 11-kb portion of the *Agrobacterium* chromosome. The expression of this region is constitutive.[36] The specificity[131] of *Agrobacterium* binding results in 100 to 300 bacteria bound per plant cell at a bacterial concentration above $1 \times 10^7/m\ell$.

In experiments on bacterial cocultivation with plant cells, since there is 100-fold excess of bacteria vs. plant cells, it is in order to know whether the frequency of transformation is limited by the number of plant cells capable of being transformed or the frequency with which effective bacterial contacts are established. This problem was approached[32] by obtaining the transformation frequencies of tobacco cells with two different selectable T-DNAs, when present either in one and the same bacterium or in two different bacteria. The first T-DNA employed was wild-type T-DNA of plasmid C58. The second T-DNA was a mini-T-

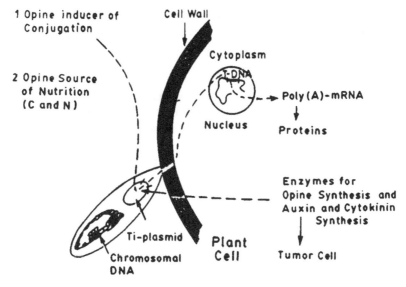

1 Opine inducer of
Conjugation

Cell Wall

Cytoplasm

T-DNA

2 Opine Source
of Nutrition
(C and N)

Poly (A)-mRNA

Proteins

Nucleus

Enzymes for
Opine Synthesis and
Auxin and Cytokinin
Synthesis

Ti-plasmid

Plant
Cell

Tumor Cell

Chromosomal
DNA

Agrobacterium tumefaciens

FIGURE 2. Diagrammatic representation of transformation of a plant cell into tumorous cell on infection by *A. tumefaciens*. T-region of plasmid integrates as T-DNA into chromosomal DNA of plant cell. Expression of T-DNA results in conversion of hormone heterotrophic plant cell into hormone autotrophic tumor cell, and also tumor cell synthesizes opines. These opines can only be catabolized by agrobacteria living in soil. (After Hooykaas, P. J. J. and Schilperoort, R. A., *Trends Biochem. Sci.*, 10, 307, 1985. With permission.)

DNA construction containing a marker gene Tn5, neomycin phosphotransferase coding sequence. These T-DNAs cotransformed plant cells at a frequency equal to the product of their independent transformation frequencies which indicated that all plant cells are equally competent. On the other hand, when these T-DNAs were located on the same Ti-plasmid vector within one bacterial strain, the cotransformation frequency was significantly higher than the products of single transformation frequencies. These results were interpreted to indicate that transformation is limited more by the establishment of effective bacterial/plant cell interaction than by the process of DNA integration or by the number of plant cells capable of being transformed.

VI. CROWN GALL — AN EXAMPLE OF GENETIC ENGINEERING IN NATURE

Crown gall is a neoplastic disease of many dicot plants incited by the soil bacterium *A. tumefaciens*. Transformation[10] of plant cells into tumorous type (Figure 2) is brought about by transfer of the tumor-inducing-principle (TIP) from bacterium to host cells. The tumor-inducing capacity of bacterium lies in the large extrachromosomal plasmid which is called Ti-plasmid.[177,193] Plasmid is essential for induction of tumors; all virulent forms of *Agrobacterium* carry a large plasmid. The bacterium can be made avirulent by high-temperature treatment and this is accompanied by loss of plasmid. The tumors formed can be made free of bacteria by exposing the infected cells to high temperature. The tumor[11] cells are characterized by their capability to grow without external supply of hormones.

These cells also synthesize a number of low-molecular-weight compounds called opines. Hormone autonomy and opine synthesis distinguish crown gall cells from normal cells. Opines characterize crown gall cells from other neoplastic cells. Due to the presence of specific opines,[65] it is possible to classify tumor cells into octopine-, nopaline-, and agropine-type tumors. Phytohormone-independent[11] growth of crown gall cells is due to an alteration of auxin-cytokinin production or release. Opines produced are utilized by bacteria as a source of energy and the bacteria benefit in this parasitic relationship. Opines released by crown gall cells can be catabolized only by agrobacteria, which aids in genetic colonization of rhizosphere only by these bacteria over other microbes. Agrobacteria are classified into three biotypes based on properties such as ketolactase production (biotype 1 and 3), growth on erythritol as carbon source (biotype 2), and growth on NaCl (biotype 3). Agrobacteria of a wide host range belong to biotypes 1 and 2 and agrobacteria isolated from grapes have a narrow host range and belong to biotype 3.

Agrobacteria are closely related to rhizobia and hence are placed in Rhizobiaceae. In addition to the common crown gall caused by *A. tumefaciens,* cane gall, also a tumorous outgrowth, is caused by *A. rubi.* Hairy root disease, which results in abundant production of roots by infected cells, is due to *A. rhizogenes.*

Ti-plasmids are subdivided[65] according to the opines they specify. The octopine-type plasmid (Type I) directs the synthesis of octopine[144] and agropine.[44] These are, respectively, arginine pyruvic acid condensation product and a carbohydrate derivative. Another opine, mannopine,[172] is found together with agropine and is considered to be an open-chain precursor in agropine synthesis. Type II plasmid directs the synthesis of octopine only and has a restricted host range, such as grape vine. Type III plasmid specifies the synthesis of nopaline, a condensation product of arginine and α-ketoglutaric acid. Other opines found in nopaline-type tumors are agrocinopine A and B and sugar phosphate derivatives. Ri-plasmids from *A. rhizogenes* have a restricted host range, but tumors induced by this plasmid on *N. glauca* have agropine and mannopine.[180]

The molecular basis of crown gall cells lies in the transfer of part of the Ti-plasmid to cells. The transformed cells maintain this alien DNA. The part of the Ti-plasmid that is transferred and maintained in plant cells is known as the T-region in bacteria and T-DNA in plant cells. T-DNA is about 10% of the plasmid and is found in DNA of transformed cells.[21,181] T-DNA is directly responsible for opine synthesis and hormone autonomy of transformed cells. The introduction of Ti-plasmid into a related bacterium, *Rhizobium trifolii,* renders this bacterium oncogenic.[76]

Crown gall cells constitute a case of genetic engineering of plant cells occurring in nature[77a] and a suitable system for the study of cell physiology and cell transformation. Hormone autonomy of crown gall cells can be utilized for the study of control of differentiation in plants by hormones and transformation of plant cells by Ti-plasmid makes it a suitable vector for transfer of foreign genes into plant cells. However, a prerequisite for effective use of this sytem is an understanding of crown gall cells and their transformation, which is discussed in the following sections.

A. Cell Transformation — Basic Aspects

Transformation of plant cells by *A. tumefaciens* into crown gall or neoplastic cells requires wounding, which is the basic stimulus for cell proliferation. Certain plant factors influence the host range and degree of tumor organization. Different species of tobacco and kalanchoe respond differently to the same strain of *A. tumefaciens.* Even on the same plant of tobacco, teratomas may arise on the upper half and tumors on the lower half. Growth regulators have an important function during induction of tumors. Even a mutation in the bacterium can be complemented by an exogenous supply of auxin. Cytokinin treatment during the induction phase by root-inducing *A. rhizogenes* results in undifferentiated tumorous growth.[5]

B. Cell Transformation — Molecular Aspects

For an understanding of the molecular aspects of crown gall cell formation, a prerequisite is familiarity with the structure of the Ti-plasmid.

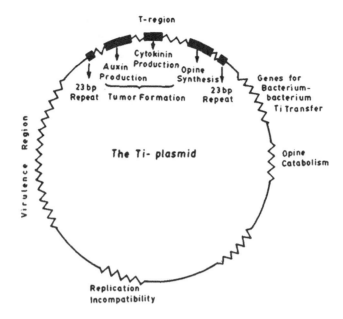

FIGURE 3. Genomic organization of tumor-inducing plasmid of *A. tumefaciens.* (From Hooykaas, P. J. J. and Schilperoort, R. A., *Trends Biochem. Sci.,* 10, 307, 1985. With permission.)

1. Ti-Plasmid

The genetic map (Figure 3) of Ti-plasmids, both octopine and nopaline types, has been prepared by transposon insertion mutogenesis and deletion mapping. The locations of insertion or deletion have been determined by changes in restriction patterns or by heteroduplex analysis. Along with genetic mapping, which allows the assignment of altered phenotype to a particular region of the Ti-plasmid, restriction maps have also been constructed.

a. Virulent Region

It has been shown by transposon insertion mutation that the region of Ti-plasmid essential for virulence is at the 10 o'clock position.[34] The vir-region displays a great degree of homology in plasmids from *A. tumefaciens* and *A. rhizogenes.*[98] This is also supported by experiments in which avirulent octopine Ti-plasmid mutants were complemented by Ri-plasmids.[77] The vir-locus is further divided into five major complementation groups (A to E); each group is separated by a region in which transposon insertion has no effect. All these groups are complemented by a corresponding wild-type region in a *trans* configuration on a compatible plasmid.[94] This indicates that the vir-region codes for functions that are expressed in a bacterial cell. Although DNA from the vir-region of a plasmid is not present in transformed plant cells, some mutation in this region results in altered tumor phenotype,[29,77] which indicates that this region is important during early events of transformation.

The genes in the vir-region are expressed in the bacterium, but only after induction directly/ indirectly by the unknown factors present in plant. The function of vir-genes is not known, but it is likely that some are essential for T-region processing and transfer. The expression of vir-region of Ti-plasmid is induced by the presence of plant products.[139] Recently, it has been found that addition of plant exudate to *Agrobacterium* causes induction of circular T-DNA copies by recombination at the 24-bp border sequences.[101] These circular T-DNA copies might be the elusive T-DNA intermediates that are transferred from *Agrobacterium* to plant cells during tumor induction. Also, it has been possible to identify[167] two compounds, acetosyringone (As, 4-acetyl-2,6-dimethoxy-phenol) and α-hydroxy acetosyringone (OH-

As) as signal molecules that activate T-DNA transfer. Each of these compounds induces vir-genes as well as helps form T-DNA intermediates. Wounding of plant cells stimulates the synthesis of these compounds. This is consistent with earlier observations that wounding is the first trigger, *Agrobacterium* recognizes wounded cells and is able to infect them.

The significance of the vir-region becomes apparent by experiments on helper effect. In a comparative study of opine synthesis in galls on *Kalanchoe daigremontiana* by two strains of *A. tumefaciens* — B 653 (octopine strain) and C 58 (nopaline strain) — the former induced a high level of octopine and the latter induced a low level of nopaline. However, a low level of nopaline synthesis induced by C 58 strain could be greatly enhanced by mixed infection with B 653. This "helper effect" was due to the activity of a 5-kb fragment from the vir-region of B 653. Incorporation of this fragment into the C 58 plasmid enabled it to induce a high level of nopaline synthesis in the absence of the B 653 strain.[141]

b. R- and Tra-Regions

Other regions identified on the plasmid are one corresponding to replication and incompatibility[42,99] and another encoding conjugative functions. Although tumor induction is a thermosensitive process[10,152,173] the transfer of genes from the Ti-plasmid does not seem to affect bacterial conjugation. This is supported by the isolation of several conjugation-negative mutants that were still oncogenic and the reverse was also true — many onc⁻ Ti-plasmid mutants were conjugative.

c. T-Region (Transferred DNA in Plant Cells)

The most significant part of the Ti-plasmid which is of interest in applied work on transformation is the T-region in a bacterium. This region is transferred to plants resulting in tumor formation and opine synthesis and is known as T-DNA in plants. The T-region of Ti-plasmid is homologous to T-DNA present in crown gall cells. This portion of the plasmid is stably incorporated into the genome of host DNA. Incorporation can occur in repetitive or unique host DNA.[59] However, all attempts to reveal homology[158] between T-DNA and plant DNA have failed.

Preliminary evidence that the Ti-plasmid is involved in gene transfer to crown gall cells concerned genetic analysis of Ti-plasmid function - synthesis of octopine or nopaline.[9] In addition, direct evidence for the transfer of T-DNA was obtained by kinetic analysis of DNA renaturation.[20] This was confirmed by DNA/DNA hybridization using the southern gel blotting technique.[21,181]

The T-region of both octopine- and nopaline-type plasmids is roughly 23 kb in size and is only a part of the entire plasmid.[26,42,111,174] Of this, only 8 to 9 kb is common to the octopine- and nopaline-type plasmids. This has been revealed by southern blotting and cross-hybridization of restriction endonuclease digests of two types of plasmids as well as heteroduplex analysis.[19,33]

The mechanism of T-DNA insertion appears to be precise because detailed analyses of some nopaline lines have indicated that the same continuous segment of Ti-plasmid is always present.[111,188,194,195] In some lines there is single T-DNA copy; others have multiple copies of T-DNA, organized in tandem array. However, octopine T-DNA is more variable. A left region (TL) containing the common or core sequence is always present; this is usually 12 kb in size,[174] but in one petunia tumor line it is shortened by about 4 kb.[26] Also present is a right T-DNA region (TR) which contains sequences which are adjacent but not contiguous. TL and TR are contiguous in only one octopine tumor line.[130] In some tumor lines TR is amplified, whereas TL is not and seems to act as an independent T-DNA.[34]

2. Insertion of T-Region

The mode of insertion of the T-region remains obscure. It is not known whether insertion is the result of plant or Ti-plasmid function and it is likely that both are involved.[34] Transfer during absence of a nuclear membrane in the host is indicated by susceptibility of the cells to infection during the proliferation phase. Also, gradual transformation and temperature sensitivity during the early phase indicate amplification and subsequent stabilization.

T-DNA is covalently linked to plant DNA. This is evidenced by analysis of clones of the T-DNA/plant-DNA junction. The T-DNA/plant junction fragments hybridize to either highly or moderately repeated or unique plant DNA sequences, indicating that there are multiple sites of integration in plant DNA. Further, the DNA sequence of right and left ends of the T-region, in comparison to the T-DNA/plant junction, indicates that "ends" of T-DNA are actively involved in T-DNA insertion.[34] There are also indications that T-DNA transfer and integration are probably coded by genes outside the T-region or by a combination of functions with different genetic locations.[89,110]

The incorporation of T-DNA occurs in a manner that it is associated with a repeat in 25 base pairs at either end of T-DNA. However, Ti-plasmid deletion spanning T-region borders has been obtained without noticeable effects. Deletion of the right octopine border region does not result in reduced tumor induction.[110] Similarly, the left border of the nopaline T-region could be deleted without any effect on tumor induction. Removal of the right border sequence abolishes and reintroduction restores tumor formation.[178a] Also, T-DNA does not have distinctly defined borders in different tumors, indicating that it does not behave as a transposon.[189,194] Sometimes many scrambled sequences are found at the left terminus.[164] In this respect T-DNA is similar to integration of virus SV-40. However, in one property T-DNA is similar to bacterial and eukaryotic transposons — it moves as a discrete unit of DNA with a capacity to integrate in nonhomologous DNA.

In a recent study of the role of border regions of T-DNA in crown gall formation by way of recombinant plasmids containing deletion and rearrangement derivatives of this region, it was found that four border regions exist in the Ti-plasmid.[154] New constructs containing as their only border region the right border, either TL-DNA or TR-DNA, were fully tumorigenic, whereas analogous constructs containing only TL-DNA left border were not tumorigenic. Furthermore TR-DNA right border can confer tumor-forming ability despite the presence of an intervening copy of TL-DNA left border.

In experiments on transformation of plant protoplasts in which segregation of T-DNA-linked function is observed, phytohormone independence is lost after some time while opine synthesis is retained, indicating an alteration in the stabilization of T-DNA.

3. Expression of T-DNA in Plant Cells

When T-DNA was considered to be responsible for tumorous properties of infected plant cells it was expected that both nuclei and polyribosomes of crown gall cells would contain RNA sequences which are complementary to T-DNA. The first evidence[40] that T-DNA is transcribed in crown gall cells was provided by the demonstration that total RNA from these cells hybridized to a specific fragment of the Ti-plasmid. Similarly, it was shown that octopine T-DNA[64] and nopaline T-DNA[56] are transcribed actively in crown gall cells. In a detailed study,[181] labeled RNA from a nuclear fraction and a poly-A polysomal fraction was hybridized to southern blots of T-region fragments. It was found that without any difference between nuclear and polysomal-RNA fractions, T-DNA is transcribed over its entire length. Further, it was found that transcription of T-DNA[184] is inhibited by a low concentration of α-amanitin. The observations that all RNAs mapped within T-DNA sequence and transcription is inhibited by α-amanitin indicate that each transcript is determined by a specific promoter site on T-DNA and transcription is directed by host RNA polymerase II.

In a further study,[183] tumor-specific RNAs were detected and mapped by hybridization of ^{32}P-labeled Ti-plasmid fragments to polyadenylated RNA which had been separated on agarose gels and then transferred to diazobenzyloxymethylcellulose, DBM paper. These results show that cells contained specific transcripts which differed in their relative abundance and size. They all bound to oligo-*dt*-cellulose, indicating that they were polyadenylated. Thus, T-DNA transferred from a prokaryotic organism provides specific poly-A addition sites.

The direction of transcription,[182] 5′→3′ polarity, was identified by hybridization to T-DNA separated single-stranded DNA. Since not all transcripts were synthesized from the same strand of DNA, the simplest model for transcription would be that there is one promoter site per group of transcripts. Accordingly, one would expect that the deletion of a 5′ proximal gene of a group would also lead to the disappearance of transcripts from 3′ distal genes. However, studies on cell lines containing T-DNA from Ti-plasmid mutants indicated that genes could not be activated by mutation lying far from the coding region.[110] Therefore, from the available evidence it can be concluded that each gene on T-DNA has its own signals for transcription in eukaryotic plant cells.

a. Translation of T-DNA-Derived mRNA in Plant Cells

In order to prove the function of T-DNA in transformed plant cells it was important to demonstrate that T-DNA-derived mRNAs are translated into proteins. An earlier study[125] indicated that the concentration of these mRNAs in transformed cells was very low. However, they were active in in vitro translation[125,161] using extract from wheat germ or rabbit reticulocyte lysate. The concentration of total T-DNA-specific RNA in an octopine tumor line was found to be between 0.0005 to 0.001%. Therefore, it was necessary to develop a hybridization selection procedure that was sensitive and specific to detect 0.001% of the total mRNA activity in the polyribosomal-RNA of transformed plant cells. This procedure was used to enrich T-DNA-derived mRNAs by hybridizing them to Ti-plasmid fragments that were covalently bound to microcrystalline cellulose. The hybridized RNAs were eluted and translated in vitro with extract from wheat germ. Using this approach,[160] at least three T-DNA-derived mRNAs could be translated into distinct proteins. One of these proteins is identical in size with the octopine-synthesizing enzyme in octopine tumors. In brief, it can be said that T-DNA contains genes that are expressed in plant cells.

An interesting aspect of this in vitro translation study[160] was that translation of all three mRNAs was inhibited by the cap analogue pm7G, indicating that they have a cap structure at the 5′-end. This is typical of eukaryotic mRNA since caps have not been described in prokaryotic RNA, which suggests that T-DNA genes are designed to function in eukaryotic rather than prokaryotic cells.

b. Sequencing of Genes for Octopine and Nopaline Synthesis

The region coding for the octopine-synthesizing enzyme has been sequenced.[30] The information indicates that the promoter sequence is more eukaryotic than prokaryotic in its recognition signals, suggesting that the gene is designed to function in eukaryotic cells. Essentially the same conclusion can be derived regarding the nopaline synthase gene.[35]

4. Functional Analysis of T-DNA

It has been possible to assign functions[89,110] to most of the different transcripts by introducing mutations in the T-DNA region of octopine or nopaline Ti-plasmids and observing the phenotypes of plant cells transformed by partially inactivated T-DNA. An important conclusion derived from these experiments is that none of the transcripts is essential for transfer and maintenance of T-DNA in a plant genome. Instead, two different functions of T-DNA transcripts are opine synthesis and tumorous growth.

In opine synthesis, the octopine plasmid contains one or two genes. One is located at the right end of TL and codes for octopine synthase, and the other is located at the right end of TR and codes for agropine and mannopine synthase (Velten, unpublished; see Schell et

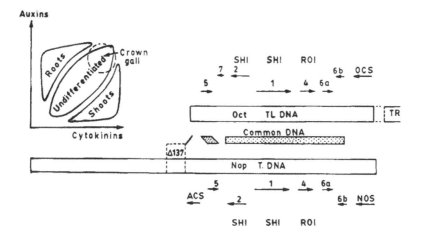

FIGURE 4. T-DNA maps of octopine- and nopaline-type plasmids. Their homology is indicated by a bar (common DNA). The transcripts corresponding to each genome are shown as SHI, ROI, OCS, and NOS (shoot and root inhibition, octopine, and nopaline synthesis, respectively). Auxin-cytokinin diagram is the representation of how hormone levels can influence organogenesis or tumorous growth in a callus culture. (From Depicker, A., Van Montagu, M., and Schell, J., in *Genetic Engineering of Plants*, Kosuge, T., Meredith, C., and Hollaender, A., Eds., Plenum Press, New York, 1983, 143. With permission.)

al.).[158] Tumorous cells with both TL and TR produce both octopine and agropine. Nopaline plasmids also contain at least two genes coding for different opines. One is located at the right end of T-DNA and codes for nopaline synthase, whereas the other is located on the left part of the T-DNA and codes for agrocinopine.

Genes responsible for tumorous growth (onc genes) are in the common or core region of T-DNA (Figure 4). Six well-defined transcripts are identified to be derived from this core region. Transcripts 1 and 2 specifically suppress shoot formation; their effect is similar to that of auxin-like plant hormone. Transcript 4 specifically prevents root formation; its effect is similar to that of plant cytokinins. Ti-plasmids from which gene 1 and/or 2, as well as gene 4, were deleted could transform the plant cells, but did not favor either root or shoot development at the site of infection. Therefore, shoot formation requires inactivity of gene 1 and/or 2 and activity of gene 4. Conversely, root formation requires inactivity of gene 4 and activity of genes 1 and 2. This interpretation of the functional aspects of transcripts 1 and 2 (auxin-like) and transcript 4 (cytokinin-like) is consistent with measurements of the endogenous level of auxin-cytokinin in teratomas and unorganized crown gall tissue.[2] Of the other three transcripts (5, 6, and 6a), transcript 5 was inhibitory to organization. Elimination of this transcript along with transcripts 1 and 2 (auxin-like function) allowed the transformed cells to organize into a teratoma, whereas transcripts 6 and 6a along with transcript 4 promoted hormone autonomy and favored continued unorganized growth.

From this information it appears that in order to obtain fully normal plants from crown gall cells containing T-DNA, it is essential that genes 1, 2, and 4 and possibly genes 5, 6, and 6a be inactivated. In fact, T-DNA genes involved in the production of auxin and cytokinin can be deleted without affecting T-DNA transfer.[110] Further, the plant cells transformed with such a disarmed T-DNA behave like normal cells and can be made to regenerate into normal fertile plants.[78]

Determinants within both T-DNA and the vir-region contribute to host specificity.[192] Within T-DNA a defective cytokinin biosynthetic gene limits host range. Nucleotide sequence analysis revealed a large deletion in the 5' coding region of this gene when compared with the homologous gene from the wide host range tumor-inducing plasmid PTiA$_6$. Introduction of the wide host range cytokinin-biosynthetic gene into T-DNA of a limited host range strain expanded its host range. Two genes from the vir-region of a wide host range plasmid, designated vir A and vir C, must also be introduced in the limited host range strain in order to make it a wide host range phenotype.

C. Cell Transformation — Physiological Aspects

Study of the biology of crown gall cells clearly indicates that T-DNA oncogenes encode plant growth regulators. This raises the question[190] of whether there are cellular homologues in plants. It has become clear from the findings[95,159] that there are at least three oncogenes in Ti-plasmid which code for specific enzymes that are involved in phytohormone synthesis. Tryptophan 2-monooxygenase and indoleacetamide hydrolase, the two enzymes that convert tryptophan to indole acetic acid, are encoded by auxin genes of T-DNA. These enzymes have not been detected in untransformed cells, but it remains to be seen whether these or similar enzymes are expressed in tissues at specific developmental stages. Also, the auxin gene codes for amidohydrolase, which converts indole-3-acetamide into indole-3-acetic acid, which is functional in crown gall cells and bacteria.[93]

Another oncogene codes for dimethylallyl pyrophosphate transferase, an enzyme that catalyzes synthesis of kinetin.[1] The activity of this enzyme has been reported in habituated tobacco tissue,[18] which grows on cytokinin-free medium, tumor tissue of genetic origin, and in cambial cells.[17]

VII. PLANT VIRUSES — ANOTHER EXAMPLE OF GENETIC ENGINEERING IN NATURE

Plant viruses are considered to be another potential vector for transfer of foreign genes since they replicate to give a very high copy number of nucleic acid molecules which are expressed in plant cells. However, there are two basic limitations: (1) virus is pathogenic and (2) no virus has been shown to integrate with plant DNA. Three types of plant viruses considered to be potential vectors are the following:

1. Single-stranded RNA virus
2. Single-stranded DNA virus
3. Double-stranded DNA virus

A. Single-Stranded RNA Virus

The genetic material of about 90% of plant viruses is single-stranded RNA. This material is positive stranded and infectious. Attempts are being made to develop tobacco rattle virus (TRV) RNA as a vector.[102] The strategy will involve the conversion of single-stranded RNA to double-stranded C-DNA using the enzymes reverse transcriptase and DNA polymerase. DNA would then be cloned into a prokaryotic plasmid or a cosmid, a combined plasmid/lambda vector. The desired gene would be inserted into the C-DNA moiety of recombinant plasmid. For infection of plant cells, C-DNA with the foreign gene can be taken if it is found to be infectious; otherwise, the viral vector and the foreign gene could be transcribed to RNA which is expected to be infectious. A diagrammatic representation of this scheme is given in Figure 5.

Recently, doubts[178] have been expressed about the ability of RNA viruses to act as possible vectors, in particular, that the low fidelity of RNA-synthesizing enzymes may limit the usefulness, but counter arguments[163] indicate that the problem may not be so serious.

B. Single-Stranded DNA Virus

This is a small group of viruses,[58] about a dozen or more, described as gemini virus.

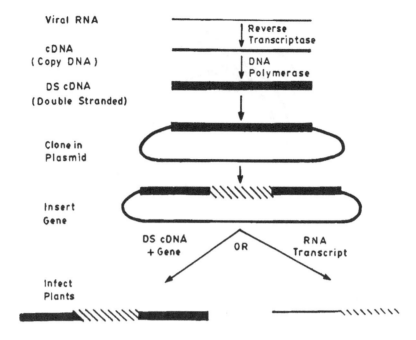

Viral RNA

Reverse Transcriptase

cDNA (Copy DNA)

DNA Polymerase

DS cDNA (Double Stranded)

Clone in Plasmid

Insert Gene

DS cDNA + Gene — OR — RNA Transcript

Infect Plants

FIGURE 5. Scheme for using viral RNA as a vector for cell transformation. (After Hull, R., in *Plant Biotechnology*, Cambridge University Press, Cambridge, 1983, 229. With permission.)

Some have a narrow host range (abutilon mosaic virus) and others have a broad host range (curley top virus). They are also known to infect legumes (bean gold mosaic virus, BGMV) and monocots (maize streak virus, MSV).

In this group of viruses, BGMV is the best known. The encapsulated DNA is either circular single-stranded DNA or a linear molecule and is infectious in both forms.

C. Double-Stranded DNA Virus

Caulimoviruses (CaMVs) are the only known double-stranded DNA viruses of plants. These viruses have a narrow host range, confined to crucifers. Among the viruses, they are the more important potential vectors,[75] because double-stranded DNA is more amenable to manipulation for recombinant DNA technology. A unique feature of CaMV is that the DNA of this virus can be directly introduced into plants. However, CaMV DNA does not integrate into host DNA, but this disadvantage can be an advantage. The large-scale multiplication of virus in plant cells provides not only an amplification of the viral genome, but also an equal chance for amplification of passenger genome, thus providing an efficient means for expression of foreign genome.

DNA isolated from CaMVs has two unusual features. First, it has discontinuities at specific sites.[85] The majority of isolates have one discontinuity in one strand (α-strand) and two in the other. The second unusual feature is that a large proportion of DNA molecules is a twisted structure when viewed in the electron microscope. Sequencing data, however, reveal that the discontinuities are in fact overlaps and DNA is apparently triple stranded in these regions.[52] The sequence data also reveal that there are (Figure 6) six possible open reading regions (I to VI). Two further reading regions (VII to VIII) have also been suggested.[75]

Only one strand of CaMV DNA, that with a single gap, α-strand is transcribed.[80,84] The map location of two major transcripts of CaMV is known. One is 19S RNA derived from coding region VI; the other is a large transcript of genome length.[134] Insertion or deletion of nucleotides in CaMV DNA usually results in loss of infectivity. However, at two sites

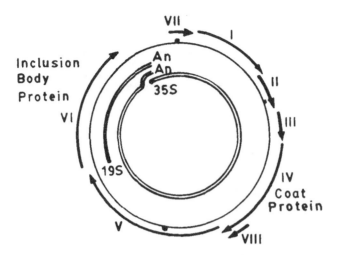

FIGURE 6.   Genomic organization of cauliflower mosaic virus. Main genome is indicated by a circle with three gaps. Marked outside the circle, identified by arrows, are eight open reading frames. Inside the circle are two principal transcripts. (After Gardner, R. C., in *Genetic Engineering of Plants*, Kosuge, T., Meredith, C., and Hollaender, A., Eds., Plenum Press, New York, 1983, 121. With permission.)

(one in a large intergenic region and the other in open reading region II) inserts made do not result in loss of infectivity.[63,82] Nonetheless, there appears to be a limit to the amount of foreign DNA which CaMV can accommodate. CaMV DNA with inserts of bacterial DNA of 60 and 250 bp was infectious and these foreign fragments were retained. When larger fragments (500 to 1200 bp) were inserted, infected plants contained viral DNA from which most of the bacterial DNA was deleted.

D. CaMV as a Vector

Contrary to earlier reports, it has been shown that the unusual secondary structure of CaMV DNA is not required for infectivity. The linearized CaMV DNA which lacks secondary structure is still infective. When linearized DNA molecules are cloned into bacterial plasmid pBR322, the cloned viral DNA infects plants if it is excised from recombinant plasmid.[81,106] Viral DNA extracted from plants infected with cloned DNA has all the unusual properties associated with CaMV DNA. Thus plants can religate a linear molecule, but they are more efficient at religating molecules with longer cohesive single-stranded ends.[75]

There are a number of limitations of CaMV virus as a vehicle for carrying passenger DNA. CaMV DNA is so tightly packed with coding regions that there is little room to insert foreign DNA. However, as pointed out earlier, modifications are possible in coding region II at the Xho I site, but there is a limit to the amount of foreign DNA which CaMV can accommodate. To circumvent the size limitation, an attempt was made to use a helper virus system.[82] This was done to replace substantial portion of the viral DNA and replace it with foreign DNA. The modified viral DNA will be noninfective, but its loss of infection could be complemented by noninfection with a helper virus DNA. It was shown that plants could be infected with two noninfective viral DNAs bearing lesions at different regions of the genome. However, the rescue was due to recombination and not complementation because in the recombination process deletions were eliminated and the normal viral genome was recovered. Therefore, to make this approach work recombination must be suppressed.

VIII. TRANSFORMATION VIA VECTOR

Attempts to transform cells or whole plants by DNA, either free or liposome coated, have not been very successful. Hence, many investigators have turned to the use of vectors in the transformation of plant cells. A prerequisite for transformation is the introduction of foreign DNA into plant cells. Following this is the problem of identification of transformed plant cells.

A vector is a molecular vehicle. To be more precise, it is recombinant DNA or an RNA molecule comprising two parts — a vehicle component and passenger component. The vehicle is used to deliver the passenger component. A vehicle could be any object, DNA or RNA, and the passenger component is the desired DNA to be introduced. For example, Ti-plasmid of *Agrobacterium* is one of the vehicles known in nature which can be used for the transformation of plant cells. In this system, a defined piece of bacterial plasmid DNA (T-region) is transferred from bacterium to plant cell, where it is integrated into the plant DNA. Another potential vector could be a virus, the foreign nucleic acid which replicates and expresses in the plant cell. Of the two vehicles, *Agrobacterium* is considered to be more promising because it is possible to identify transformed cells by their hormone autonomy and opine synthesis. Plant pathogens — bacteria and viruses — are recognized as vectors for their natural ability to introduce foreign DNA into plant cells. In plant cells, an alternative could also be native DNA. This category includes unstable components of the plant genome; transposon elements, stable components of the plant genome, and even extrachromosomal pieces of DNA in plant cells.

Plant genetic engineering using recombinant DNA technology is in its infancy. Progress is dependent on deployment of a proper vehicle and methods for identification of transformed plant cells. Simultaneously, one has to bear in mind the limitations, the foremost being that a transformed cell does not always mean a transformed plant. Despite its infancy, the field of plant genetic engineering is so fascinating that it is worth a review. The progress is summarized in the following section.

A. Cell Transformation Using T-DNA as a Vector

The first demonstration of cell transformation by recombinant DNA was in the not-too-distant past when transposon Tn7 as a passenger was carried through the vector T-37, nopaline plasmid of *A. tumefaciens* and introduced into tobacco cells.[68] This success served as a milestone in the transformation of higher plant cells using recombinant DNA. The introduction of the Tn7 gene was used as a marker to select transformed cells. This in turn also demonstrated the expression of the introduced Tn7 gene. Among other functions, Tn7 codes for activity of bacterial dihydrofolate reductase (dHFR) which makes the cells resistant to methotrexate. Normal cells of tobacco are sensitive to $2\mu g/m\ell$ of methotrexate. However, transformed cells could grow at this concentration and were isolated. This clearly indicates that bacterial gene dHFR could be transferred to plant cells, conferring them resistance to methotrexate. The main result of this work is that if a foreign gene is inserted into T-DNA of *Agrobacterium*, it is carried to the plant as a package.

1. Design of Intermediate or Shuttle Plasmid

Several attempts, in different laboratories, for site-specific insertion of foreign DNA into Ti-plasmids were unsuccessful. This was due to two reasons: (1) Ti-plasmid is large (180 to 230 kb) and difficult to handle in vitro and (2) it does not have any unique restriction site. To overcome these limitations, construction of intermediate or shuttle plasmids was attempted.[27,124]

The steps in the design of shuttle plasmids are briefly as follows:

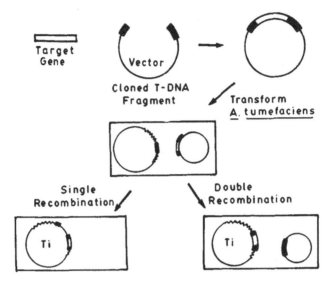

FIGURE 7. Possibilities of insertion of foreign DNA into T-DNA of *A. tumefaciens.* (After DeFramond, A. J., Bevan, M. W., Barton, K. A., Flavell, R., and Chilton, M.-D., in *Advances in Gene Technology: Molecular Genetics of Plants and Animals,* Vol. 20, Miami Winter Symp., Downey, K., Vollmy, R., Ahmed, F., and Schultz, J., Eds., Academic Press, New York, 1983, 159. With permission.)

1.   A restriction fragment from T-DNA of Ti-plasmid is prepared.
2.   This is cloned into pBR 322.
3.   A restriction fragment of foreign DNA or target DNA is prepared; preferably it should also carry a marker such as antibiotic resistance.
4.   Target DNA is then inserted, by recombinant DNA technology, into the cloned T-DNA fragment.
5.   The resulting recombinant plasmid is introduced into *Agrobacterium* by transformation or conjugation.
6.   Selection is made for recombinants between the engineered fragment and its normal counterpart in the Ti-plasmid.

Depending on whether it is single recombination (cointegration) or double recombination (homogenotization), different products are possible (Figure 7). In the former, a complex T-DNA is possible in which engineered and normal fragments are present in tandem separated by vector DNA sequences. In the latter, it is possible to have a more simple T-DNA in which the "engineered fragment" has replaced the normal T-DNA fragment.

Essentially, the vector system described above has two components:

1.   An intermediate vector which can be one of the commonly used cloning plasmids such as pBR322 into which foreign genes can be cloned and which can be easily mobilized into *Agrobacterium.*
2.   An acceptor Ti-plasmid which carries active virulence region and "borders" involved in recognition specificity. In between these border sequences should be a DNA sequence homologous to cloning vector and also a dominant selectable marker gene for plant cells.

By a single homologous recombination event the vector carrying foreign genes can be inserted next to the marker gene and between the border sequences of acceptor Ti-plasmid.

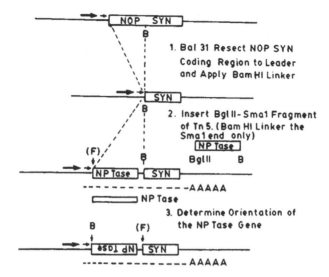

FIGURE 8. Diagrammatic representation of chimeric gene reconstruct. (After DeFramond, A. J., Bevan, M. W., Barton, K. A., Flavell, R., and Chilton, M.-D., in *Advances in Gene Technology: Molecular Genetics of Plants and Animals*, Vol. 20, Miami Winter Symp., Downey, K., Vollmy, R., Ahmed, F., and Schultz, J., Eds., Academic Press, New York, 1983, 159. With permission.)

Such a vector system was shown to transfer and integrate foreign DNA into plant cells.[196] However, a number of foreign genes of bacterial, plant, and animal origin introduced into tobacco[196] or sunflower[92] via Ti-plasmids were not expressed. This was in sharp contrast to opine synthase genes of Ti-plasmid that are expressed on transfer to a wide variety of plant cells. These genes are therefore very useful for study of the expression of foreign genes into plant cells and help design chimeric gene construct.

2. Chimeric Gene Construct

Chimeric gene construct (Figure 8) and transfer of bacterial genes into plant cells has been possible by a number of research groups. One of the genes transferred which is expressed in plant cells[7] is neomycin phosphotransferase-II (NEO or NPT), a Tn5-encoded kanamycin detoxification enzyme. The strategy involved was that instead of using plant gene "signals", the signals of T-DNA gene-encoding nopaline synthase (NOS) were employed. This gene is expressed in a wide range of dicot plants using the following steps:

1. Bgl II/Sm I fragment of Tn-5 was prepared carrying the coding region of NPT, but not its prokaryotic promoter.
2. Sm I site was converted to a Bam HI site with molecular linkers.
3. Deletion of the front half of the NOS gene was possible by resection with Bal 31 exonuclease and replacing it with a unique Bam HI site.
4. At the junction between the leader and coding sequence of NOS, a Bam HI molecular linker was applied, thus allowing the possibility of linking the NPT gene in two reverse orientations.
5. The resulting chimeric genes were engineered into nopaline Ti-plasmid T37, with the chimeric gene replacing the wild-type NOS gene.

Tumors incited by *Agrobacterium* containing the chimeric gene construct, in the correct orientation, were able to grow on G418, an aminoglycoside antibiotic that inhibited the growth of normal tobacco cells.

Using the same strategy of having a unique Bam HI cloning site at the end of 5′ untranslated leader sequence of the NOS gene, coding sequences derived from a number of bacterial genes were inserted individually to make a chimeric gene construct. The specific genes introduced into plant cells are (1) NPT from Tn5, (2) methotrexate-insensitive dHFR of plasmid R67, and (3) chloramphenicol acetyltransferase from pBR322. All these genes were expressed in plant cells.[69]

In a similar approach,[51] the gene NPT from Tn5 was joined to the 5′ and 3′ regulatory region of the NOS gene and this chimeric gene construct was cloned into an intermediate vector pMON 120. It was then inserted into octopine-type Ti-plasmid by recombination. Plant cells infected with the bacterium were able to grow on kanamycin medium.

From the above results it can be concluded that using the natural gene transfer system of *A. tumefaciens* it is possible to introduce bacterial genes into plant cells. A chimeric gene must be constructed in which bacterial coding sequences are placed under the control of transcriptional signals of NOS genes of T-DNA. When this chimeric gene construct is introduced into plant cells using Ti-plasmid of *Agrobacterium,* the bacterial sequences are expressed as functional proteins.

Following this strategy it has also been possible to introduce native plant gene 1,5-bisphosphate carboxylase into tobacco cells. The gene was linked to the bacterial gene chloramphenicol acetyltransferase (CAT).[70] This chimeric gene construct was introduced into plant cells using Ti-plasmid of *A. tumefaciens.* The chimeric gene construct (SS-CAT) comprised the promoter of *Pisum sativum* small subunit (SS1) of 1,5-bisphosphate carboxylase and the coding sequence of bacterial gene CAT. The SS-CAT chimeric gene construct was linked to the 3′ end of the NOS gene from Ti-plasmid. The expression of the CAT and SS genes in transformed plant cells was demonstrated by the presence of CAT activity and poly-A RNA homologous to the SS gene. The activity was further accentuated in chloroplasts containing green tissue in light compared to dark. This was possible in the NOS-SS-CAT gene and not in the NOS-CAT gene.

From these experiments it can be concluded that genes cloned between promoter and terminator regions of T-DNA are chimeric genes and are able to be expressed in plant cells. From these transformed cells it is possible to regenerate plants. The success of the system can be seen in a spate of reports appearing in one year (1985) from different laboratories. These include transfer to tobacco cells of genes from (1) soybean (lectin gene, kunitz-trypsin inhibitor gene, and nonseed protein genes), (2) pea (ribulose bisphosphate carboxylase small subunit gene),[130b] (3) wheat (gene Wh AB87 for chlorophyll a/b binding protein),[103a] (4) glyphosphate-resistant[22a] gene from *Salmonella typhimurium,* (5) phaseolin gene[130a] from french bean, (6) pea (chlorophyll a/b binding protein gene)[163a] and (7) heat shock genes from *Drosophila.*[166a]

Similarly, it has also been possible to obtain expression of maize zein genes in transformed sunflower cells.[57a] Genes encoding maize seed storage proteins of 19 and 15 kd were inserted at various sites in T-DNA of Ti-plasmid pTiA6 of *A. tumefaciens.* The bacteria harboring these plasmids were employed to incite tumors on sunflower stem sections. Some of the tumors formed were found to contain mRNAs transcribed from zein genes. Recently even the expression of NOS and human growth hormone chimeric genes has been shown in transformed tobacco and sunflower cells.[4a]

3. Problems and Prospects of T-DNA-Mediated Transfer of Genes

Although transfer, integration, and expression of bacterial as well as higher plant genes have been possible using T-DNA as a vector, there are limitations to this transfer system. Transformation by T-DNA results in a tumorous cell which does not regenerate easily and

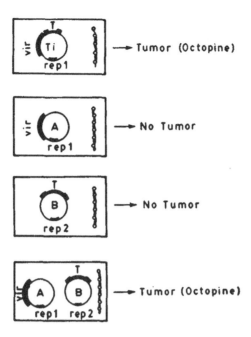

FIGURE 9. The binary plant-vector strategy. T-region is active even after separation from Ti-plasmid; it can be manipulated in vitro. Transfer of this plasmid can be mediated to plant cells by *Agrobacterium* with virulent plasmid. (From Hooykaas, P. J. J. and Schilperoort, R. A., *Trends Biochem. Sci.*, 10, 307, 1985. With permission.)

fails to form normal plants. This is due to the high levels of phytohormones produced by crown gall cells. A number of T-DNA genes are involved in oncogenicity. However, analysis of spontaneous deletion mutants indicates that it is possible to cure T-DNA of *onc* function.

In an experiment on cells transformed by mutated T-DNA, differentiation of plants was possible. One plant was studied in great detail; its T-DNA was found to have undergone a large deletion, having lost all but the right part of T-DNA[31] which codes for octopine synthase gene. This raised the prospects of growing normal plants, and T-DNA can become an ideal vehicle when cured of its tumorogenic function.

In earlier literature there appeared suggestions that T-DNA could not pass through meiosis.[191] However, there is growing evidence that genes introduced in plants through T-DNA can be sexually inherited.[133,140,186]

A new approach in vector construction is a mini-Ti-plasmid and a binary vector system[27] (Figure 9). This new approach is the simplification of procedures used earlier, in the sense that it does not encompass the recombination step with Ti-plasmid. This was possible due to splitting the Ti-plasmid into two independent replicons, one carrying the T-DNA and the other the vir-genes, and remained tumorogenic by *trans* complementation. The T-DNA replicon of a small size can be manipulated readily by foreign gene insertions and is thus an attractive vector for plant genetic engineering. A mini-Ti-plasmid containing full-length T-DNA from a nopaline-type plasmid pTi T37 has been constructed. A cointegrate form of this plasmid, designated mini-Ti-pRK, contains wide host range plasmid pRK 290 and can therefore replicate in *Agrobacterium*. This plasmid is unable to induce tumors. However, when mini-Ti-pRK is introduced into an *Agrobacterium* strain carrying octopine-type Ti-plasmid with virulent genes, tumors which are octopine- as well as nopaline-positive form. In order to determine that T-DNA of mini-plasmid does not combine with helper plasmid and functions in *trans* to the vir-genes of helper plasmid, two methods were used to block recombination. First, a *rec⁻* mutant of *A. tumefaciens* was used. The two plasmids helped produce tumors characteristic of wild pTi T37. A second method of eliminating recombination was to eliminate all homology between mini-Ti-pRK and helper plasmid. Again, in the presence of the helper plasmid, the mini-Ti-pRK induced tumors typical of pTi T37. From this it can be concluded that mini-Ti-pRK functions in *trans* to vir functions on helper plasmid. The significance of the new vector system is that it requires minimal genetic manipulation.

Using this strategy on deleting oncogenes from mini-Ti, a micro-Ti was produced which contained the NOS gene and the ampicillin resistance gene and origin of replication of pBR322 flanked by left and right T-DNA borders. This micro-Ti was recloned into wide host range plasmid pCK 290 and transferred into the *A. tumefaciens* strain carrying a helper plasmid that could supply vir-genes in *trans*. Using the octopine Ti-plasmid pTi B6-306 as a helper, transformed tobacco cells were obtained which produced both nopaline and oc-topine. Two cloned cell lines producing both opines were hormone dependent and produced fertile tobacco plants. When selfed, these two opine markers segregated in F₁ progeny in Mendelian fashion. This indicated that two T-DNAs were not linked in transformed plant genome and two unlinked T-DNAs could transform the same cell and segregate in F₁ progeny.[28] Employing a binary Ti-vector system, a high-frequency transformation of tobacco was achieved.[3a] Later, this was also possible with tomato, potato, and *Arabidopsis*.[3b]

B. Cell Transformation Using CaMV as a Vector

As mentioned earlier, in CaMV genome open reading frame II can be deleted or expanded with small inserts of foreign DNA without loss of function. An attempt made to insert bacterial gene R67 plasmid-encoded dHFR was successful.[12] The chimeric viral DNA could be stably propagated in turnip plants expressing the dHFR gene, producing a functional enzyme.

More recently,[143b] the protein coding region of gene VI of CaMV was replaced by coding part of the bacterial gene for aminoglycoside phosphotransferase II from Tn5. This hybrid genome was expected to be expressed under the control of gene VI expression signals. The altered modified viral genome was found to be infectious for both entire plants as well as protoplasts of *Brassica napus* var. *rapa*. Stable genetically transformed cell lines were recovered. Significantly, DNA of the hybrid CaMV genome was also found to be integrated into the high-molecular-weight host genomic DNA. Transformation was, however, achieved only when modified viral genome was supplied together with wild viral DNA as helper molecule.

C. Potential Vectors for Cell Transformation

Instead of using exotic DNA as a vector for carrying the foreign genome, parts of the plant genome itself can also be used. At present, three principal candidates are mobile genetic elements, chloroplast, and mitochondrial extrachromosomal elements. Recent reports indicate transfer of mobile genetic elements,[158a] transformation of chloroplast DNA,[26a] and mitochondrial[26a] transformation.

1. Mobile Genetic Elements

In her studies on genetics of corn plant in the early 1940s, Barbara McClintok realized that some of the bizarre variegation effects could be explained by mobile gene segments — a startling discovery at a time when genes were thought to be permanently aligned on a chromosome, like beads on a chain. Further, it was postulated that when these genes jumped from one place to another on the chromosome they turned other genes on and off, providing a possible explanation for the varied gene expression that occurs during development. At that time it was difficult to find evidence to support this hypothesis of jumping genes and this discovery was simply discounted. At present, efforts are being made to redescribe controlling elements in molecular terms[168] in analogy to transposable elements of bacteria.

If mobile genetic elements of higher organisms are similar to transposable elements,[15] then they are potential molecular vehicles to mobilize a foreign genome into the plant genome. At present, attention is focused on two elements — Ac DS controlling element system and its effect on shrunken locus (Sh) in maize. The Sh locus in maize encodes structural components of the enzyme sucrose synthetase which is the critical enzyme in starch synthesis in maize endosperm. Mutant kernels produce less starch than normal and shrink upon drying. In lines of maize in which DS has been transposed from its normal position between the waxy locus and the centromere on chromosome 9 to points in and around the Sh locus, one finds large insertions of DNA from about 11 to 21 kb.[13] These lines are shrunken mutable in the presence of transacting Ac element. DS may transpose away from the locus, leaving either a stably altered or restored allele.

2. Mitochondrial Extrachromosomal Elements

Cytoplasmic male sterility (cms) is one of the essential elements in the production of seeds in hybrid corn. The molecular basis of cms lies in extrachromosomal DNA elements in maize mitochondria. Various cms lines of maize have been classified into four groups:[41] N (normal), T, C, and S, based on their response to nuclear restorer genes. This correlates well with restriction digest profiles of the mitochondrial genome of four classes, which are similar within but differ between the four groups.[147] Further, it has been found that mitochondria of the cms-S group contain two small DNAs[148] in addition to mitochondrial DNA. These two elements, S_1 and S_2, are linear molecules of 6.2 and 5.2 kb, respectively. Following this, it has been shown that all four cytoplasms could be characterized by their extrachromosomal pieces of DNA.[90,91]

The elements S_1 and S_2 in the cms-S line seems to behave like transposon elements in that they appear to jump in and out of the mitochondrial genome. This is inferred from an analysis of mitochondrial DNA in the cms line reverting to fertility. The reversion to fertility was correlated with loss of free S_1 and S_2 elements.[112] However, it has yet to be determined whether S_1 and S_2 code for specific proteins because mitochondria of cms-T cytoplasm shut down the synthesis of a 21,000-mol wt polypeptide produced in the normal line and instead synthesize a new 13,000-mol wt polypeptide. In the T-restored cytoplasm synthesis of the 13,000-mol wt polypeptide is suppressed and synthesis of the 21,000-mol wt protein is restored.[45,46]

Further studies on S_1 and S_2 elements are likely to reveal their utility as carriers of foreign DNA.

IX. TRANSFORMATION BY DIRECT GENE TRANSFER

Both the successful vector systems — Ti-plasmid and CaMV DNA — suffer from major disadvantages. Their host range is limited and, specifically for CaMV, there is a limitation on the size of foreign DNA which can be inserted. In view of these drawbacks several research groups have attempted to look for alternatives. Taking the lead from recent pioneering attempts of animal cell biologists[151] showing that animal cells in culture could be transformed with foreign DNA without any specialized vector, attempts made to transform plants cells by direct gene transfer have been successful and are enumerated below.

The first success[143a] was achieved at Friedrich-Meischer Institute, Basel, Switzerland. The bacterial transposon gene-encoding aminoglycoside phosphotransferase, which inactivates kanamycin, was ligated to promoter and termination sequences of CaMV gene VI. These control sequences were selected because CaMV DNA is transcribed by the RNA polymerase

II of its host, and gene VI, which codes for inclusion of body protein, is strongly expressed in host cells. The resulting hybrid gene was inserted into *E. coli* plasmid pUCB. Tobacco mesophyll protoplasts were then exposed to these recombinant plasmids in the presence of PEG 6000 which promoted the uptake of plasmids into protoplasts. To select for transformed cells, the protoplasts were grown on a medium containing kanamycin. In the transformed cell the enzyme aminoglycoside phosphotransferase was detected. Molecular evidence of transformation was provided by southern blot analysis. Recovery of plants resistant to kanamycin and about 50% resistant plants from anther culture and of segregation ratios 3:1 and 1:1 for resistant vs. sensitive seedings in the first self-cross and the first back-cross seedling populations indicated the transmission of a transformed phenotype in a Mendelian manner.

This result on direct transfer of gene was confirmed[66] when protoplast of *N. tabacum* were transformed by plasmid pBR322 containing a chimeric gene consisting of the NOS promoter, the coding region of the aminoglycoside phosphotransferase gene of Tn5, and the polyadenylation signal region of the octopine synthase gene. This chimeric gene conferred resistance to kanamycin to the transformed cells. The transformation frequencies were about 0.01%. It is not clear how transformation occurs. The input DNA appears to be integrated into high-molecular-weight cellular DNA.

The main outcome of these results is that there is no need for specialized gene vectors, thus avoiding some of the associated problems. A real step forward in this direction is the transformation of monocot protoplasts of *Lolium multiflorum*[146] and *Triticum monococcum*.[113]

Direct transformation of plant protoplasts was possible on incubating the protoplasts and engineered plasmids in the presence of PEG or PVA. For this, one must wait weeks or months in order to pick up the transformed colonies. However, a quick assessment of transformation is possible[54] if plasmid-DNA and plant protoplasts are given a high-voltage pulse. DNA molecules are taken during or shortly after the electric pulse, presumably through pores generated in the cell membrane (electroporation). After 1 to 2 days of electroporation, protoplasts of carrot, maize, and tobacco were assayed for expression of bacterial gene-encoding CAT.

Using electroporation, a stable transformation[104] of carrot protoplasts has been possible by taking naked DNA from Ti-plasmid of *A. tumefaciens*. Protoplast regeneration, somatic embryogenesis, and plantlet regeneration were unaffected by high-voltage electric pulse at a field strength of 0.5 to 3.5 kV/cm. The transformation resulted in hormone-independent growth of carrot protoplasts. Somatic embryos were detected in these cultures after 45 days of electroporation. The transformed embryos developed into teratomas which synthesized opine.

The same group[132] has reported electroporation-mediated infection of mesophyll protoplasts of tobacco with RNAs of TMV and cucumber mosaic virus (CMV). Approximately 30 to 40% of the protoplasts survived electroporation and under optimal conditions 75% of the protoplasts were infected with TMV-RNA. The conditions for optimal infection were (1) several DC pulses of 90 μsec at a field strength of 5 to 10 kV/cm, (2) changing position of protoplasts within the chamber between electric pulses, and (3) RNA concentration of about 10 μg/mℓ in a solution of 0.5*M* mannitol without any buffer.

These results indicate that electroporation is a promising technique and more exciting results are awaited. On treatment of protoplasts with DNA in the presence of PEG it is also possible to follow transient gene expression. This strategy can also be employed for studying plant promoters.[179] The promoter of the shrunken gene isolated from *Zea mays* was fused to the bacterial gene neomycin phosphotransferase II from Tn5 and this plasmid construction ps KAN I was used to transform protoplasts of *T. monococcum*. The expression of the chimeric gene was analyzed in terms of activity of neomycin phosphotransferase II activity. Maximum activity was seen after 4 days, but it could be detected up to 10 days. These

FIGURE 10.   Microinjection of DNA into plant protoplast. (From Crossway, A., Oakes, J. Y., Irvine, J. M., Ward., B., Knauf, V. C., and Shewmaker, C. K., *Mol. Gen. Genet.*, 202, 179, 1986. With permission.)

results show that a promoter fragment from Z. *mays* can promote initiation of transcription in *T. monococcum* cells.

X. TRANSFORMATION FOLLOWING MICROINJECTION OF DNA

Transformation of animal cells by microinjection of DNA has been successful for many years.[86,155] However, attempts at plant cell microinjection were hampered by either a rigid plant cell wall or the fragile nature of the plant protoplast. Another difficulty encountered in microinjection of genetic material into plant protoplast employing a glass micropipette was the presence of large central vacuoles. The vacuoles make a protoplast turgid and it is difficult to insert a glass micropipette into the protoplast unless it is held firmly in place. Vacuoles also function as a lysosomal compartment in plant cells and hence genetic material should be microinjected into the intracytoplasmic or intranuclear compartment of protoplasts. Thus, several factors affect success. For example, frequency of transformation following nuclear injection of mammalian cells or eggs has been higher than cytoplasmic injections.[14]

Recently, methods for immobilizing plant protoplasts for microinjection[62,105,127,169] have been devised; also, using a DNA fluorochrome complex, localization of injected DNA in the protoplast compartment could be directly monitored.[127]

More recently, efficient transformation[23] of N. *tabacum* protoplasts has been demonstrated on microinjection, using the holding pipette method (Figure 10), of DNA from *A. tumefaciens*. The frequency of integration averaged 14% with intranuclear injection compared to 6% with cytoplasmic injection. A diagrammatic representation of this scheme is given in Figure 11.

XI. TRANSFORMATION FOLLOWING TRANSPLANTATION OF FOREIGN NUCLEUS

Transfer of genetic information by nuclear transplantation is well documented in animal cells.[100] However, success with plant material using protoplasts and foreign nuclei was limited[114] and there was no proof of transfer and expression of biological information. Nonetheless, in a recent attempt, [157] uptake of isolated nuclei from *Vicia hajastana* cells into protoplasts of an auxotrophic cell line of *Datura innoxia* has been demonstrated. Fol-

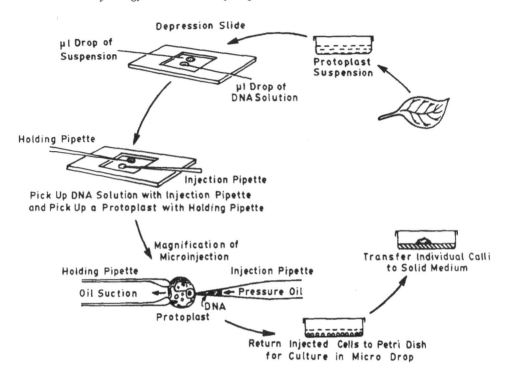

FIGURE 11.   Summary diagram of microinjection of plant protoplasts for cell transformation. (After Crossway, A., Oakes, J. Y., Irvine, J. M., Ward, B., Knauf, V. C., and Shewmaker, C. K., *Mol. Gen. Genet.*, 202, 179, 1986. With permission.)

lowing uptake of foreign nuclei, the protoplasts could grow on minimal medium and the presence of the *Vicia* genome in induced prototrophic clones of *Datura* was confirmed by dot-blot analysis. In some transformed clones recovery of prototrophy was also accompanied by restoration of morphogenic potential.

XII. PROSPECTS

The feasibility of transfer of foreign genes into plant cells using recombinant DNA technology has been demonstrated. Diverse genes have been shown to integrate into the plant genome and express themselves in an alien environment. The first result was achieved using T-DNA from *A. tumefaciens* as a vehicle, followed by CaM virus. The use of other potential vectors has yet to be demonstrated. Cell transformation has also been accomplished by direct gene transfer and also following microinjection of DNA and transplantation of a foreign nucleus.

Although plant genetic engineering was being attempted for quite some time, a breakthrough was possible in 1980[68] when an unequivocal demonstration of uptake, integration, and expression of bacterial gene was possible in plant cells. Since then, transfer of many select foreign genes and a native gene has been demonstrated by a number of groups using novel techniques. These strides in a matter of a few years are the promise for the future, and there is room for enthusiasm, and results of practical importance are awaited. An achievement in this direction[22a] is the transfer of herbicide (glyphosate) resistance from *Salmonella typhimurium* "via Ti-plasmid" to tobacco cells and, ultimately, to plants.

REFERENCES

1. **Akiyoshi, D. E., Klee, H., Amasino, R. M., Nester, E. W., and Gordon, M. P.,** T-DNA of *Agrobacterium tumefaciens* encodes an enzyme of cytokinin biosynthesis, *Proc. Natl. Acad. Sci. U.S.A.,* 81,, 5994, 1984.
2. **Amasino, R. M. and Miller, C.,** Hormonal control of tobacco crown gall tumour morphology, *Plant Physiol.,* 69, 389, 1982.
3. **An, G.,** High efficiency transformation of cultured tobacco cells, *Plant Physiol.,* 79, 568, 1985.
3a. **An, G.,** Development of plant promoter expression vectors and their use for analysis of differential activity of nopaline synthase promoter in transformed tobacco cells, *Plant Physiol.,* 81, 86, 1986.
3b. **An, G., Watson, B. D., and Chang, C. C.,** Transformation of tobacco, tomato, potato, and *Arabidopsis thabania* using a binary Ti-vector system, *Plant Physiol.,* 81, 301, 1986.
4. **Aoki, S. and Takebe, I.,** Infection of tobacco mesophyll protoplasts by tobacco mosaic virus ribonucleic acid, *Virology,* 39, 439, 1969.
4a. **Barta, A., Sommergruber, K., Thompson, D., Hartmuth, K., Matzke, M. A., and Matzke, A. J. M.,** The expression of nopaline synthase — human growth hormone chimeric gene in transformed tobacco and sunflower callus tissue, *Plant Mol. Biol.,* 6, 347, 1986.
5. **Beiderbeck, R.,** Wurzelinduktion an Blattern von *Kalanchoe daigremontiana* durch *Agrobacterium rhizogenes* und der Einfluss von kinetin auf diesen Prozess, *Z. Pflanzenphysiol.,* 68, 440, 1973.
6. **Bendich, A. J. and Filner, P.,** Uptake of exogenous DNA by pea seedlings and tobacco cells, *Mutat. Res.,* 13, 199, 1971.
7. **Bevan, M. W., Flavell, R. B., and Chilton, M. D.,** A chimaeric antibiotic resistance gene as a selectable marker for plant cell transformation, *Nature (London),* 304, 184, 1983.
8. **Binachi, F. and Walet-Foederer, H. G.,** An investigation into the anatomy of shoots apex of *Petunia hybrida* in connection with the results of transformation experiments, *Acta Bot. Neerl.,* 23, 1, 1974.
9. **Bomhoff, G., Klapwijk, P. M., Kester, B. C. M., Schilperoot, R. A., Hernalsteens, J. P., and Schell, J.,** Octopine and nopalene synthesis and breakdown are genetically controlled by a plasmid of *Agrobacterium tumefaciens,* Mol. Gen. Genet., 145, 177, 1976.
10. **Braun, A. C.,** Thermal studies on the factors responsible for tumour initiation in crown gall, *Am. J. Bot.,* 34, 234, 1947.
11. **Braun, A. C.,** The activation of two growth substance systems accompanying the conversion of normal to tumour cells in crown gall, *Cancer Res.,* 16, 53, 1956.
12. **Brisson, N., Paszkowski, J., Penswick, J. R., Gronenborn, B., Potrykus, I., and Hohn, T.,** Expression of a bacterial gene in plants by using a viral vector, *Nature (London),* 310, 511, 1984.
13. **Burr, B. and Burr, F. A.,** Detection of maize controlling element events at the shrunken locus, *Genetics,* 98, 143, 1981.
14. **Capecchi, M. R.,** High efficiency transformation by direct microinjection of DNA into cultured mammalian cells, *Cell,* 22, 479, 1980.
15. **Carlos, M. P. and Miller, J. H.,** Transposable elements, *Cell,* 20, 579, 1980.
16. **Cassells, A. C.,** Uptake of charged lipid vesicles by isolated tomato protoplasts, *Nature (London),* 275, 760, 1978.
17. **Chen, C. M., Ertl, J. R., Leisner, S. M., and Chang, C. C.,** Localization of cytokinin biosynthetic sites in pea plants and carrot roots, *Plant Physiol.,* 78, 510, 1985.
18. **Chen, C. M. and Melitz, D. K.,** Cytokinin biosynthesis in a cell-free system from cytokinin-autotrophic tobacco tissue cultures, *FEBS Lett.,* 107, 15, 1979.
19. **Chilton, M. D., Drummond, H. J., Merlo, D. J., and Sciaky, D.,** Highly conserved DNA of Ti-plasmids overlaps T-DNA maintained in plant tumours, *Nature (London),* 275, 147, 1978.
20. **Chilton, M. D., Drummond, M. H., Merlo, D. J., Sciaky, D., Montoya, A. L., Gordon, M. P., and Nester, E. W.,** Stable incorporation of plasmid DNA into higher plant cells: the molecular basis of crown gall tumorigenesis, *Cell,* 11, 263, 1977.
21. **Chilton, M. D., Saiki, R. K., Yadav, N., Gordon, M. P., and Quetier, P.,** T-DNA from *Agrobacterium* Ti-plasmid is in the nuclear DNA fraction of crown gall tumour cells, *Proc. Natl. Acad. Sci. U.S.A.,* 77, 4060, 1980.
22. **Christen, A. A. and Lurquin, P. F.,** Infection of cowpea mesophyll protoplasts with cowpea chlorotic mottler virus (CCMV) RNA encapsulated in large liposomes, *Plant Cell Rep.,* 2, 43, 1983.
22a. **Comai, L., Facciotti, D., Hiatt, W. R., Thompson, G., Rose, R. E., and Stalker, D. M.,** Expression in plants of a mutant gene from *Salmonella typhimurium* confers tolerance to glyphosate, *Nature (London),* 317, 741, 1985.
23. **Crossway, A., Oakes, J. V., Irvine, J. M., Ward, B., Knauf, V. C., and Shewmaker, C. K.,** Integration of foreign DNA following microinjection of tobacco mesophyll protoplasts, *Mol. Gen. Genet.,* 202, 179, 1986.

24. **Culter, A. J., Shargool, P. D., Kurz, W. G. W., and Constabel, F.**, Liposomes: a vehicles for transferring low molecular weight compounds into periwinkle protoplasts, *J. Plant Physiol.*, 117, 29, 1985.

25. **Davey, M. R., Cocking, E. C., Freeman, J., Pearce, N., and Tudor, I.**, Transformation of *Petunia* protoplasts by isolated Agrobacterium plasmid, *Plant Sci. Lett.*, 18, 307, 1980.

26. **DeBeuckeleer, M., Lemmers, M., De Vos, G., Willmitzer, L., Van Montagu, M., and Schell, J.**, Further insight on the transferred-DNA of octopine crown gall, *Mol. Gen. Genet.*, 193, 283, 1981.

26a. **DeBlock, M., Schell, J., and Montagu, M.**, Chloroplast transformation by *Agrobacterium tumefaciens*, *Eur. Mol. Biol. Organ. J.*, 4, 1367, 1985.

27. **DeFramond, A. J., Bevan, M. W., Barton, K. A., Flavell, R., and Chilton, M. D.**, Mini Ti-plasmid and a chimeric gene construct: new approaches to plant gene vector construction, in *Advances in Gene Technology: Molecular Genetics of Plants and Animals*, Vol. 20, Miami Winter Symp., Downey, K., Vollmy, R., Ahmed, F., and Schultz, J., Eds., Academic Press, New York, 1983, 159.

28. **DeFramond, A. J., Back, E. W., Chilton, W. S., Kayes, L., and Chilton, M-D.**, Two unlinked T-DNAs can transform the same tobacco plant and segregate in the F_1 generation, *Mol. Gen. Genet.*, 202, 125, 1986.

29. **DeGreve, H., Decraemer, H., Seurinck, J., Van Montagu, M., and Schell, J.**, The functional organization of the octopine *Agrobacterium tumefaciens* plasmid pTiB6S3, *Plasmid*, 6, 235, 1981.

30. **DeGreve, H., Dhaese, P., Seurinck, J., Lemmers, M., Van Montagu, M., and Schell, J.**, Nucleotide sequence and transcription map of the *Agrobacterium tumefaciens* Ti-plasmid encoded octopine synthase gene, *J. Mol. Appl. Genet.*, 1, 499, 1982.

31. **DeGreve, H., Leemans, J., Hernalsteens, J. P., Thia-Toong, L., De Beuckeleer, M., Willmitzer, L., Otten, L., Van Montagu, M., and Schell, J.**, Regeneration of normal and fertile plants that express octopine synthase from tobacco crown galls after deletion of tumour controlling functions, *Nature (London)*, 1982.

31a. **Dellaporta, S. and Fraley, R.**, unpublished.

32. **Depicker, A., Herman, L., Jacobs, A., Schell, J., and Van Montagu, M.**, Frequencies of simultaneous transformation with different T-DNAs and their relevance to the *Agrobacterium* plant cells interaction, *Mol. Gen. Genet.*, 201, 477, 1985.

33. **Depicker, A., Van Montagu, M., and Schell, J.**, Homologous DNA sequences in different Ti-plasmids are essential for oncogenicity, *Nature (London)*, 275, 150, 1978.

34. **Depicker, A., Van Montagu, M., and Schell, J.**, Plant cell transformation by *Agrobacterium* plasmids, in *Genetic Engineering of Plants*, Kosuge, T., Meredith, C., and Hollaender, A., Eds., Plenum Press, New York, 1983, 143.

35. **Depicker, A., Stachel, S., Dhaese, P., Zambryski, P., and Goodman, H. M.**, Nopaline synthase: transcript mapping and DNA sequence, *J. Mol. Appl. Genet.*, 1, 561, 1982.

35a. **Deshayao, A., Herrera-Estrella, L., and Caboche, M.**, Liposome mediated transformation of tobacco mesophyll protoplasts by an *E. coli* plasmid, *Eur. Mol. Biol. Organ. J.*, 4, 2731, 1985.

36. **Douglas, C. J., Staneloni, R. J., Rubin, R. C., and Nester, E. W.**, Identification and genetic analysis of an *Agrobacterium tumefaciens* chromosomal virulence region, *J. Bacteriol.*, 161, 850, 1985.

37. **Doy, C. H.**, The transfer and expression (transgenosis) of foreign genes in plant cells, reality and potential, in *The Eukaryote Chromosome*, Peacock, W. J. and Brock, R. D., Eds., Australian National University Press, Canberra, 1975, 447.

38. **Doy, C. H., Gresshoff, P. M., and Rolfe, B. G.**, Time course of phenotypic expression of *Escherichia coli* gene Z following transgenosis in haploid *Lycopersicon esculentum* cells, *Nature (London), New Biol.*, 244, 90, 1973.

39. **Draper, J., Davey, M. R., Freeman, J. P., Cocking, E. C., and Cox, B. J.**, Ti-plasmid homologous sequences present in tissues from *Agrobacterium* plasmid transformed Petunia protoplasts, *Plant Cell Physiol.*, 23, 255, 1982.

40. **Drummond, M. H., Gordon, M. P., Nester, E. W., and Chilton, M.-D.**, Foreign DNA of bacterial plasmid origin is transcribed in crown gall tumours, *Nature (London)*, 269, 535, 1977.

41. **Duvick, D. N.**, Cytoplasmic pollen sterility in corn, *Adv. Genet.*, 13, 1, 1965.

42. **Engler, G., Depicker, A., Maenhaut, R., Villarnoel-Mandiola, R., Van Montagu, M., and Schell, J.**, Physical mapping of DNA base sequence homologies between an octopine and a nopaline Ti-plasmid of *Agrobacterium tumefaciens*, *J. Mol. Biol.*, 152, 183, 1981.

43. **Fernandez, S. M., Lurquin, P. F., and Kado, C. I.**, Incorporation and maintenance of recombinant-DNA plasmid vehicle PBR 313 and PCRI in plant protoplasts, *FEBS Lett.*, 87, 277, 1978.

44. **Firmin, J. L. and Fenwick, G. R.**, Agropine — a major new plasmid-determined metabolite in crown gall tumours, *Nature (London)*, 276, 842, 1978.

45. **Forde, B. G. and Leaver, C. J.**, Nuclear and cytoplasmic genes controlling synthesis of variant mitochondrial polypeptides in male sterile maize, *Proc. Natl. Acad. Sci. U.S.A.*, 77, 418, 1980.

46. **Forde, B. G., Oliver, R. J. C., Leaver, C. J., Gunn, R. E., and Kemble, R. J.,** Classification of normal and male sterile cytoplasm in maize. I. Electrophoretic analysis of variation in mitochondrially synthesized proteins, *Genetics,* 95, 443, 1980.

47. **Fraley, R., Dellaporta, S., and Papahadjopoulos, D.,** Liposome-mediated delivery of tobacco mosaic virus RNA into tobacco protoplasts: a sensitive assay for monitoring liposome-protoplast interaction, *Proc. Natl. Acad. Sci. U.S.A.,* 79, 1859, 1982.

48. **Fraley, R. T., Fornari, C. S., and Kaplan, S.,** Entrapment of bacterial plasmid in phospholipid vesicles: Potential for gene transfer, *Proc. Natl. Acad. Sci. U.S.A.,* 76, 3348, 1979.

49. **Fraley, R. T. and Horsch, R. B.,** In vitro plant transformation system using liposomes and bacterial cocultivation, in *Genetic Engineering of Plants,* Kosuge, T., Meredith, C., and Hollaender, A., Eds., Plenum Press, New York, 1983, 177.

50. **Fraley, R. T., Horsch, R. B., Matzke, A., Chilton, M. D., Chilton, W. S., and Sanders, P. R.,** In vitro transformation of petunia cells by an improved method of cocultivation with *A. tumefaciens* strains, *Plant Mol. Biol.,* 1, 171, 1984.

51. **Fraley, R. T., Rogers, S. G., and Horsch, R. B.,** Use of chimeric gene to confer antibiotic resistance to plant cells, in *Advances in Gene Technology: Molecular Genetics of Plants and Animals,* Vol. 20, Miami Winter Symp., Downey, K., Vollmy, R., Ahmed, F., and Schultz, J., Eds., Academic Press, New York, 1983, 211.

52. **Franck, A., Jonard, G., Richard, K., Hirth, L., and Guilley, H.,** Nucleotide sequence of cauliflower mosaic virus DNA, *Cell,* 21, 285, 1980.

53. **Freeman, J. P., Draper, J., Davey, M. R., Cocking, E. C., Garland, K. M. A., Harding, K., and Pental, D.,** A comparison of methods for plasmid delivery into plant protoplasts, *Plant Cell Physiol.,* 25, 1353, 1984.

54. **Fromm, M., Taylor, L. P., and Walbot, Y.,** Expression of genes transferred into monocot and dicot plant cells by electroporation, *Proc. Natl. Acad. Sci. U.S.A.,* 82, 5824, 1985.

55. **Fukunaga, V., Nagata, T., and Takebe, I.,** Liposome-mediated infection of plant protoplasts with tobacco mosaic virus DNA, *Virology,* 113, 752, 1981.

56. **Gelvin, S. B., Gordon, M. P., Nester, E. W., and Aronson, A. I.,** Transcription of *Agrobacterium* Ti-plasmid in the bacterium and in crown gall tumours, *Plasmid,* 6, 17, 1981.

57. **Goldberg, R. B., Okamura, J. K., and Jafaku, D.,** Soybean seed protein genes are developmentally regulated in tobacco plants in *Abstr. 1st Int. Congr. Plant Mol. Biol.,* Georgia University, Georgia, 1985, 1.

57. **Goldsbrough, P. B., Gelvin, S. B., and Larkins, B. A.,** Expression of maize zein genes in transformed sunflower cells, *Mol. Gen. Genet.,* 202, 374, 1986.

58. **Goodman, R. M.,** Gemini viruses, *J. Gen. Virol.,* 54, 9, 1981.

59. **Gordon, M. P., Amasino, R., Garfinkel, D., Huffman, G., Klee, H., Knauf, V., Kwok, W., Lichtenstein, C., Montoya, A., Nester, E., Powell, A., Ream, L., Rubin, R., Stachel, S., Taylor, B., Watson, B., White, F., and Yanofsky, M.,** Current developments in transformation of plants, in *Advances in Gene Technology: Molecular Genetics of Plants and Animals,* Vol. 20, Miami Winter Symp., Downey, K., Vollmy, R., Ahmed, F., and Schultz, J., Eds., Academic Press, New York, 1983, 37.

60. **Grandmann-Rebel, W. and Hemleben, V.,** Incorporation T4 phage DNA into a specific DNA fraction from the higher plant *Matthiola incana, Z, Naturforsch.,* 31, 558, 1976.

61. **Greierson, D., McKee, R. A., Attridge, T. H., and Smith, H.,** Studies on the uptake and expression of foreign genetic material by higher plant cells, in *Modification of the Information Content of Plant Cells,* Markham, R., Davies, D. R., Hopwood, D. A., and Horne, R. W., Eds., North-Holland, Amsterdam, 1975, 91.

62. **Griesbach, R. J.,** Protoplast microinjection, *Plant Mol. Biol. Rep.,* 1, 32, 1983.

63. **Gronenborn, B., Gardner, R. C., Schaefer, S., and Shepard, R. J.,** Propagation of foreign DNA in plants using cauliflower mosaic virus as vector, *Nature (London),* 294, 773, 1981.

64. **Gurley, W. B., Kemp, J. D., Alber, M. J., Sulton, D. W., and Gallis, J.,** Transcription of Ti-plasmid derived sequences in three octopine type crown gall tumour lines, *Proc. Natl. Acad. Sci. U.S.A.,* 76, 2828, 1979.

65. **Guyon, P., Chilton, M.-D., Petit, A., and Tempe, J.,** Agropine in null type crown gall tumours: evidence for the generality of opine concept, *Proc. Natl. Acad. Sci. U.S.A.,* 77, 269, 1980.

66. **Hain, R., Stabel, P., Czernilofsky, A. P., Steinbiss, H. H., Herrera-Esterella, L., and Schell, J.,** Uptake, integration expression and genetic transmission of a selectable chimeric gene by plant protoplasts, *Mol. Gen. Genet.,* 199, 161, 1985.

67. **Hanold, D.,** In vitro transformation of protoplast-derived *Hyoscyamus muticus* cells by *Agrobacterium tumefaciens, Plant Sci. Lett.,* 30, 177, 1983.

68. **Hernalsteens, J. P., Var Vilet, F., de Beuckeleer, M., Depicker, A., Engler, G., Leemers, M., Holsters, M., Van Montagu, M., and Schell, J.,** The *Agrobacterium tumefaciens* Ti-plasmid as a host vector system for introducing foreign DNA in plant cells, *Nature (London),* 287, 654, 1980.

69. Herreara-Esterella, L., DeBlock, M., Messens, E., Hernalsteens, J. P., Van Montagu, M., and Schell, J., Chimeric genes as dominant seletable markers in plant cells, *Eur. Mol. Biol. Organ. J.*, 2, 987, 1983.

70. Herreara-Esterella, L., Van den Broeck, G., Maenhaut, R., Van Montagu, M., Schell, J., Timko, M., and Cashmore, A., Light-inducible and chloroplast associated expression of a chimeric gene introduced into *Nicotiana tabacum* using Ti-plasmid vector, *Nature (London)*, 310, 115, 1984.

71. Hess, D., Versuche zur Transformation an hoheren Pflanzen: Wiederholung der Anthocyan-Induktion bei Petunia und erste Charakterisierung des transformierenden Prinzips, *Z. Pflanzenphysiol.*, 61, 286, 1969.

72. Hess, D., Uptake of DNA and bacteriophage into pollen and genetic manipulation, in *Genetic Manipulation with Plant Material*, Ledoux, L., Ed., Plenum Press, New York, 1975, 519.

73. Hess, D., Gresshoff, P. M., Fielitz, U., and Gleiss, D., Uptake of protein and bacteriophage into swelling and germinating pollen of *Petunia hybrida*, *Z. Pflanzenphysiol.*, 74, 371, 1974.

74. Heyn, R. F. and Schilperoort, R. A., The use of protoplasts to follow the fate of *Agrobacterium tumefaciens* DNA on incubation with tobacco cells, *Colloq. Int. C.N.R.S.*, 212, 385, 1971.

75. Hohn, T., Richards, K., and Lebeuvier, G., Cauliflower mosaic virus on its way to becoming a useful plant vector, *Curr. Top. Microbiol. Immunol.*, 96, 193, 1982.

76. Hooykaas, P. J. J., Klapwijk, P. M., Nuti, M. P., Schilperoort, R. A., and Rorsch, A., Transfer of the *Agrobacterium tumefaciens* Ti-plasmid to avirulent Agrobacteria and to Rhizobium explant, *J. Gen. Microbiol.*, 98, 477, 1977.

77. Hooykaas, P. J. J., Ooms, G., and Schiperroort, R. A., Tumours induced by different strains of *Agrobacterium tumefaciens*, in *Molecular Biology of Plant Tumours*, Kahl, G. and Schell, J., Eds., Academic Press, New York, 1982, 374.

77a. Hooykaas, P. J. J. and Schilperoort, R. A., The Ti-plasmid of *Agrobacterium tumefaciens*: a natural genetic engineer, *Trends Biochem. Sci.*, 10, 417, 1985.

78. Horsch, R. B., Fraley, R. T., Rogers, S. G., Sanders, P. R., Lloyd, A., and Hoffmann, N., Inheritance of functional foreign genes in plants, *Science*, 223, 496, 1984.

79. Hotta, Y. and Stern, H., Uptake and distribution of heterologous DNA in living cells, in *Informative molecules in Biological Systems*, Ledoux, L., Ed., North-Holland, Amsterdam, 1971, 176.

80. Howell, S. H. and Hull, R., Replication of cauliflower mosaic virus and transcription of its genome in turnip leaf protoplasts, *Virology*, 86, 468, 1978.

81. Howell, S. H., Walker, L. L., and Dudley, R. K., Cloned cauliflower mosaic virus DNA infects turnips, *Brassica rapa*, *Science*, 208, 1265, 1980.

82. Howell, S. H., Walker, L. L., and Walden, R. M., Rescue of in vitro generated mutants of cloned cauliflower mosaic virus genome in infected plants, *Nature (London)*, 293, 483, 1981.

83. Hughes, B. G., White, F. G., and Smith, M. A., Fate of bacterial plasmid DNA during uptake by barley protoplasts, *FEBS Lett.*, 79, 80, 1977.

84. Hull, R., Covey, S. N., Stanley, J., and Davies, J. W., The polarity of cauliflower mosaic virus genome, *Nucl. Acid Res.*, 7, 669, 1979.

85. Hull, R. and Howell, S. H., Structure of the cauliflower mosaic virus genome. II. Variation in DNA structure and sequence between isolates, *Virology*, 86, 482, 1978.

86. Jaenisch, R. and Mintz, B., Simian virus 40 DNA sequences in DNA of healthy adult mice derived from perisimplantation blastocytes injected with viral DNA, *Proc. Natl. Acad. Sci. U.S.A.*, 71, 1250, 1974.

87. Jia, J. F., Shillito, R. D., and Potrykus, I., Crown gall transformation of regenerating protoplasts of haploid and diploid *Petunia hybrida* var. Mitchell by *Agrobacterium tumefaciens*, *Z. Pflanzephysiol.*, 112, 1, 1983.

88. Johnson, C. B., Grierson, D., and Smith, H., Expression of λp lac 5 DNA in cultured cells of higher plant, *Nature (London), New Biol.*, 244, 105, 1973.

89. Joos, H., Inze, D., Caplan, A., Sormann, M., Van Montagu, M., and Schell, J., Genetic analysis of T-DNA transcripts in nopaline crown galls, *Cell*, 32, 1057, 1983.

90. Kemble, R. J. and Bedbrook, J. R., Low molecular weight circular and linear DNA in mitochondria from normal and male sterile *Zea mays* cytoplasm, *Nature (London)*, 284, 365, 1980.

91. Kemble, R. J., Gunn, R. E., and Flavell, R. B., Classification of normal and male sterile cytoplasms in maize. II. Electrophoretic analysis of DNA species in mitochondria, *Genetics*, 95, 451, 1980.

92. Kemp, J. D., Sutton, D. W., Fink, C., Barker, R. F., and Hall, T. C., *Agrobacterium* -mediated transfer of foreign genes into plants, in *Genetic Engineering, Application to Agriculture*, Beltsville Symp. Agric. Res., Owens, L. D., Ed., Rowman and Allanhall, London, 1983, 215.

93. Kemper, E., Waffenschmidt, S., Weiler, E. W., Rausch, T., and Schroder, J., T-DNA-encoded auxin formation in crown gall cells, *Planta*, 163, 257, 1985.

94. Klee, H. J., Gordon, M. P., and Nester, E. W., Complementation analysis of *Agrobacterium tumefaciens* Ti-plasmid mutation affecting oncogenicity, *J. Bacteriol.*, 150, 322, 1982.

95. Klee, H. J., Montoya, A., Horodyski, F., Lichtenstein, C., Garfinkel, D., Fuller, S., Flores, C., Peschon, J., Nester, E. W., and Gordon, M. P., *Proc. Natl. Acad. Sci. U.S.A.*, 81, 1728, 1984.

96. **Kleinhofs, A.**, DNA-hybridization studies of the fate of bacterial DNA in plants, in *Genetic Manipulation with Plant Material*, Ledoux, L., Ed., Plenum Press, New York, 1975, 461.
97. **Kleinhofs, A. and Behki, R.**, Prospects for plant genome modification by non-conventional methods, *Annu. Rev. Genet.*, 11, 79, 1977.
98. **Knauf, V. C.**, Ph.D. thesis, University of Washington, Seattle, 1982.
99. **Koekman, B. P., Hooykaas, P. J. J., and Schilperoort, R. A.**, Localization of the replication control region of the physical map of the octopine Ti-plasmid, *Plasmid*, 4, 184, 1980.
100. **Kondorosi, E. and Duda, E.**, Introduction of foreign genetic material into cultured mammalian cells by liposomes loaded with isolated nuclei, *FEBS Lett.*, 120, 37, 1980.
101. **Koulikova-Nicola, Z., Shilloto, R. D., Hohn, B., Wang, K., Van Montagu, M., and Zambryski, P.**, Involvement of circular intermediates in the transfer of T-DNA from *Agrobacterium tumefaciens* to plant cells, *Nature (London)*, 313, 191, 1985.
102. **Koziel, M. and Siegel, A.**, Cloning of a DNA copy of tobacco-rattle virus RNA-2, in *Proc. 5th Int. Congr. Virology, Int. Union of Mol. Biol., Strasbourg*, 1981, 257.
103. **Krens, F. A., Molendijk, L., Wullems, G. J., and Schilperoort, R. A.**, In vitro transformation of plant protoplasts with Ti-plasmid DNA, *Nature (London)*, 296, 72, 1982.
103a. **Lamppa, G., Nagy, F., and Chua, N. H.**, Light regulated and organ specific expression of a wheat cab gene in transgenic tobacco, *Nature (London)*, 316, 750, 1985.
104. **Langridge, W. H. R., Li, B. J., and Szalay, A. A.**, Electric field mediated stable transformation of carrot protoplasts with naked DNA, *Plant Cell Rep.*, 4, 155, 1985.
105. **Lawrence W. A. and Davies, D. R.**, A method for the microinjection and culture of protoplasts at very low densities, *Plant Cell Rep.*, 4, 33, 1985.
106. **Lebeurier, G., Hirth, L., Hohn, T., and Hohn, B.**, Infectivities of native and cloned DNA of cauliflower mosaic virus, *Gene*, 12, 139, 1980.
107. **Ledoux, L. and Huart, R.**, Integration and replication of DNA of *M. lysodeikticus* in DNA of germinating barley, *Nature (London)*, 218, 1256, 1968.
108. **Ledoux, L., Huart, R., and Jacobs, M.**, Fate of exogenous DNA in *Arabidopsis thaliana*. Translocation and integration, *Eur. J. Biochem.*, 23, 96, 1971.
109. **Ledoux, L., Huart, R., and Jacobs, M.**, DNA mediated genetic correction of thamineless *Arabidopsis thaliana*, *Nature (London)*, 249, 17, 1974.
110. **Leemans, J., Deblaere, R., Willmitzer, L., DeGreve, H., Hernalsteens, J. P., Van Montagu, M., and Schell, J.**, Genetic identification of functions of TL-DNA transcripts in octopine crown galls, *Eur. Mol. Biol. Organ. J.*, 1, 142, 1982.
111. **Lemmers, M., De Beuckeleer, M., Holsters, M., Zambryski, P., Depicker, A., Hernalsteens, J. P., Van Montagu, M., and Schell, J.**, Internal organization, boundaries and integration of Ti-plasmid. DNA in nopaline crown gall tumours, *J. Mol. Biol.*, 144, 353, 1980.
112. **Levings, C. S., Kim, B. D., Pring, D. R., Conde, M. F., Mans, R. J., Laughnan, J. R., and Gabay-Laughnan, S. J.**, Cytoplasmic reversion of *cms* in maize associated with a transpositional event, *Science*, 209, 1021, 1980.
113. **Lorz, H., Baker, B., and Schell, J.**, Gene transfer to cereal cells mediated by protoplast transformation, *Mol. Gen. Genet.*, 199, 178, 1985.
114. **Lorz, H. and Potrykus, I.**, Investigations on the transfer of isolated nuclei into plant protoplasts, *Theor. Appl. Genet.*, 53, 251, 1978.
115. **Lurquin, P. F.**, Entrapment of plasmid DNA by liposomes and their interactions with plant protoplasts, *Nucl. Acid Res.*, 6, 3773, 1979.
116. **Lurquin, P. F. and Hotta, Y.**, Reutilization of bacterial DNA by *Arabidopsis thaliana* cells in tissue culture, *Plant Sci. Lett.*, 5, 103, 1975.
117. **Lurquin, P. F. and Kado, C. I.**, *Escherichia coli* plasmid pBR313 insertion into plant protoplasts and into their nuclei, *Mol. Gen. Genet.*, 154, 113, 1977.
118. **Lurquin, P. F. and Sheehy, R. E.**, Binding of large liposomes to plant protoplasts and delivery of encapsulated DNA, *Plant Sci. Lett.*, 25, 133, 1982.
119. **Lurquin, P. F. and Sheehy, R. E.**, Effects of conditions of incubation on the binding of DNA-loaded liposomes to plant protoplasts, *Plant Sci. Lett.*, 28, 49, 1983.
120. **Malhotra, K., Rashid, A., and Maheshwari, S. C.**, Transgenosis in higher plant cells — a reinvestigation, *Z. Pflanzenphysiol.*, 95, 21, 1979.
121. **Marton, L., Wullems, G. J., Molendijk, L., and Schilperoort, R. A.**, In vitro transformation of cultured cells from *Nicotiana tabacum* by *Agrobacterium tumefaciens*, *Nature (London)*, 277, 129, 1979.
122. **Mathews, B. F. and Cress, D. F.**, Liposome-mediated delivery of DNA to carrot protoplasts, *Planta*, 153, 90, 1981.
123. **Mathews, B. F., Dray, S., Widholm, J., and Ostro, M.**, Liposome-mediated transfer of bacterial RNA into carrot protoplasts, *Planta*, 145, 37, 1979.

124. **Matzke, A. J. M. and Chilton, M.-D.,** Site-specific insertion of genes into T-DNA of *Agrobacterium* tumour-inducing plasmid: an approach to genetic engineering of higher plant cells, *J. Mol. Appl. Genet.,* 1, 39, 1981.

125. **McPherson, J. C., Nester, E. W., and Gordon, M. P.,** Proteins encoded by *Agrobacterium tumefaciens* Ti-plasmid DNA (T-DNA) in crown gall tumors, *Proc. Natl. Acad. Sci. U.S.A.,* 77, 2666, 1980.

126. **Meyer, P., Walgenbach, E., Bussmann, K., Hombrecher, G., and Saedlar, H.,** Synchronized tobacco protoplasts are efficiently transformed by DNA, *Mol. Gen. Genet.,* 201, 513, 1985.

127. **Morikawa, H. and Yamada, Y.,** Capillary microinjection into protoplasts and intranuclear localization of injected materials, *Plant Cell Physiol.,* 26, 229, 1985.

128. **Motoyoshi, F., Bancraft, J. B., Watts, J. W., and Burgess, J.,** The infection of tobacco protoplasts with cowpea chlorotic mottle virus, *J. Gen. Virol.,* 20, 177, 1973.

129. **Mukherjee, A. B., Orloff, S., Butler, J. D. B., Triche, T., Lalley, P., and Schulman, J. P.,** Entrapment of mataphase chromosomes into phospholipid vesicles (lipochromosomes): carrier potential in gene transfer, *Proc. Natl. Acad. Sci. U.S.A.,* 75, 1361, 1978.

130. **Murai, N. and Kemp, J. D.,** T-DNA of pTi-15955 from *Agrobacterium tumefaciens* is transcribed into a minimum of seven polyadenylated RNAs in a sunflower crown gall tumour, *Nucl. Acid Res.,* 10, 1679, 1982.

130a. **Murai, N., Ohshima, M., and Sakamoto, M.,** Plant transformation with T-plasmid derived binary vector, in *Abstr. 1st Int. Congr. Plant Mol. Biol.,* Georgia University, Athens, 1985, 115.

130b. **Nagy, F., Morelli, G., Fraley, R. T., Rogers, S. G., and Chua, N. H.,** Photoregulated expression of a pea rbe S gene in leaves of transgenic plants, *Eur. Mol. Biol. Organ. J.,* 4, 3063, 1985.

131. **Neff, N. T. and Binns, A. N.,** *Agrobacterium tumefaciens* interaction with suspension cultured tomato cells, *Plant Physiol.,* 77, 35, 1985.

132. **Nishiguchi, M., Langridge, W. H. R., Szalay, A. A., and Zaitlin, M.,** Electroporated-mediated infection of tobacco leaf protoplasts with tobacco mosaic virus RNA and cucumber mosaic virus RNA, *Plant Cell Rep.,* 5, 57, 1986.

133. **Norton, R. A. and Towers, G. H. N.,** Transmission of nopaline crown gall tumour markers through meiosis in regenerated whole plants of *Bidens alba, Can. J. Bot.,* 62, 408, 1984.

134. **Odell, J. T., Dudley, K., and Howell, S. H.,** Structure of 19S RNA transcript encoded by the cauliflower mosaic virus genome, *Virology,* 111, 377, 1981.

135. **Ohyama, K.,** Genetic transformation in plants, in *Handbook of Plant Cell Culture, Vol. 1,* Evans D. A., Sharp, W. R., Ammirato, P. V., and Yamada, Y., Eds., Macmillan, New York, 1983, 501.

136. **Ohyama, K., Gamborg, O. L., and Miller, R. A.,** Isolation and properties of DNA from protoplasts of cell suspension cultures of *Ammi visnage* and *Daucus carota, Plant Physiol.,* 50, 319, 1972.

137. **Ohyama, K., Gamborg, O. L., Shyluk, J. P., and Miller, R. A.,** Studies on transformation: uptake of exogenous DNA by plant protoplasts, in *Protoplastes et Fusion de Cellules Somatiques Vegetales,* Tempe, J., Ed., Centre National de la Recherche Scientifique, Paris, 1973, 212, 423.

138. **Ohyama, K., Pelcher, L. E., and Horn, D.,** DNA binding and uptake by nuclei isolated from plant protoplasts, *Plant Physiol.,* 60, 98, 1977.

139. **Okker, R. J. H., Spaink, H., Hille, J., Van Brussel, T. A. N., Lugtenberg, B., and Schilperoort, R. A.,** Plant-inducible virulence promoter of the *Agrobacterium tumefaciens* Ti-plasmid, *Nature (London),* 312, 564, 1984.

140. **Otten, L., De Greve, H., Hernalsteens, J. P., Van Montagu, M., Schieder, O., Straub, J., and Schell, J.,** Mendelian transmission of genes introduced into plants by the Ti-plasmids of *Agrobacterium tumefaciens, Mol. Gen. Genet.,* 183, 209, 1981.

141. **Otten, L., Piotrowiak, G., Hooykaas, P., Dubois, M., Szegedi, E., and Schell, J.,** Identification of an *Agrobacterium tumefaciens* pTiB653 vir region fragment that enhances the virulence of pTiC58, *Mol. Gen. Genet.,* 199, 189, 1985.

142. **Owens, L. D.,** Binding of Col E-Kan plasmid DNA by tobacco protoplasts. Non-expression of plasmid gene, *Plant Physiol.,* 63, 683, 1979.

143a. **Paszkowski, J., Shillito, R. D., Saul, M., Mandak, V., Hohn, T., Hohn, B., and Potrykus, I.,** Direct gene transfer to plants, *Eur. Mol. Biol. Organ. J.,* 3, 2717, 1984.

143b. **Paszkowski, J., Pisan, B., Shillito, R. B., Hohn, T., Hohn, B., and Potrykus, I.,** Genetic transformation of *Brassica campestris* var rapa protoplasts with an engineered cauliflower mosaic virus, *Plant Mol. Biol.,* 6, 303, 1986.

144. **Petit, A., Delhaye, S., Tempe, J., and Morel, G.,** Recherches sur les guanidines des Tissus de crown gall. Mise en evidence d'une relation biochemique specifique entre les souches d'*Agrobacterium* et les tumeurs qu'elles indusient, *Physiol. Veg.,* 8, 205, 1970.

145. **Pollock, K., Barfield, O. G., Robinson, S. J., and Shields, R.** Transformation of protoplast-derived cell colonies and suspension cultures by *Agrobacterium tumefaciens, Plant Cell Rep.,* 4, 202, 1985.

146. **Potrykus, I., Saul, M., Petruska, J., Paszkowski, J., and Shillito, R. D.,** Direct gene transfer into protoplasts of graminaceous monocot, *Mol. Gen. Genet.,* 199, 183, 1985.

147. **Pring, D. R. and Levings, C. S.,** Heterogeneity of male cytoplasmic genome among male sterile cytoplasm, *Genetics,* 80, 121, 1978.

148. **Pring, D. R., Levings, C. S., Hu, W. W., and Timothy, D. H.,** Unique DNA associated with mitochondria in the "S" type cytoplasm of male sterile maize, *Proc. Natl. Acad. Sci. U.S.A.,* 74, 2904, 1977.

149. **Rebel, W., Hemleben, V., and Seyffert, W.,** Fate of T4 phage DNA in seedlings of *Matthiola incana, Z. Naturforsch.,* 28, 473, 1973.

150. **Redei, G. P., Acedo, G., Weingarten, H., and Kier, L. D.,** Has DNA corrected genetically thiamineless mutants of *Arabidopsis,* in *Cell Genetics in Higher Plants, Proc. Int. Training Course, Szeged, Hungary,* Dudits, D., Farkas, G. L., and Maliga, P., Eds., Budapest Akademie, Kiado, 1977, 91.

151. **Robins, D. M., Ripley, S., Henderson, A. S., and Axel, R.,** Transforming DNA integrates into the host chromosome, *Cell,* 23, 29, 1981.

152. **Rogler, C. E.,** Plasmid-dependent temperature sensitive phase in crown gall tumorigenesis, *Proc. Natl. Acad. Sci. U.S.A.,* 77, 2688, 1980.

153. **Rouze, P., Deshayes, A., and Caboche, M.,** Use of liposomes for the transfer of nucleic acids: optimization of the method for tobacco mesophyll protoplasts with tobacco mosaic virus RNA, *Plant Sci. Lett.,* 31, 55, 1983.

154. **Rubin, R. A.,** Genetic studies on the role of octopine T-DNA border regions in crown gall tumor formation, *Mol. Gen. Genet.,* 202, 312, 1986.

155. **Rubin, G. M. and Spradling, A. C.,** Genetic transformation of *Drosophila* with transposable element vectors, *Science,* 218, 348, 1982.

156. **Sanford, J. C., Skubik, K. A., and Reisch, B. I.,** Attempted pollen-mediated plant transformation employing genomic donor DNA, *Theor. Appl. Genet.,* 69, 571, 1985.

157. **Saxena, P. K., Mii, M., Crosby, W. L., Fowke, L. C., and King, J.,** Transplantation of isolated nuclei into plant protoplasts: a novel technique for introducing foreign DNA into plant cells, *Planta,* 168, 29, 1986.

158. **Schell, J., Van Montagu, M., Holsters, M., Zambryski, P., Joos, H., Inze, D., Herrera-Estrella, L., Depicker, A., De Block, M., Caplan, A., Dhaese, P., Van Haute, E., Hernalsteens, J. P., De Greeve, H., Leemans, J., Deblaere, R., Willmitzer, L., Schröder, J., and Otten, L.,** Ti-plasmids as experimental gene vectors for plants, in *Advances in Gene Technology: Molecular Genetics of Plant and Animals,* Vol. 20, Miami Winter Symp., Downey, K., Vollmy, R., Ahmed, F., and Schultz, J., Eds., Academic Press, New York, 1983, 191.

158a. **Schell, J., Kreuzaler, F., Kaulen, H., Willmitzer, L., Eckes, P., Roshal, S., Spena, A., Baker, B., and Fedoroff, N.,** Use of Ti-plasmid vectors to study the transfer and regulation and expression of chimeric genes in plants, in *Abstr. 1st Int. Congr. Plant Mol. Biol., Georgia University,* Athens, 1985, 5.

159. **Schröder, G., Klipp, W., Hillebrandt, A., Ehring, R., Koncz, C., and Schröder, J.,** The conserved part of the T-region in Ti-plasmid expresses four proteins in bacteria, *Eur. Mol. Biol. Organ. J.,* 2, 403, 1983.

160. **Schröder, J. and Schröder, G.,** Hybridization, selection and translation of T-DNA encoded mRNAs from octopine tumours, *Mol. Gen. Genet.,* 185, 51, 1982.

161. **Schröder, J., Schröder, G., Huisman, H., Schilperoort, R. A., and Schell, J.,** The mRNA of lysopine dehydrogenase in plant tumours is complimentary to a Ti-plasmid fragment, *FEBS Lett.,* 129, 166, 1981.

162. **Sheehy, R. and Lurquin, P. F.,** Targeting of large liposomes with lectins increases their binding to plant protoplasts, *Plant Physiol.,* 72, 386, 1983.

162a. **Shillito, R. D., Saul, M. W., Paszkowski, J., and Potrykus, I.,** Direct gene transfer to plants, *IAPTC Newslett.,* 5, 1986.

163. **Siegel, A.,** Plant-virus-based vectors for gene transfer may be of considerable use despite a preserved high error frequency during RNA synthesis, *Plant Mol. Biol.,* 4, 327, 1985.

163a. **Simpson, J., Tinko, M. P., Cashmore, A. R., Schell, J., Van Montagu, M., and Herrera-Estrella, L.,** Light inducible and tissue specific expression of a chimeric gene under control of 5'-flanking sequence of a pea chlorophyll a/b binding protein, *Eur. Mol. Biol. Organ. J.,* 4, 2723, 1985.

164. **Simpson, R. B., O'Hara, P. J., Kwok, W., Montoya, A. L., Lichtenstein, C., Gordon, M. P., and Nester, E. W.,** DNA from the A65/2 crown gall tumor contains scrambled Ti-plasmid sequence near its junctions with plant DNA, *Cell,* 29, 1005, 1982.

165. **Smith, H., McKee, R. A., Atridge, T. H., and Grierson, D.,** Studies on the use of transducing bacteriophages as vectors for the transfer of foreign genes to higher plants, in *Genetic Manipulation with Plant Material,* Ledoux, L., Ed., Plenum Press, New York, 1975, 551.

166. **Soyfer, V. N.,** Hereditary variability of plants under the action of exogenous DNA, *Theor. Appl. Genet.,* 58, 225, 1980.

166a. **Spena, A., Hain, R., Ziervogel, V., Saedler, H., and Schell, J.,** Construction of a heat-inducible gene for plants. Demonstration of heat inducible activity of *Drosophila* hsp 70 promoter in plants, *Eur. Mol. Biol. Organ. J.,* 4, 2739, 1985.

167. **Stachel, S. E., Messens, E., Van Montagu, M., and Zambryski, P.,** Identification of signal molecules produced by wounded plant cells that activate T-DNA transfer in *Agrobacterium tumefaciens, Nature (London),* 318, 624, 1985.

168. **Starlinger, P.,** A reexamination of McClintock's "controlling elements" in maize in view of recent advances in molecular biology, in *Genome Organization and Expression in Plants,* Leaver, C. J., Ed., Plenum Press, New York, 1980, 537.

169. **Steinbiss, H. H. and Stable, P.,** Protoplast derived tobacco cells can survive capillary microinjection of the fluorescent dye, Lucifer yellow, *Protoplasma,* 116, 223, 1983.

170. **Suzuki, M. and Takebe, I.,** Uptake of single-stranded bacteriophage DNA by isolated tobacco protoplasts, *Z. Pflanzenphysiol.,* 78, 421, 1976.

171. **Szoka, F. and Papahadjopoulos, D.,** Procedure for preparation of liposomes with large internal aqueous space and high capture by reverse phase evaporation, *Proc. Natl. Acad. Sci. U.S.A.,* 75, 4194, 1978.

172. **Tate, M. E., Ellis, J. C., Kerr, A., Tempe, J., Murray, K., and Shaw, K.,** Agropine: a revised structure, *Carbohydr. Res.,* 104, 105, 1982.

173. **Tempe, J., Petit, A., Holsters, M., Van Montagu, M., and Schell, J.,** Thermosensitive step associated with transfer of the Ti-plasmid during conjugation, a possible relation to transformation in crown gall, *Proc. Natl. Acad. Sci. U.S.A.,* 74, 2848, 1977.

174. **Thomashow, M. F., Nutter, R., Montoya, A. L., Gordon, M. P., and Nester, E. W.,** Integration and organization of Ti-plasmid sequences in crown gall tumours, *Cell,* 19, 729, 1980.

175. **Uchimiya, H.,** Parameters influencing the liposome-mediated insertion of fluorescein diacetate into plant protoplasts, *Plant Physiol.,* 67, 629, 1981.

176. **Uchimiya, H. and Murashige, T.,** Quantitative analysis of the fate of exogenous DNA in *Nicotiana* protoplasts, *Plant Physiol.,* 59, 301, 1977.

177. **Van Larebecker, N., Engler, G., Holsters, M., Elsacker, S., Zaenen, I., Schilperoort, R., and Schell, J.,** Large plasmid in *Agrobacterium tumefaciens* is essential for crown gall-inducing ability, *Nature (London),* 252, 169, 1974.

178. **Van Vloten-Doting, L., Bol, J. F., and Cornelissen, B.,** Plant virus based vectors for gene transfer will be of limited use because of high error frequency during viral RNA synthesis, *Plant Mol. Biol.,* 4, 323, 1985.

178a. **Wang, K., Herrera-Estrella, L., Van Montagu, M., and Zambryski, P.,** Right 25 bp terminus sequence of nopaline T-DNA is essential for determination of DNA transfer from *Agrobacterium* to plant genome, *Cell,* 38, 455, 1984.

179. **Werr, W. and Lörz, H.,** Transient gene expression in a Gramineae cell line. A rapid procedure for studying plant promotors, *Mol. Gen. Genet.,* 202, 471, 1986.

180. **White, F., Ghidossi, G., Gordon, M., and Nester, E. W.,** Tumor-induction by *Agrobacterium rhizogenes* involves the transfer of plasmid DNA to the plant genome, *Proc. Natl. Acad. Sci. U.S.A.,* 79, 3193, 1982.

181. **Willmitzer, L., de Beuckeleer, M., Leemers, M., Van Montagu, M., and Schell, J.,** DNA from Ti-plasmid is present in the nucleus and absent from plastids of plant crown gall cells, *Nature (London),* 287, 359, 1980.

182. **Willmitzer, L., Dhaese, P., Shreier, J., Schmalenbach, M., Van Montagu, M., and Schell, J.,** Size, location and polarity of T-DNA encoded transcripts present in nopaline crown gall tumours, *Cell,* 32, 1045, 1983.

183. **Willmitzer, L., Otten, L., Simons, G., Schmalenbach, W., Schröder, J., Schröder, G., Van Montagu, M., De Vos, G., and Schell, J.,** Nuclear and polysomal transcripts of T-DNA in octopine crown gall suspension and callus cultures, *Mol. Gen. Genet.,* 182, 255, 1981.

184. **Willmitzer, L., Schmalenbach, W., and Schell, J.,** Transcription of T-DNA in octopine and nopaline crown gall tumours is inhibited by low concentration of α-amanitin, *Nucl. Acid Res.,* 9, 4801, 1981.

185. **Wullems, G. J., Krens, F. A., Ooms, G., and Schilperoort, R. A.,** Crown gall: a model system for genetic manipulations of higher plants, in *Plant Cell Culture in Crop Improvement,* Sen, S. K. and Giles, K. L., Eds., Plenum Press, New York, 1983, 269.

186. **Wullems, G. J., Molendijk, L., Ooms, G., and Schilperoort, R. A.,** Retention of tumour markers in F_1 progeny plants from in vitro induced octopine and nopaline tumour tissues, *Cell,* 24, 719, 1981.

187. **Wullems, G. J., Molendijk, L., and Schilperoort, R. A.,** The expression of tumor markers in intraspecific somatic hybrids of normal and crown gall cells from *N. Tabacum, Theor. Appl. Genet.,* 56, 203, 1980.

188. **Yadav, N. S., Postel, K., Saiki, R. K., Thomashow, M. F., and Chilton, M.-D.,** T-DNA of crown gall teratoma is covalently joined to host plant DNA, *Nature (London),* 287, 458, 1980.

189. **Yadav, N. S., Vanderleyden, J., Bonnett, D. R., Barnes, W. M., and Chilton, M.-D.,** Short direct repeats flank the T-DNA on a nopaline T-plasmid, *Proc. Natl. Acad. Sci. U.S.A.,* 79, 6322, 1982.

190. **Yang, N. V.,** *Agrobacterium* T-DNA oncogenes encode plant growth regulators: are there cellular homologues in plants, *Trends Biotechnol.,* 3, 197, 1985.

191. **Yang, F., Montoya, A. L., Merlo, D. J., Drummond, M. H., Chilton, M.-D., Nester, E. W., and Gordon, M. P.,** Foreign DNA sequences in crown gall teratomas and their fate during the loss of the tumorous traits, *Mol. Gen. Genet.,* 177, 707, 1980.
192. **Yanofsky, M., Lowe, G., Montoya, A., Rubin, R., Krul, W., Gordon, M., and Nester, E.,** Molecular and genetic analysis of factors controlling host range in *Agrobacterium tumefaciens, Mol. Gen. Genet.,* 201, 237, 1985.
193. **Zaenen, I., Van Larebecke, N., Teuchy, H., Van Montagu, M., and Schell, J.,** Supercoiled circular DNA in crown gall-inducing *Agrobacterium* strains, *J. Mol. Biol.,* 86, 109, 1974.
194. **Zambryski, P., Depicker, A., Kruger, K., and Goodman, H. M.,** *J. Mol. Appl. Genet.,* 1, 361, 1982.
195. **Zambryski, P., Holsters, M., Kruger, K., Depicker, A., Schell, J., Van Montagu, M., and Goodman, H. M.,** Tumour DNA structure in plant cells transformed by *A. tumefaciens, Science,* 209, 1385, 1980.
196. **Zambryski, P., Joos, H., Genetello, C., Van Montagu, M., and Schell, J.,** Ti-plasmid vector for the introduction of DNA into plant cells without alteration of their normal regeneration capacity, *Eur. Mol. Biol. Organ. J.,* 2, 2143, 1983.
197. **Zhengkaixu, Luciano, C. S., Bellard, S. T., Rhoad, R. E., and Shaw, J. G.,** Infection of tobacco protoplasts with liposome-encapsulated polyviral RNA, *Plant Sci. Lett.,* 36, 137, 1985.

Chapter 5

CELL PRESERVATION

I. INTRODUCTION

In order to maintain the dynamics of any crop species it is essential to have an inflow of genes into breeding populations from other resources. The required pool of genetic variability has been provided in the past by the primitive cultivars and wild relatives of crop plants. Until recently these plants have been preserved in primitive agricultural systems around the world as well as in their natural habitats. However, it is now realized that these are endangered plants because this supply is supplanted by the introduction of new varieties required to meet the dietary requirement of an increasing world population. Also, due to deforestation and increasing industralization, the changes in natural habitats of the wild relatives of crop plants have led to their loss of adaptation and consequent erosion. The danger of erosion of genetic resources all over the world was recognized in 1974 with the formation of an International Board for Plant Genetic Resources (IBPGR) under the authority of the Food and Agriculture Organization of the United Nations with the mandate " . . . to create and coordinate a worldwide collaborative network for genetic resources conservation and to mobilize financial support for such a programme".[28,29]

Although the concern for conservation is considerable, the task is equally complex and difficult. There are two possible ways to approach the problem. Conservation is possible either *in situ* or *ex situ*. In the former, there is a need to preserve primitive agricultural systems and the natural habitats of wild relatives of crop plants. This approach can be particularly beneficial for forest plants. However, this is feasible only at the national scale because of many considerations such as type, location, nature of germplasm to be preserved, and whether it can be exchanged internationally. By contrast, conservation *ex situ*, in gene banks, can be taken up at the international scale.

Storage of germplasm as seed has been practiced because large quantities of seeds can be dried and stored in a small space. This is an acceptable method for preservation by the IBPGR and conditions for storage and management of seedbanks have been specified.[22] However, this method cannot be applied to plants producing recalcitrant seeds which remain viable for a short period only and particularly to those plants which grow vegetatively and do not set seed. Plants producing recalcitrant seeds are predominantly woody plants, which set seed after many years of growth and these seeds lose viability quickly. Hence, such plants must be preserved in their natural habitats, which require large areas of land. Also, in vegetatively propagated crops there is annual storage and replanting of perishable germplasm. This involves the risk of losing germplasm as well as crop due to a possible pathogen attack. The vegetative material is also not suitable for international exchange because of quarantine considerations. Therefore, these two conventional means of preservation of germplasm as seeds and propagules, although practiced, are unable to meet all the requirements. Instead, an alternative approach, preservation of plant cells/organs in vitro, is gaining ground.

Regeneration of plants from meristem and even free cells, somatic as well as gametic, in an ever-increasing number of plants has led to consideration of cell preservation as an alternative method for germplasm preservation. The main advantages of this method are that (1) in a relatively small space a large stock of cells can be preserved, which can be retrieved at will to regenerate into plants, and (2) the material is free from pests and pathogens and is therefore suitable for international exchange.

II. STORAGE OF PLANT CELLS

A number of methods have been used for storage of plant cells and every method has its own limitations. The different methods used are given below.

A. Low-Oxygen Storage or Mineral Oil Overlay

Similar to the storage of microbial cultures, the callus tissue of *Daucus carota*[13] could be stored in culture tubes, overlaid with about 0.5 cm of mineral oil.[13] This tissue was maintained for 3 years at 25°C by subculturing at a 5-month interval. In this method of storage evaporation from agar surface is reduced. The tissue, however, continues to grow at a reduced rate. Slow growth is due to the reduced supply of oxygen. A callus tissue cannot survive if immersed in liquid medium, but can continue to grow at a slow rate when immersed in mineral oil due to greater solubility of oxygen.

B. Cold Storage

Tissue cultures can also be preserved by maintaining them at low temperature (1 to 10°C). At this temperature the growth is slowed down, but not completely stopped. Grape plants[58] could be stored for 15 years at 9°C by yearly transfer to fresh medium. It has been estimated that 800 cultivars of grape, each replicated six times, could be maintained in 2 m², whereas 1 ha was required in the field. Cold storage has been practiced for fruit and foliage plants, *Fragaria*,[64] *Prunus*,[53] *Musa*,[4] *Malus*,[51] and *Trifolium*.[19] Strawberry plants could be maintained for 6 years at 4°C by adding a few drops of liquid medium every third month.

Cold storage can be especially suitable for ornamentals, for commercial purposes, and for research material. It appears that tissues from all plants cannot be stored in the cold. Cold storage at 4°C for 3 months or more was effective for tissue derived from chilling tolerant plants such as *Atractylodes, Atropa, Bupleurum, Dioscorea, Lithospermum*, and *Phytolacca*, whereas tissue from the cold-sensitive plants *Datura* and *Perilla* could not survive cold storage.[43] Storage of *Lotus* callus at 4°C for up to 130 days had little effect on differentiation potential, but longer periods (150 to 300 days) of low-temperature storage were inhibitory.[71] Cold storage also affected secondary metabolite production. Nicotine and betalain production suffered after cold storage of tissue of *Nicotiana* and *Phytolacca*, respectively. However cold storage is good for retention of morphogenic potential. Root-forming ability was retained for more than 1 year in cold-stored callus of *Bupleurum* against the controls which lost it rapidly.

Survival during cold storage of cultures of *Prunus*[53] was found to be dependent on interactions of storage temperature, light, and age of subculture. Complete darkness appeared more suitable than a 16-hr photoperiod for successful storage at −3°C for up to 10 months; 1 or 2 weeks of growth at normal temperature before storage enhanced the survival.

To further reduce the growth of a tissue during cold storage special chemicals, such as B-9 (succinic acid 2,2-dimethyl hydrazide), can be used, as was done with potato.[57]

C. Dry Storage

The callus tissue of *D. carota* could be dried and stored[68] in a desiccated state, after pregrowth in a medium containing an elevated level of sucrose (0.15 *M*) and abscisic acid (ABA). The dried tissue survived up to 14 days and could tolerate freezing to −80°C. The dried callus could be stored for 1 year.[69] Drying of cells can be done in a sterile capsule under nonsterile conditions or it can be done on a sterile filter paper placed in a laminar air flow cabinet. Storage of dried tissue is possible at room temperature. However, low humidity is necessary and low temperature is beneficial. Temperature above 0°C should be used. Pretreatment of tissue with ABA and 0.15 *M* sucrose is decidedly beneficial for increased survival. The specific role of ABA in this process is not clear. Both ABA and rifampicin decrease RNA polymerase acitvity, but rifampicin is of no help in survival of dried cells. Sucrose is reported[70] to act in altering the structure of cell water or increase in dry matter content which helps in the survival of cells during drying.

D. Low-Pressure Storage

Low-pressure storage is practiced for increasing the shelf-life of meat, poultry, fish, vegetables, cut flowers, and potted plants. The system operates on the following principles: (1) controlled temperature, (2) reduced atmospheric pressure which results in increased gaseous exchange, (3) air exchange to flush away any toxic vapors released into the storage area, and (4) high humidity which prevents shrinkage, loss in weight, and desiccation.

Successful storage of tobacco tissue[8] at different partial pressures of oxygen indicates the feasibility of this system in preservation of plant cells. However, it is not clear how low-pressure storage is effective in reducing the growth of cells. In analogy with horticulture plants which are maintained by florists under low-pressure storage, low-pressure storage may be able to prolong the life of a tissue by delaying senescence. By decreasing the partial pressure of oxygen in the atmosphere the amount of carbon dioxide evolved is reduced[47] and with low temperature the respiration is also reduced.

III. AIMS IN PRESERVATION

In the common methods — mineral oil overlay, cold storage, dry storage, and low-pressures storage — employed for storage of tissues and cells, growth does not come to a stop by instead is slowed down. Since growth is slowed down, genetic changes in a tissue cannot be ruled out. Genetic stability is the prime aim and *sine qua none* for any cell preservation program. The possibility of storage for an indefinite period is another requirement of any preservation program. Economy is the third consideration, and finally feasibility of international exchange must be considered. A system of preservation that is able to meet most of these requirements is cryopreservation of plant cells.

IV. CRYOPRESERVATION OF PLANT CELLS

Cryopreservation (Gr. kryos = frost) is literally preservation in the frozen state. However, in practice it refers to preservation at ultra low temperature: $-79°C$ (dry ice), $-80°C$ (low-temperature deep freezers), $-140°C$ (vapor phase), and $-180°C$ (liquid nitrogen). At these ultra low temperatures, all intracellular, biological, and chemical processes are brought to a halt and the material is in a state of suspended animation, technically described as stabilate. In this state there is no chance of genetic drift. Cryopreservation is also economical. Once established, it brings considerable savings in terms of cost of equipment and personnel and reduces the risk of losing invaluable material through contamination, human error, or equipment failure.

The two technological advances which made cryopreservation a possibility are (1) liquification of air and its constituent gases, occurring at the end of 19th century, which helped achieve ultra low temperatures and (2) substances such as glycerol[74] and dimethylsulfoxide (DMSO)[50] which could prevent cells against frost injury in cryopreservation.

A. Principle and Terminology

Before discussing the details of cryopreservation of plant cells, it is appropriate to present the principle and terminology of the process of supercooling of the tissue and its preservation.

1. Supercooling of Water and Ice Nucleation

Although pure water becomes ice at 0°C, cell water needs a much lower temperature because of freezing point depression due to salts and organic molecules present in a cell sap

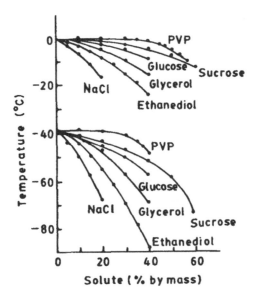

FIGURE 1.   Effect of various solutes on the depression of equilibrium freezing point of water (upper curves) and temperature of homogeneous nucleation of ice (lower curves). PVP, polyvinylpyrrolidone. (From Mackenzie, A. P., *Philos. Trans. R. Soc. London,* 8, 278 and 167, 1977. With permission.)

(Figure 1, upper curves). The lowest temperature at which liquid water has been demonstrated[83] in living cells is −68°C. Therefore, the storage temperature of cells should be lower than −70°C. Pure water can be supercooled to −38°C,[1] which is known as the homogenous nucleation temperature. The homogenous nucleation temperature can be further brought down by addition of solutes such as sodium chloride and other substances (Figure 1, lower curves). Therefore, in a cell one should expect many heterogenous ice nucleation sites due to the presence of many solutes.

Nucleation and growth of ice crystal require time to develop and the rate at which these processes proceed determines the morphological structure of ice crystal. The extent and rate of supercooling, the number of potential nucleation sites, the viscocity of the medium, and the type of solutes present all interact to determine the rate of nucleation and type of crystals to be formed. When a sample is cooled slowly at the site of nucleation, hexagonal ice is formed. At faster cooling rates ice forms into irregular dendrites, and at very fast cooling rates small ice spherules are formed.[52] There is a tendency for spherule formation when high concentrations of solutes and viscous additives are present.

2. *Evolution of Latent Heat During Cooling*

Evolution of latent heat occurs during cooling of any biological material. This must be taken into consideration; otherwise it can result in considerable damage. The evolution of heat is due to the fact that nucleation of ice involves the conversion of randomly oriented molecules with a high level of free energy to an ordered structure with a low level of free energy, the excess being liberated as heat. In practice, as soon as ice begins to form, the temperature rises rapidly and then falls rapidly to the temperature of the cooling chamber.

3. *Vitrification and Devitrification*

Theoretically, when an aqueous solution is cooled rapidly there is insufficient time for ice crystals to grow and the solution remains amorphous. Below a certain temperature (glass temperature) this condition becomes stable and the solution is referred to as vitrified. Therefore, in principle, if cellular contents can be vitrified, formation of intracellular ice can be avoided and the material can be saved from freeze injury. Extremely rapid cooling rate is required for vitrification. Vitrification of pure water is almost impossible; it requires a cooling rate of 10^{10}°Cp/sec.[10] However, it is possible to reduce the cooling rate for vitrification by using either high-molecular-weight substances such as glycol and ethanediol as cryoprotectants or by increasing the barometric pressure. With an increase in concentration of cryoprotectant or an increase of pressure, the cooling rate required for vitrification is reduced. Although pressure can prevent the material from freezing damage, it can be harmful in itself. Cryoprotectants can also be a safeguard against pressure and this approach forms the basis of a method under investigation for cryopreservation of whole animal organs.[23]

For recovery of preserved material, at some point during slow warming of tissue there is spontaneous evolution of heat corresponding to sudden crystallization or devitrification. Devitrification is not harmful to cells; however, on further warming above the recrystallization point the minute ice crystals start to melt and refreeze resulting in the formation of larger crystals which can be damaging to cells.

4. Dehydration and Substitution of Water

The transfer of water to the gaseous state results in dehydration, which is one of the causes of cell damage. However, in nature dehydration is a method of preservation. Some seeds can remain alive for 100 years in dehyrated state.[2] Tiller buds of *Vitis ripara*[73] dehydrate and overcome freezing damage during winter.

In a cell, water can be replaced by organic solvents. Pollen[44] of *Camellia japonica* could be stored using 19 different organic solvents. However, little is known about the biochemical basis of this mode of preservation.

B. Frost Damage

Of the many reasons for freeze injury, the significant ones are given below.

1. Cold Shock or Chilling Injury

The term cold shock or chilling injury or stress is applied to describe the response of a tissue to a reduction in temperature. A recent response study to chilling injury of plant cells[103] revealed that inhibition of cyclosis and disappearance of transvacuolar strands in the cytoplasm were prime changes in chilling-sensitive plants. These changes were not present in chilling-resistant plants. Following cold shock, an important metabolic change is leakage of cytoplasmic enzymes and solvents from cells, suggesting that there is damage to the cell membrane. It is accounted for in terms of differential contraction of membrane components[49] and irreversible protein denaturation.[7] Also, electrolyte concentrations increase following freezing and metabolic processes become out of step because active ionic pumps cannot keep pace with passive diffusion.[24] In particular, a study of freezing injury of sugarbeet cells frozen at -5°C indicated that freezing affects the vacuolar membrane and increases its permeability for sucrose. Freezing increases membrane fragility and modifies transtonoplast electrical potential difference.[5] In this case tonoplast was the main site of freezing injury.

A more direct effect is the modification of lipid constituents of the cell membrane. Upon reduction in temperature, membrane lipids undergo a phase change. On exposure of cells to cold, solidification of membrane lipids takes place, which is damaging because the membrane should remain sufficiently fluid for normal enzyme activity and transmembrane transport of water. The temperature at which this phase change occurs can be further lowered by increasing the degree of unsaturation of phospholipids. This is seen in *Medicago*[31] and *Chlorella*.[72] Injury can be avoided or reduced by including cryoprotectants, specific phospholipids,[12] or even by a change in concentration of specific anions.[59] During cold acclimation of mulberry cells,[106] there was a significant increase in phospholipids and unsaturation of their fatty acids in the plasma membrane. The fluidity of the plasma membrane also increased with hardiness.

Studies on the molecular aspects of freezing injury have revealed that plasma membranes lose semipermeability immediately after cooling through a critical temperature, suggesting a membrane structural transition.[79] Also, lipid bilayers may undergo an amorphous change after freezing at a lethal temperature.[87] This is possible in every membrane in the cell.[9] Other suggested irreversible alterations in membrane structure include changes in lipid-protein interaction and protein conformation.[107]

2. Cell Shrinkage

One of the effects of frost damage is cell shrinkage, which becomes progressively apparent as the temperature is lowered. In plant cells in particular, due to the rigid cell wall the shrinkage of protoplast away from the cell wall leaves enough space for ice formation and the consequent collapse of the cell.[62] However, some plant cells, instead of being deformed, retain their shape, but at the expense of shedding of membrane to the surrounding medium.[35] This renders the cells susceptible to injury on recovery of preserved material when a return to normal conditions results in melting of ice and enables the water to flow into the cells.

It has been suggested that there is a critical limit of cell shrinkage, and shrinkage below this level may cause the structural proteins to come close together and become denatured by the formation of disulfide bonds.[48] During the freeze-thaw cycle, volumetric contraction and subsequent expansion must be successfully contended with if the cells are to survive. In protoplasts isolated from nonacclimated rye leaves, sufficiently large volumetric contractions were irreversible.[88,89] During osmotic contraction, membrane material is deleted from the plasma membrane[32,104,105] of isolated protoplasts. Sufficiently large contractions and deletion of the plasma membrane were irreversible, and upon expansion nonacclimated protoplasts lysed before regaining their original size.

3. Ice Formation

An important cause of injury is ice formation. Formation of ice initiates in intercellular space and this is followed by intracellular ice formation. The latter is always damaging to cells.[55] Injury is due to physical disruption produced by crystals and lesions produced in membranes.[30]

At extremely rapid rates of cooling the ice crystals formed are very small and less damaging. During recovery, such a tissue must be warmed very rapidly to prevent further growth of ice crystals.

C. Prevention of Frost Damage

In nature, several plant species acquire tolerance to cold — cold acclimation/frost hardiness — in a number of ways, such as:

1. Increased ability to remain supercooled
2. Manufacture of cryoprotectant compounds
3. Ability to sustain ice formation
4. Ability to sustain phase changes in membrane lipids
5. Synthesis of a range of compounds such as fatty acids, proteins, and carbohydrates, etc. which have the ability to bind to water and depress its melting point.

These adaptations are dependent on genotype, environment, physiological conditions, and degree of vacuolation of tissue.

Growth temperature is an important factor ensuing survival on freezing. This is evidenced from the hardening effect on some crops when grown at low temperature. For example, cabbage grown at 20 to 30°C dies when exposed to −2°C. When grown at 5°C for 1 week it dies at −7°C and if it is grown at 0°C for 1 week it can withstand up to −12°C. However, this is not true for sunflower; it is killed at −2°C irrespective of growth temperature.

Preculture of cells at lower temperature significantly affects the metabolic rate and subsequent survival on cryopreservation. Cells of *Chlorella*[60] after preculture at 4°C for different durations had correspondingly increased survival rates compared to growth at 20°C. A similar effect was possible on reducing nutrient concentration and including metabolic inhibitors in cell culture at 20°C.[63] In higher plants, many species cold harden in response to low temperature. However, it is not clear what elicits this hardening process. ABA has been shown to play an important role in plant-water balance and in the adaptation of plants to stress conditions including cold hardiness. When cell suspensions of winter wheat, winter rye, and brome grass were treated with 7.5×10^{-5} *M* ABA for 4 days at 20°C they could tolerate a temperature of −30°C, whereas control cultures could tolerate only −7 to −8°C.[15] Of the ten species tested, ABA was effective only on those cultures which responded to cold hardening on exposure to low temperature. These results suggests that ABA triggers the genetic system responsible for induction of the hardening process.

Degree of vacuolation is a determining factor in cell preservation,[61] and it is considered to be a function of growth temperature and age of culture. In fact, stationary phase sycamore cells did not stand cryopreservation as compared to exponential phase cells.[61,102] Cells pregrown for 3 to 4 days in medium containing osmotically active additives, such as mannitol (6%) or proline (10%), showed a reduction in cell size, and a single large vacuole was replaced by smaller vesicles.[77,95] Proline is also reported to remove excess intracellular water by the osmotic gradient.[40] Also, cells pregrown in mannitol were different in behavior during early stages of freezing as seen in the cryomicroscope.[76]

D. Cryoprotectants

Cryoprotectants are those chemicals which protect cells from frost damage during cryopreservation. They are a heterogeneous group of compounds which appear to act in a variety of different mechanisms. For the sake of convenience, cryoprotectants can be classified into two types: penetrating and nonpenetrating. However, this classification is not critical, particularly when considering the temperature for addition of cryoprotectants. Glycerol, the first substance to be recognized as a cryoprotectant,[74] is a penetrating compound when added at room temperature and is nonpenetrating at 0°C. The main site of damage in cryopreservation is the membrane, and it is generally considered that nonpenetrating cryoprotectants act on membranes. Another recognized and commonly used cryoprotectant is dimethylsulfoxide (DMSO).[50] It is reported to cause redistribution of intramembranous particles[20] which may enhance membrane permeability.

The main function of cryoprotectants are

1. Depression of freezing point and supercooling point of water.
2. Elevation of devitrification point.
3. Reduction of the amount of water to be removed in the form of ice.
4. Prevention from dehydration of cells by maintaining structural water.
5. Reduction of concentration of solutes in cell sap.
6. Increase the viscosity of cell sap.
7 Retardation of the growth of ice crystals, and facilitation of vitrification.

Cryoprotectants, however, present certain problems associated with their addition and removal. A permeating cryoprotectant initially causes cell shrinkage and then, depending

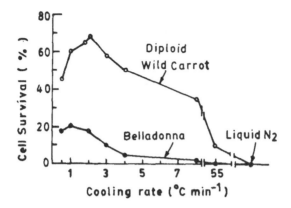

FIGURE 2.   Effect of cooling rate on cell survival of *Daucus carota* and *Atropa belladonna*. (From Nag, K. K. and Street, H. E., *Physiol. Plant.*, 34, 261, 1975. With permission.)

upon the temperature as the cryoprotectant enters the cell, the cell regains its original shape. This process is reversed during the removal of cryoprotectants; the cell swells on return of isotonic condition and then gradually returns to normal. These volume changes may be damaging to cells. Therefore, in order to minimize damage, cryoprotectants should be added gradually and removed stepwise. The stress of dilution is more than that of addition.

E. Process of Cryopreservation

The process of cryopreservation of plant cells comprises two essential steps, freezing and storage of cells. The main consideration during preservation is that a cell is to be saved from cryogenic injury. In order to accomplish this, formation of intracellular ice (which is the main cause of injury) and leakage of cellular membranes must be avoided. Since plant materials are variable in their sensitivity to freezing injury, the various methods employed for freezing are

1.   Slow cooling, with a temperature decrease of 0.1 to 10°C/min.
2.   Rapid cooling, with a temperature decrease of 50 to 1200°C/min.
3.   Stepwise cooling, a combination of slow and rapid cooling.
4.   Dry freezing, dehydration followed by freezing.

One of the reasons for the various methods of cooling is that no single method has been found to be universally applicable. Ideally, a cryopreservation schedule should have a cooling rate which is slow enough to dehydrate the cells and avoids the formation of intracellular ice, but is fast enough to prevent damage from increased solute concentration.

1. Slow Cooling

In this method the cooling rate is 0.1 to − 10°C/min, from 0°C to − 100°C, and then the material is transferred to liquid nitrogen. Slow cooling involves the beneficial effects of dehydration which minimizes the amount of intercellular water and consequent intercellular ice. The slow cooling method is commonly used, especially for preservation of cells taken from a cell suspension. However, a cooling rate must be determined for each tissue because cell survival is dependent on the cooling rate (Figure 2). For controlled cooling, ready-made instruments and even computer programed freezers are available.

During slow cooling, the phase change in membrane lipids as a result of lowered temperature occurs slowly, and protein components are pushed to regions where the membrane is still fluid.[86] The slow cooling method saves the tissue from postthaw deplasmolysis injury, which is quite common in the rapid cooling method.[96]

2. Rapid Cooling

In rapid cooling the material in storage vials is directly plunged into liquid nitrogen. In this way a cooling rate of $-300°C$ to $1000°C/min$ or faster is possible. A relatively slower decrease (-10 to $-70°C/min$) is possible when the vial containing plant material is suspended in the atmosphere over the liquid nitrogen and is then gradually lowered.

In rapid freezing smaller intracellular ice crystals are formed and injury is minimized. In rapid cooling there is no lateral movement of membranous proteins; instead they are fixed.

3. Stepwise Cooling

Two-step cooling can be done in two ways: slow cooling followed by rapid cooling or rapid cooling followed by another rapid cooling with a gap between. The first method, slow cooling followed by rapid cooling, combines the advantages of both methods described above. A slow cooling process down to -20 to $-40°C$ permits protective dehydration of cells; following it, rapid cooling prevents the formation of large ice crystals. In rapid cooling followed by rapid cooling prefreezing of cells occurs by cooling rapidly to subzero temperature, holding at this temperature for some time, and further cooling rapidly to storage temperature (liquid nitrogen). During prefreezing a high concentration of extracellular solutes produced during the initial cooling step induces sufficient shrinkage which protects the cells against damage produced during second rapid cooling.[26]

4. Dry Freezing

Dehydrated tissue is considerably resistant to cryogenic injury. This is consistent with the situation in seeds; dry seeds are resistant to freezing injury but are susceptible after imbibing water. Dehydration of cells is possible in an oven or under vacuum. The latter is better for survival of cells when exposed to the storage temperature of liquid nitrogen.[97] However, there is a dehydration optimum which varies from tissue to tissue.

F. Factors Affecting Cryopreservation

A number of factors determine the success of cryopreservation; the significant factors are given below.

1. Nature of Material

Cell cytology and physiology are determining factors in preservation. In general, small richly cytoplasmic cells (resembling meristematic cells) have better survival rates compared to older vacuolated cells. Therefore, cells in suspension in an actively growing phase are able to withstand freezing injury better than stationary phase cells.[66,98] This is also true for preservation of shoot apices, embryos, or whole plantlets where regrowth occurs only in the meristematic region after preservation and older vacuolated cells die.

2. Tissue Pretreatment

A number of tissue pretreatments are helpful in cryopreservation. A cold preconditioning of tissue by transfer of plants to 4°C for 3 days helps in increasing the rate of survival of tissues on preservation.[85]

Preculture of tissue is also helpful in cryopreservation. Shoot apices of potato could withstand cryopreservation when precultured for 48 hr in the presence of a cryoprotectant (5% DMSO).[33] However, it was not clear whether the effect was due to preculture alone or to the cryoprotectant. This was clarified later when a preculture alone, for at least 48 hr, was found to be helpful.[92]

Dehydration of tissue is also helpful in survival during freezing and thawing. When water content has been adequately removed, the thawing rate is less critical.[93,98] Hydrated tissues alone survive rapid thawing.

3. Cryprotectant

The choice and concentration of cryoprotectant are determining factors in cryopreservation. For plant tissues, DMSO is the most effective cryoprotectant. The other cryoprotectant employed either alone or in combination with DMSO is glycerol.[66] However, it was quite ineffective for shoot tips of pea[45] and inferior to DMSO for shoot tips of strawberry[46] and carnation.[93]

A new cryoprotectant for plants is proline.[39] For cell suspensions of maize[100] proline proved to be a more effective cryoprotectant than DMSO or glycerol. It is quite interesting that cell suspensions grown in the presence of 10% proline for 3 to 4 days could survive cryopreservation even without further addition of any cryoprotectant. Low-temperature preservation may be a condition of stress because plants under stress, caused either by drought or salt, accumulate proline, which is regarded as a natural protectant for enzyme and other activities. Proline accumulates at cold-hardening temperature and is suggested to act as a membrane stabilizer.[39] Yet another cryoprotectant recently discovered for plant cell cryopreservation is trehalose.[6] Cells of *Daucus carota* and *Nicotiana plumbaginifolia* were successfully cryopreserved with trehalose. Trehalose is accumulated in high concentrations by anhydrobiotic organisms such as yeast, some fungi, and nematodes, which survive a high degree of desiccation. Also, in studies with artificial membranes, trehalose substitutes for water molecules in the membrane during dehydration and thus helps maintain membrane integrity.

The concentration of cryoprotectant used varies from tissue to tissue. About 5 to 10% DMSO is used for cell suspension[66,80] as well as shoot apices. However, some apices may tolerate up to 20% DMSO;[33,81] 10% proline is optimal for cell suspensions. The cryoprotectant should be added gradually, over 30 to 60 min. This period should be prolonged when using glycerol[102] because of its low permeability, while with DMSO it is preferable to maintain a low temperature (around 0°C)[37,66,95,97] in order to minimize its toxic effect. A common practice is to dissolve cryoprotectants in a culture medium. However, it has been suggested that cryoprotectants should be dissolved in water with or without sucrose.[93]

A new trend is to use more than one cryoprotectant, to obtain an additive effect, and to dissolve them in a sucrose solution. Cells of *Zea mays* cryoprotected with DMSO, glycerol, and proline or DMSO, glycerol, and sucrose, each mixture prepared either in water or nutrient medium, differed in their postthaw performance. Cells cryoprotected with DMSO, glycerol, and proline remained alive after thawing, but were incapable of recovery growth.[98] However, when a cryoprotectant mixture was prepared in a nutrient medium, cells retained recovery potential on cryopreservation. This difference was not noticed in the sucrose-containing cryoprotectant mixture, suggesting that sucrose substitutes for medium components. This was further shown in experiments on using either a single cryoprotectant or a mixture of two cryoprotectants prepared either in water or nutrient medium.

Although either 5 to 10% DMSO or 10% proline is effective as a cryoprotectant, a more effective cryoprotectant is a three-component mixture of DMSO, glycerol, and sucrose. This mixture is able to protect tissue from under- or overdehydration and ice formation. This is demonstrated[98] by comparing the performance of cells frozen under different conditions and cryoprotected with this mixture as compared to DMSO and glycerol. For cells treated with DMSO and glycerol, direct quenching in liquid nitrogen is lethal, whereas cells cryoprotected with the three-component mixture are buffered against the deleterious effect of freezing in liquid nitrogen once the early stages of cooling have been passed.

G. Storage and Recovery

Frozen samples are stored at the temperature of liquid nitrogen ($-196°C$) because a temperature above $-130°C$ may allow the growth of intracellular crystals. For long-term storage a regular supply of liquid nitrogen is essential. About 20 to 25 ℓ of liquid nitrogen is required for a storage flask holding 4000 sample ampoules, each having 2 mℓ material.[102]

1. Warming

For successful cryopreservation, the warming rate required to recover the material is as important as the cooling rate. The change occurring in a tissue during slow warming as seen in a cryomicroscope is a darkening or flashing that corresponds to sudden formation of intracellular ice. Finally, this darkening disappears when intracellular ice melts. Slow warming is injurious to tissue because it allows the growth of small ice crystals, formed during freezing. This phenomenon is known as migratory recrystallization.[54] Hence, rapid warming which does not permit migratory recrystallization gives better survival rates than slow warming.[24]

For rapid warming, the ampoule containing the tissue is plunged directly in a water bath maintained at 37°C. This gives a warming rate of 500 to 750°C/min. An advantage of rapid warming is that the tissue is moved quickly through the damaging temperature range, just below the melting point.

A common form of damage in the process of recovery is the dilation of cytoplasmic organelles.[96]

2. Postthaw Treatment

Careful handling of thawed material is essential to retain viability. The important factors determining the viability of postthawed cells are the temperature and speed at which dilution of the cryoprotectant is performed. When slowly penetrating cryoprotectants are used, warming to only 0°C may not allow rapid diffusion of the cryoprotectant out of the cell. Instead, cryoprotectants are more mobile at room temperature and their diffusion is faster. However, the toxicity of cryoprotectants increases with a rise in temperature and this disadvantage may outweigh the advantage of exit.

Fortunately, it has recently been found that it is not essential to remove the cryoprotectants because any washing serves to traumatize the cells.[99] Therefore, it is advisable not to carry out postthaw washing. Following thawing many cells continue to degenerate and only a minority of cells in a population are recovered.[98] Hence, in this technique performance of a population of cells rather than individual cells should be taken into consideration.

V. ACHIEVEMENTS

Cryopreservation has been attempted using different cell and tissue systems. Some representative examples are given below.

A. Callus

Although a callus culture is well suited for cryopreservation, the prevalence of genetic instability in this system makes it unsuitable for considerations of germplasm and its preservation. Calli of several species have been cryopreserved successfully. Stepwise freezing is done, but warming must be rapid. For cryoprotectants, the three-component mixture — PEG (6000) 10% + glucose 8% + DMSO 10% — was found to be suitable for preservation of rice, alfalfa, date palm, and sugarcane tissues.[25-27,91,94]

The morphogenic potential of callus cells is retained during cryopreservation, because plant regeneration has been possible after thawing. It appears that it is not possible to cryopreserve shoot primordial structures along with callus cells. When differentiating, callus of *Saccharum* was preserved in liquid nitrogen; the primordial structures were destroyed, but cells could be preserved.[94]

A recent interesting observation concerning cryopreservation of *Oryza sativa*[28] cells is that the best preservation was possible when cryoprotectants were applied at 0°C and removed by washing with medium.

B. Cell Suspension

Cell suspensions are relatively more suitable for preservation than callus cultures. Here it is possible to obtain a uniform population of growing cells. The first reported storage was of *Linum usitatissimum*[78] cells at −50°C, followed by the first report of cryopreservation of carrot[65] cells at the temperature of liquid nitrogen. The cells preserved were capable of embryogenesis. An increased frequency of preservation was possible while cells were in their exponential phase of growth. The successful protocol for cryopreservation is slow freezing, rapid warming, and washing in fresh liquid medium to return to culture condition.[65,82,102] New developments in this field are (1) pregrowth of cells, (2) three-component mixture of cryoprotectants, (3) combination of slow cooling and stepwise freezing, and (4) postthaw condition.

Preculture of cells in medium containing osmotically active substances conditions the cells to cryopreservation and helps them acquire freeze tolerance. Cells are cultured for 3 to 4 days[77,100] on medium supplemented with either 6% mannitol or 10% proline. This results in reduction in cell size and the large vacuole is replaced by smaller ones. The observed increase in survival following freezing has been attributed to a reduction in mean cell size and an increased cytoplasm to vacuole ratio.[98] However, in a comparative study of osmotically stressed cells of sycamore and soybean on culture on 6% mannitol solution, only sycamore cells became less vacuolate and no such change occurred in stressed soybean cells.[75,76]

DMSO, found to be suitable for pregrowth of shoot tips, was unsuitable for cells.[99] However, incorporation of DMSO into liquid growth medium had a significant effect in preservation of periwinkle cells.[16] Compared to control, less water crystallized at temperatures below −30°C in DMSO-treated cells. Similar results were obtained with sorbitol, but it was less effective. A combination of DMSO and sorbitol was most effective. Further, a close correlation was found between percent of unfrozen water at −40°C and percent of cell survival after freezing for 1 hr in liquid nitrogen. Pregrowth of *Dioscorea* cells in amino acid medium and *Penax ginseng* cells in 7 to 25% sucrose was helpful in cryopreservation.[11] However, for pregrowth of cells of *Digitalis lantana*,[21] the same group of workers used 3% mannitol. ABA was recently introduced to this list of pregrowth compounds.[17] Treatment of wheat cells with ABA improved survival on preservation in liquid nitrogen.

It is of interest to know that biosynthetic and differentiation potentials did not change after cell storage in liquid nitrogen. This has been shown in *Dioscorea* and *Penax*[11] cells and *Catharanthus*[18] cells.

The three-component mixture of cryoprotectants[98] used is comprised of DMSO, glycerol, and sucrose.

Slow cooling is at the rate of 1 to 2°C/min.[102] A linear rate of cooling can be achieved using preprogrammed freezing apparatus or nonlinear cooling at 0.1°C/min can be achieved using a deep freeze running at −70°C, by placing the material ampoule in a vacuum flask and in turn placing this flask in the deep freeze.[82] An inexpensive method of achieving slow and nonlinear cooling is suspension of the ampoule over the atmosphere of liquid nitrogen.[65] In stepwise freezing or prefreezing, the specimen is first taken to subzero temperature (−35°C) and held at this temperature for about 30 to 40 min. This is possible in an alcohol bath cooled by a dip-cooling coil. Holding temperature can be maintained by a thermostat. Finally, the specimen is quenched in liquid nitrogen. This method has been practiced routinely for the preservation of many plant cells.[100,101]

For postthaw conditions, it has recently been found that it is not necessary to remove cryoprotectants and washing is not needed. Rather, continued presence of cryoprotectants is helpful in recovery.[99] Also, liquid culture of thawed material is deleterious. Instead, layering of cells over agar medium is better for recovery. Rice[14] cells on preservation in liquid nitrogen and return to in vitro culture were found to be altered in a number of physiological characters. These included a reduction in respiration and glucose uptake, loss of intracellular potassium, and a decrease in the level of the key metabolites ATP, glucose-6-phosphate, and pyruvate. Because of this, after thawing, periwinkle[18] cells were transferred to filter paper discs over an agar nutrient medium for 4 to 5 hr. These discs, with the cells, were transferred to fresh medium for growth. It appears that a period of equilibration is required for the resumption of normal growth.[21,99]

C. Shoot Tip

Cells in culture are not ideal for germplasm.[41] Therefore, emphasis on cryopreservation is shifting from cells to meristematic structures such as shoot apices and young embryos. This is because of the following.

1. Genetic instability of cell cultures
2. Decline in morphogenic potential of cells on progressive culture
3. Ploidy, which is unstable in cell cultures, can be maintained though shoot apices and axillary meristem.

The first successful cryopreservation of shoot apices involved direct quenching of shoot apices of *Dianthus caryophyllus*[84] into liquid nitrogen. Although rapid freezing is the least demanding approach, for recovery of material it requires very rapid thawing rates, and this may result in damage and make the material brittle. Later, it was found that the best cooling rate for this material was 50°C/min.[85] Also, in a reinvestigation[93] it was found that direct quenching of shoot apices of *Dianthus* in liquid nitrogen gave no survival; instead, if the shoot apices were frozen gradually to −15°C and then plunged in liquid nitrogen, all survived. However, interestingly, shoot apices of strawberry[81] and *Solanum goniocalyx*[34] did not survive slow (1°C/min) or intermediate (60°C/min) cooling, but survived ultrarapid cooling by direct quenching in liquid nitrogen.

A short preculture of shoot apices on basal medium or medium supplemented with 5% DMSO has been found to be essential for survival on cryopreservation in pea,[45] *Solanum* spp.,[42,92] and strawberry.[46] Preculture on basal medium may relate to the healing of excision injury. However, more interesting is preculture on DMSO. For shoot apices of *Brassica napus*[99] preculture for 1 day on callus-inducing medium lacking DMSO and 2 days on basal medium containing DMSO may be optimal. Another cryoprotectant such as proline may be as effective as DMSO in promoting freeze tolerance. Incidentally, on recovery, the cryopreserved shoot apices have a tendency to undergo callusing, even on basal medium, which could be overcome by inclusion of gibberellic acid.[36]

D. Somatic Embryos

Somatic embryos of carrot[97] could not be preserved following the protocol worked out for cryopreservation of carrot cells. However, a modification of the technique involving dry freezing was quite successful. The specimens were treated with 5 to 10% DMSO and were blotted dry. Freezing at a slow rate (10°C/min to −100°C) was followed by quenching in liquid nitrogen. Thawing was relatively slow and recovery was on a semisolid medium. Globular and heart-shaped embryos resumed growth, but growth in late torpedo-stage embryos and plantlets was possible from meristematic regions.

E. Pollen Embryos

Pollen embryos of *N. tabacum* and *Atropa belladonna*[3] could be cryopreserved with DMSO or glycerol. The greater freeze tolerance was possible by early-stage embryos; globular embryos showed 31% survival, early heart-shaped embryos 9%, and late heart-shaped embryos only 2%. Embryos in the cotyledonary stage did not survive.

F. Protoplasts

Recently it has been possible to cryopreserve protoplasts of *Bromus inerimis* and *Daucus carota*.[56] Freezing was carried out in four sequential steps ($+4$, -20, -60, and $-196°C$) with the cryoprotectant DMSO or glycerol, each at 0.7 *M*. In another study,[90] protoplasts of carrot were preserved using various cryoprotectants, of which 10% DMSO or 5% DMSO + 10% glucose were most effective. Freezing was either slow (1°C/min to $-196°C$) or fast (-30 to $-40°C$), followed by quenching in liquid nitrogen. Thawing was rapid. It is interesting that no special pregrowth or postthaw treatments, as needed for plant cells, were required. Of protoplasts of *Datura innoxia*, *N. tabacum*, and *Daucus carota* preserved with 10% DMSO and 10% glycerol, only carrot protoplasts[38] regenerated cell walls and underwent cell division.

REFERENCES

1. **Angell, C. A., Shuppert, J., and Tucker, J. C.,** Cooperative behaviour in supercooled water: heat capacity, expansivity and PMR chemical shift anomalies from 0 to 38°C, *J. Physiol. Chem.*, 77, 3092, 1973.
2. **Aufhammer, G. and Fischbeck, G.,** Ergebnisse von Gefaeβ und Feldversuchen mit dem Nachbau Keimfaehiger Gersten und Haferkorner aus dem Grundstein des 1832 errichteten Nuernberger Stadtthaeters, *Z. Pflanzenzuecht.*, 51, 354, 1964.
3. **Bajaj, Y. P. S.,** Effects of superlow temperature on excised anthers and pollen embryos of *Atropa*, *Nicotiana* and *Petunia*, *Phytomorphology*, 28, 171, 1978.
4. **Banerjee, N. and deLanghe, E.,** A tissue culture technique for rapid clonal propagation and storage under minimal growth conditions of *Musa* (banana and plantain), *Plant Cell Rep.*, 4, 151, 1985.
5. **Barbier, H., Nalin, F., and Guern, J.,** Freezing injury in sugarbeet root cells: sucrose leakage and modification of tonoplast properties, *Plant Sci. Lett.*, 26, 75, 1982.
6. **Bhandal, I. S., Hauptmann, R. M., and Widholm, J. M.,** Trehalose as cryoprotectant for the freeze preservation of carrot and tobacco cells, *Plant Physiol.*, 78, 430, 1985.
7. **Brandts, J. F.,** Heat effects on proteins and enzymes, in *Thermobiology*, Rose, A. H., Ed., Academic Press, New York, 1967, 25.
8. **Bridgen, M. P. and Staby, G. L.,** Low pressure and low oxygen storage of *Nicotiana tabacum* and *Chrysanthemum × morifolium* tissue cultures, *Plant Sci. Lett.*, 22, 177, 1981.
9. **Briggs, S. P., Haug, A. R., and Scheffer, R. P.,** Localization of spin labels in oat leaf protoplasts, *Plant Physiol.*, 70, 662, 1982.
10. **Bruggeller, P. and Mayer, E.,** Complete vitrification in pure liquid water and dilute aqueous solution, *Nature (London)*, 288, 569, 1980.
11. **Butenko, R. G., Popov, A. S., Volkova, L. A., Chernyak, N. D., and Nosov, A. M.,** Recovery of cell cultures and their biosynthetic capacity after storage of *Dioscorea deltoidea* and *Panax ginseng* cells in liquid nitrogen, *Plant Sci. Lett.*, 33, 285, 1984.
12. **Butler, W. J. and Roberts, T. K.,** Effects of some phosphatidyl compounds on boar spermatozoa following cold stock or slow cooling, *J. Reprod. Fertil.*, 43, 183, 1975.
13. **Caplin, S. M.,** Mineral oil overlay for conservation of plant tissue culture, *Am. J. Bot.*, 46, 324, 1959.
14. **Cella, R., Colombo, R., Galli, M. G., Nielson, E., Rollo, F., and Sala, F.,** Freeze-preservation of rice cells. A physiological study of freeze-thawed cells, *Physiol. Plant.*, 55, 279, 1982.
15. **Chen, T. H. H. and Gusta, L. V.,** Abscisic acid-induced freezing resistance in cultured plant cells, *Plant Physiol.*, 73, 71, 1983.
16. **Chen, T. H. H., Kartha, K. K., Constabel, F., and Gusta, L. V.,** Freezing characteristics of cultured *Catharanthus roseus* cells treated with dimethylsulfoxide and sorbitol in relation to cryopreservation, *Plant Physiol.*, 75, 720, 1984.

17. **Chen, T. H. H., Kartha, K. K., and Gusta, L. V.,** Cryopreservation of wheat suspension culture and regenerable callus, *Plant Cell Tissue Organ Culture*, 4, 101, 1985.

18. **Chen, T. H. H., Kartha, K. K., Leung, N. L., Kurz, W. G. W., Chatson, K. B., and Constabel, F.,** Cryopreservation of alkaloid producing cell cultures of periwinkle *(Catharanthus roseus), Plant Physiol.*, 75, 726, 1984.

19. **Chenye, V. A. and Dale, P. J.,** Shoot-tip culture in forage legumes, *Plant Sci. Lett.*, 19, 303, 1980.

20. **DeGroot, C. and Van Leene, W.,** The influence of cryoprotectants, temperature, divalent cations and serum proteins on the structure of plasma membrane in rabbit peripheral blood lymphocytes, *Eur. J. Cell Biol.*, 19, 19, 1979.

21. **Diettrich, B., Haack, U., Popov, A. S., Butenko, R. G., and Luckner, M.,** Long-term storage in liquid nitrogen of an embryogenic cell strain of *Digitalis lantana, Biochem. Physiol. Pflanzen.*, 180, 33, 1985.

22. **Ellis, R. H., Roberts, E. H., and Whitehead, J.,** A new, more economic and accurate approach to monitoring the viability of accession during storage in seed banks, *Plant Genet. Resour. Newsl.*, 41, 3, 1980.

23. **Fahy, G. M. and Hirsch, A.,** Prospects for organ preservation by vitrification, in *Organ Preservation, Present and Future*, Pegg, D. E., Jacobsen, I. A., and Halast, N. A., Eds., MTP Press, Lancaster, 1982, 399.

24. **Farrant, J.,** General observations on cell preservation, in *Low Temperature Preservation in Medicine and Biology*, Ashwood-Smith, M. J. and Farrant, J., Eds., Pitman Medical, Tunbridge-Wells, England, 1980, 1.

25. **Finkle, B. J., Ulrich, J. M., Rains, D. W., Tisserat, B. B., and Schaeffer, G. W.,** Survival of alfalfa *Medicago sativa*, rice *Oryza sativa* and date palm *Phoenix dactylifera* callus after liquid nitrogen freezing, *Cryobiology*, 16, 583, 1979.

26. **Finkle, B. J., Ulrich, J. M., Schaeffer, G. W., and Sharpe, F.,** Cryopreservation of cells of rice and other plant species, in *Potentials of Cell and Tissue Culture Techniques in the Improvement of Cereal Plants*, Academia Sinica and International Rice Research Institute, Science Press, Beijing, 1983.

27. **Finkle, B. J., Ulrick, J. M., Tisserat, B., and Rains, D. W.,** Regeneration of date palm and alfalfa plants after freezing callus tissue to −196°C in a combination of cryoprotective agents, *Cryobiology*, 17, 625, 1980.

28. **Frankel, O. H. and Bennett, E., Eds.,** *Genetic Resources in Plants Their Exploration and Conservation*, International Biological Program Handbook No. 11, Blackwell Scientific, Oxford, 1970.

29. **Frankel, O. H. and Hawkes, J. G.,** *Crop Genetic Resources for Today and Tomorrow*, Cambridge University Press, Cambridge, 1975.

30. **Fujikawa, S.,** Morphology evidence of membrane damage caused by intracellular ice crystals, *Cryobiology*, 15, 707, 1978.

31. **Gerloff, E. D., Richardson, T., and Stahlmann, M. A.,** Changes in fatty acids of alfalfa roots during cold hardening, *Plant Physiol.*, 41, 1280, 1966.

32. **Gordon-Kamm, W. J. and Steponkus, P. L.,** The behaviour of plasma membrane following osmotic contraction of isolated protoplasts: implications in freezing injury, *Protoplasma*, 123, 83, 1984.

33. **Grout, B. W. W. and Henshaw, G. G.,** Freeze preservation of potato shoot tip cultures, *Ann. Bot.*, 42, 1227, 1978.

34. **Grout, B. W. W. and Henshaw, G. G.,** Structural observations on the growth of potato shoot tip cultures after thawing from liquid nitrogen, *Ann. Bot.*, 46, 243, 1980.

35. **Grout, B. and Morris, G. J.,** Membrane as a factor in cryoinjury, in *Proc. 4th Symp. Problem of Low Temperature Preservation of Cells, Tissues and Organs*, Humboldt University, Berlin, 1981, 63.

36. **Grout, B. W. W., Westcott, R. J., and Henshaw, G. G.,** Survival of shoot-meristems of tomato seedlings frozen in liquid nitrogen, *Cryobiology*, 15, 478, 1978.

37. **Haskins, R. H. and Kartha, K. K.,** Freeze-preservation of pea meristems cell survival, *Can. J. Bot.*, 58, 833, 1980.

38. **Hauptmann, R. M. and Widholm, J. M.,** Cryostorage of cloned amino acid analog resistant carrot and tobacco suspension cultures, *Plant Physiol.*, 70, 30, 1982.

39. **Heber, U., Tyankova, L., and Stantarius, K. A.,** Stabilization and inactivation of biological membranes during freezing in the presence of amino acids, *Biochim. Biophys. Acta*, 241, 578, 1971.

40. **Hellengren, J. and Li, F. H.,** Survival of *Solanum tuberosum* suspension cultures to −14°C, the mode of action of proline, *Physiol. Plant.*, 52, 449, 1981.

41. **Henshaw, G. G. and O'Hara, J. F.,** In vitro approaches to the conservation and utilization of global plant genetic resources, in *Plant Biotechnology*, Mantell, S. H. and Smith, H., Eds., Cambridge University Press, Cambridge, 1983, 219.

42. **Henshaw, G. G., Stamp, J. A., and Westcott, R. J.,** Tissue culture and germplasm storage, in *Plant Cell Cultures: Results and Perspectives*, Sala, F., Parisi, B., Cella, R., and Cifferi, O., Eds., Elsevier/North-Holland, Amsterdam, 1980, 277.

43. **Hiraoka, N. and Kodama, T.,** Effects of non-frozen cold storage on growth, organogenesis and secondary metabolism of callus cultures, *Plant Cell Tissue Organ Culture,* 3, 349, 1984.
44. **Iwanami, Y.,** Retaining the viability of *Camellia japonica* pollen in various organic solvents, *Plant Cell Physiol.,* 13, 1139, 1972.
45. **Kartha, K. K., Leung, M. L., and Gamborg, O. L.,** Freeze-preservation of pea meristems in liquid nitrogen and subsequent plant regeneration, *Plant Sci. Lett.,* 15, 7, 1979.
46. **Kartha, K. K., Leung, N. L., and Pahl, K.,** Cryopreservation of strawberry meristems and mass propagation of plantlets, *J. Am. Soc. Hortic. Sci.,* 105, 481, 1980.
47. **Kessel, R. H. J. and Carr, A. H.,** The effect of dissolved oxygen concentration on growth and differentiation of carrot *(Daucus carota)* tissue, *J. Exp. Bot.,* 28, 996, 1972.
48. **Levitt, J.,** Winter hardiness in plants, in *Cryobiology,* Meryman, H. T., Ed., Academic Press, New York, 1966.
49. **Lovelock, J. E.,** Haemolysis by thermal shock, *Br. J. Haematol.,* 1, 117, 1955.
50. **Lovelock, J. E. and Bishop, M. W. H.,** Prevention of freezing damage to living cells by dimethyl sulfoxide, *Nature (London),* 183, 1394, 1959.
51. **Lundergan, C. and Janick, J.,** Low temperature storage of in vitro apple shoots, *HortScience,* 14, 514, 1979.
52. **Luyet, B. J.,** On the possible biological significance of some physical changes encountered in the cooling and the rewarming of aqueous solutions, in *Cellular Injury and Resistance in Freezing Organism,* Proc. Int. Conf. Low Temp. Sci., Bunyeido Printing Corporation, Sapporo, Japan, 1966, 1.
53. **Marino, G., Rosati, P., and Sagrati, F.,** Storage of in vitro cultures of *Prunus* rootstocks, *Plant Cell Tissue Organ Culture,* 3, 73, 1985.
54. **Mazur, P.,** Physical and chemical basis of injury in single celled microorganisms subjected to freezing and thawing, in *Cryobiology,* Merymen, H. T., Ed., Academic Press, New York, 1966, 214.
55. **Mazur, P.,** Cryobiology: the freezing of biological systems, *Science,* 168, 939, 1970.
56. **Mazur, R. A. and Hartmann, J. X.,** Freezing of plant protoplasts, in *Plant Cell and Tissue Culture: Principles and Application,* Sharp, W. R., Larsen, P. O., Paddock, E. F., and Raghavan, V., Eds., Ohio State University Press, Columbus, 1979, 876.
57. **Mix, G.,** Kartoffelsorten aus dein Reagenzglas-Bedingungen zur Langzeitlagerung, *Der Kartoffelbau,* 32, 198, 1981.
58. **Morel, G.,** Meristem culture techniques for the long term storage of cultivated plants, in *Crop Genetic Resources for Today and Tomorrow,* Frankel, O. H. and Hawkes, J. G., Eds., Cambridge University Press, Cambridge, 1975, 327.
59. **Morris, G. J.,** Lipid loss and haemolysis by thermal shock, lack of correlation, *Cryobiology,* 12, 192, 1975.
60. **Morris, G. J.,** The cryopreservation of *Chlorella.* II. Effect of growth temperature on freezing tolerance, *Arch. Microbiol.,* 107, 309, 1976.
61. **Morris, G. J.,** Plant cells, in *Low Temperature Preservation in Medicine and Biology,* Ashwood-Smith, M. J. and Farrant, J., Eds., Pitman Medical, Tunbridge-Wells, England, 1980, 253.
62. **Morris, G. J.,** Cryopreservation, An Introduction to Cryopreservation in Culture Collections, Institute of Terrestrial Ecology, Cambridge, 1981.
63. **Morris, G. J. and Clarke, A.,** The cryopreservation of *Chlorella.* IV. Accumulation of lipid as a protective factor, *Arch. Microbiol.,* 119, 153, 1978.
64. **Mullin, R. H. and Schlegel, D. E.,** Cold storage maintenance of strawberry meristem plantlets, *HortScience,* 11, 100, 1976.
65. **Nag, K. K. and Street, H. E.,** Carrot embryogenesis from frozen cultured cells, *Nature (London),* 245, 270, 1973.
66. **Nag, K. K. and Street, H. E.,** Freeze-preservation of cultured plant cells. I. The pre-treatment phase, *Physiol. Plant.,* 34, 254, 1975.
67. **Nag, K. K. and Street, H. E.,** Freeze-preservation of cultured plant cells. II. The freezing and thawing phase, *Physiol. Plant.,* 34, 261, 1975.
68. **Nitsche, W.,** Erhaltung der Lebensfachigkeit in getrocknetem Kallus, *Z. Pflanzenphysiol.,* 87, 469, 1978.
69. **Nitzsche, W.,** One year storage of dried callus, *Z. Pflanzenphysiol.,* 100, 269, 1980.
70. **Nitzsche, W.,** Germplasm preservation, in *Handbook of Plant Cell Culture — Techniques for Propagation and Breeding,* Evans, D. A., Sharp, W. R., Ammirato, P. V., and Yamada, Y., Eds., Macmillan, New York, 1983, 782.
71. **Orshinsky, B. R. and Tomes, D. T.,** Effect of long-term culture and low temperature incubation on plant regeneration from callus line of birdsfoot trefoil *(Lotus corniculatus),* *J. Plant Physiol.,* 119, 389, 1985.
72. **Patterson, G. W.,** Effect of culture temperature on fatty acid composition of *Chlorella sorokiniana,* *Lipids,* 5, 597, 1970.
73. **Pierquet, P. and Stushoff, C.,** Relationship of low temperature exotherms to cold injury in *Vitis riparia,* *Am. J. Enol. Vitic.,* 31, 1, 1980.

74. **Polge, C., Smith, A. U., and Parkes, A. S.,** Revival of spermatozoa after vitrification and dehydration at low temperatures, *Nature (London)*, 164, 666, 1949.
75. **Pritchard, H. W., Grout, B. W. W., and Short, K. C.,** Osmotic stress as a pregrowth procedure for cryopreservation. I. Growth and ultrastructure of sycamore and soybean cell suspensions, *Ann. Bot.*, 57, 41, 1986.
76. **Pritchard, H. W., Grout, B. W. W., and Short, K. C.,** Osmotic stress as a pregrowth procedure for cryopreservation. III. Cryobiology of sycamore and soybean cell suspensions, *Ann. Bot.*, 57, 379, 1986.
77. **Pritchard, H. W., Grout, B. W. W., Reid, D. S., and Short, K. C.,** The effect of growth under water stress on the structure, metabolism and cryopreservation of cultured sycamore cells, in *Biophysics of Water*, Franks, F. and Mathias, S. F., Eds., John Wiley & Sons, Chichester, 1982, 315.
78. **Quatrano, R. S.,** Freeze-preservation of cultured flax cells using DMSO, *Plant Physiol.*, 43, 2057, 1968.
79. **Rajashekher, C., Gusta, L. V., and Burke, M. J.,** Membrane structural transitions: probable relation to frost damage in hardy herbaceous species, in *Low Temperature Stress in Crop Plants, The Role of Membranes*, Lyons, J. M., Graham, D., and Raison, J. K., Eds., Academic Press, New York, 1979, 255.
80. **Sakai, A. and Sugawara, Y.,** Survival of poplar callus at superlow temperature after cold acclimation, *Plant Cell Physiol.*, 14, 1201, 1973.
81. **Sakai, A., Yamakawa, M., Sakato, D., Harada, T., and Yakuwa, T.,** Development of a whole plant from an excised strawberry runner apex frozen to −196°C, *Low Temp. Sci. Ser. B*, 36, 31, 1978.
82. **Sala, F., Cella, R., and Rollo, F.,** Freeze-preservation of rice cells, *Physiol. Plant.*, 45, 170, 1979.
83. **Scheuermann, E. A.,** Fluessiges Wasser bei Minusgraden, *Kosmos (Stockholm)*, 63, 331, 1967.
84. **Seibert, M.,** Shoot initiation from carnation shoot apices frozen to −196°C, *Science*, 191, 1178, 1976.
85. **Seibert, M. and Wetherbee, P. J.,** Increased survival and differentiation of frozen herbaceous plant organ cultures through cold treatment, *Plant Physiol.*, 59, 1043, 1977.
86. **Singer, S. J. and Nicolson, G. L.,** The fluid mosaic model of the structure of cell membranes, *Science*, 175, 720, 1972.
87. **Singh, J. and Miller, R. W.,** Spin-probe studies during freezing of cells isolated from cold-hardened and non-hardened winter rye, *Plant Physiol.*, 69, 1423, 1982.
88. **Steponkus, P. L. and Wiest, S. C.,** Plasma membrane alterations following cold acclimation and freezing, in *Plant Cold Hardiness and Freezing Stress — Mechanism and Crop Implications*, Li, P. H. and Sakai, A., Eds., Academic Press, New York, 1979, 75.
89. **Steponkus, P. L. and Wiest, S. C.,** Freeze-thaw induced lesions in plasma membrane, in *Low Temperature Stress in Crop Plants, The Role of Membranes*, Lyons, J. M., Graham, S. G., and Raison, J. K., Eds., Academic Press, New York, 1979, 231.
90. **Takeuchi, M., Matsuschima, H., and Sugawara, Y.,** Long term freeze preservation of protoplasts of carrot and *Marchantia*, *Cryog. Lett.*, 1, 519, 1980.
91. **Tisserat, B., Ulrich, J. M., and Finkle, B. J.,** Cryogenic preservation and regeneration of date palm tissue, *HortScience*, 16, 47, 1981.
92. **Towill, L. E.,** *Solanum tuberosum*, a model for studying the cryobiology of shoot-tips of the tuber-bearing *Solanum* species, *Plant Sci. Lett.*, 20, 315, 1981.
93. **Uemura, M. and Sakai, A.,** Survival of carnation (*Dianthus caryophyllus*) shoot apices frozen to the temperature of liquid nitrogen, *Plant Cell Physiol.*, 21, 85, 1980.
94. **Ulrich, J. M., Finkle, B. J., Moore, P. H., and Ginoza, H.,** Effect of a mixture of cryoprotectants in attaining liquid nitrogen survival of callus cultures of a tropical plant, *Saccharum* cv. H-50-7209, *Cryobiology*, 16, 550, 1979.
95. **Withers, L. A.,** Freeze-preservation of synchronously dividing cultured cells of *Acer pseudoplatanus*, *Cryobiology*, 15, 87, 1978.
96. **Withers, L. A.,** A fine structural study of the freeze-preservation of plant tissue cultures. II. The thawed state, *Protoplasma*, 94, 235, 1978.
97. **Withers, L. A.,** Freeze preservation of somatic embryos and clonal plantlets of carrot (*Daucus carota*), *Plant Physiol.*, 63, 460, 1979.
98. **Withers, L. A.,** Low temperature storage of plant tissue cultures, in *Advances in Biochemical Engineering Plant Cell Cultures*, Vol. 2, Flechter, A., Ed., Springer-Verlag, Berlin, 1980, 18.
99. **Withers, L. A.,** Germplasm storage in plant biotechnology, in *Plant Biotechnology*, Mantell, S. H. and Smith, H., Eds., Cambridge University Press, Cambridge, 1983, 187.
100. **Withers, L. A. and King, P. J.,** Proline, a novel cryoprotectant for the freeze preservation of cultured cells of *Zea mays*, *Plant Physiol.*, 64, 657, 1979.
101. **Withers, L. A. and King, P. J.,** A simple freezing unit and routine cryopreservation method for plant cell cultures, *Cryog. Lett.*, 1, 213, 1980.
102. **Withers, L. A. and Street, H. E.,** Freeze-preservation of cultured plant cells. III. The pregrowth phase, *Physiol. Plant.*, 39, 171, 1977.
103. **Woods, C. M., Reid, M. S., and Patterson, B. D.,** Response to chilling stress in plants cells. I. Changes in cyclosis and cytoplasmic structure, *Protoplasma*, 121, 8, 1984.

104. **Wolfe, J. and Steponkus, P. L.,** The stress-strain relation of the plasma membrane of isolated plant protoplasts, *Biochim. Biophys. Acta,* 643, 663, 1981.
105. **Wolfe, J. and Steponkus, P. L.,** Cryobiology of isolated protoplasts, mechanical properties of the plasma membrane, *Plant Physiol.,* 71, 276, 1983.
106. **Yoshida, S.,** Studies on freezing injury of plant cells. I. Relation between thermotropic properties of isolated plasma membrane vesicles and freezing injury, *Plant Physiol.,* 75, 38, 1984.
107. **Yoshida, S.,** Chemical and biophysical changes in the plasma membrane during cold acclimation of mulberry bark cells, *Plant Physiol.,* 76, 257, 1984.

INDEX

Q

R

Printed and bound by CPI Group (UK) Ltd, Croydon, CR0 4YY

22/10/2024

01777633-0008